Postharvest Technology
and Food Process
Engineering

Postharvest Technology and Food Process Engineering

Amalendu Chakraverty
R. Paul Singh

CRC Press
Taylor & Francis Group
Boca Raton   London   New York

CRC Press is an imprint of the
Taylor & Francis Group, an informa business

CRC Press
Taylor & Francis Group
6000 Broken Sound Parkway NW, Suite 300
Boca Raton, FL 33487-2742

First issued in paperback 2016

Version Date: 20130111

ISBN 13: 978-1-138-19885-2 (pbk)
ISBN 13: 978-1-4665-5320-0 (hbk)

Library of Congress Cataloging-in-Publication Data

Chakraverty, Amalendu.
    Post harvest technology and food process engineering / Amalendu Chakraverty, R. Paul Singh.
        p. cm.
    Includes bibliographical references and index.
    ISBN 978-1-4665-5320-0 (hardcover : alk. paper)
    1. Food crops--Postharvest technology. I. Singh, R. Paul. II. Title.

SB175.C42 2013
664'.02--dc23   2012050931

Visit the Taylor & Francis Web site at
http://www.taylorandfrancis.com

and the CRC Press Web site at
http://www.crcpress.com

Dedicated to

our parents

Contents

PART II GRAIN STORAGE

PART III PARBOILING AND MILLING

PART IV BY-PRODUCTS/BIOMASS UTILIZATION

PART V FOOD PROCESS ENGINEERING

Preface

This book originates from *Postharvest Technology of Cereals and Pulses* (published in 1981), which was considered to be the first of its kind. Since then, students and professionals in the field of agricultural and food engineering have felt the need for a consolidated book on postharvest technology and food process engineering.

This comprehensive book deals with grain properties, engineering principles, numerical problems, designs, and testing and provides illustrations and descriptions of the operations of various commercial grain dryers, milling machines, and furnaces, as well as utilization of by-products/biomass for producing energy, chemicals, food, feed, and other value-added products. Adequate emphasis has been placed on postharvest management, food chemistry, preservation and processing of fruits and vegetables, and relevant food engineering operations, namely, fluid mechanics, heat transfer, drying, and associated machines.

The major aim of this book is to serve as a text or as a reference book for students, professionals, and others engaged in agricultural science and food engineering, food science, and technology in the field of primary processing of cereals, pulses, fruits, and vegetables.

I would like to acknowledge my coauthor, Dr. R. Paul Singh, for contributing Chapter 17 entitled "Postharvest Management of Fruits and Vegetables." I am also indebted to my wife, Sushmita Chakraverty, and my sons, Krishnendu and Soumendu, for their painstaking assistance in the preparation of the manuscript.

I would like to extend my gratitude to Stephen Zollo, senior editor at Taylor & Francis Group, CRC Press, for his persistence and cooperation without which this publication would not have been possible.

Sincere thanks are also due to the publisher, Taylor & Francis Group, CRC Press, and to Prof. T.K. Goswami and other faculty members and scholars in the Department of Agriculture and Food Engineering at IIT Kharagpur, India, for their cooperation in the improvement of the manuscript.

Amalendu Chakraverty
(Former Professor)
Indian Institute of Technology
Kharagpur, India

R. Paul Singh
University of California, Davis
Davis, California

Introduction

Cereals, legumes, oilseeds, fruits, and vegetables are the most important food crops in the world. The need to increase food production and supply an adequate quantity of grains and other food in order to meet the energy and nutritional requirements of the growing world population is widely recognized. Cereals include edible grains such as rice, wheat, corn, barley, rye, oats, or sorghum. Cereal grains contribute the bulk of food calories and proteins worldwide and are consumed in various forms. They are also fed to livestock and are thereby converted into meat, milk, or eggs.

Rice and wheat are two of the most important types of staple food. Corn is mainly used as an ingredient of feed in the United States, though it has numerous uses in food items as well. Generally, cereals are composed of about 10%–15% moisture, 55%–71% carbohydrate, 8%–11% protein, 2%–5% fat, and 2%–9% fiber; while milling hull, bran and germ of cereal grains are separated, removing indigestible fiber as well as fat. The removal of fatty bran is necessary to avoid rancidity and to improve shelf life as well as the functional properties of starchy endosperm of food products.

Legumes are characterized by their high protein and low fat contents. Soybean contains a high percentage of both protein and fat, though it is mainly considered as oilseed.

Fruits and vegetables are clubbed together because of their many similarities with respect to their compositions and methods of harvest as well as postharvest operations. Fruits are the mature ovaries of plants with their seeds. Usually, fruits and vegetables contain a very high percentage of water and low percentage of protein and fat. Their water content normally varies from 70% to 85%. Fruits and vegetables are common sources of digestible starches, sugars, certain minerals, vitamins A and C, and indigestible fibers, which are important constituents of a diet. Citrus fruits, some green leafy vegetables, and tomatoes are good sources of vitamin C.

Generally, the supply of grains and other food can be enhanced in two ways: by increasing production and by reducing postharvest losses. Food production has increased significantly during the last few decades with the use of improved high-yielding cultivars, suitable fertilizers, water, as well as crop management practices.

Wheat and paddy production has increased spectacularly in many countries since the mid-1960s. Table 1 shows the production of these two grains in 1996. The production of pulses and fruits and vegetables in 1996 is presented in Tables 2 and 3, respectively.

It is recognized that hunger and malnutrition can exist despite adequate food production owing to uneven distribution, losses, and deterioration of available food resources during traditional postharvest operations. Therefore, maximum utilization of available food and minimization of postharvest losses are essential. Postharvest losses of cereals and fruits and vegetables are generally estimated to be 5%–20% and 20%–50%, respectively. A country can become self-sufficient in food if it minimizes colossal postharvest losses.

Commercial food preservation methods, as a whole, include drying/dehydration, refrigeration/cold storage or freezing, canning/pasteurization, chemical addition, and other special methods such as use of microwave, infrared rays, radiations, etc. The grain PHT, in particular, may involve drying, storage, parboiling/conditioning, milling operations, and by-product utilization. Apart from these, various other conversion technologies, namely, thermal, thermochemical, chemical, and biochemical processing, are also employed to convert biomass/by-products into energy, food, feed, and chemicals. Hence, PHT covers a wide range of diversified subjects.

Table 1   Wheat and Paddy Production (1000 Tons) in Some Countries

Country	Wheat Production, 1996	Paddy Production, 1996
India	62,620.0	120,012.0
China	109,005.0	190,100.0
Russia	87,000.0	2,100.0
United States	62,099.0	7,771.0
Canada	30,495.0	—
France	35,946.0	116.0
Australia	23,497.0	951.0
Pakistan	16,907.0	5,551.0
Argentina	5,200.0	974.0
World	584,870.0	562,260.0

Source: FAO Production Year Book, Vol. 50, FAO, Rome, Italy, 1996.

Table 2   Pulses Production (1000 Tons) in Some Continents/Countries

Continent/Country	Production, 1996
Asia	28,222
Africa	7,651
Europe	9,380
N. America	5,541
S. America	3,770
Australia	2,186
India	14,820[a]
China	4,979
Brazil	2,862
France	2,636
World	56,774

Source: FAO Production Year Book, Vol. 50, FAO, Rome, Italy, 1996.

[a] *Food and Agricultural Organisation (FAO) Production Year Book (1995).*

Table 3   Fruit and Vegetable Production (Million Tons) in Some Countries

Country	Production, 1996			
	Apple	Orange	Mango	Potato
China	16.00	2.26	1.21	46.03
India	1.20	2.00	10.00	17.94
Russia	1.80	—	—	38.53
Poland	1.70	—	—	22.50
Brazil	0.65	21.81	0.44	2.70
Mexico	0.65	3.56	—	1.20
France	2.46	—	—	6.46
Germany	1.59	—	—	13.60
United States	4.73	10.64	—	22.55
World	53.67	59.56	19.22	294.83

Source: FAO Production Year Book, Vol. 50, FAO, Rome, Italy, 1996.

GRAIN PROPERTIES, DRYING, AND DRYERS

I

Chapter 1

Properties of Grains

A grain is a living biological product that germinates as well as respires. The respiration process in the grain is externally manifested by the decrease in dry weight, utilization of oxygen, evolution of carbon dioxide, and release of heat. The rate of respiration is dependent upon moisture content and temperature of the grain. The rate of respiration of paddy increases sharply (at 25°C) from 14% to 15% moisture content, which may be called the critical point. On the other hand, the rate of respiration increases with the increase of temperature up to 40°C. Above this temperature, the viability of the grain as well as the rate of respiration decreases significantly.

Structure

Wheat and rye consist mainly of pericarp, seed coat, aleurone layer, germ, and endosperm, whereas oats, barley, paddy, pulses, and some other crops consist not only of the aforementioned five parts but also an outer husk cover. The husk consists of strongly lignified floral integuments. The husk reduces the rate of drying significantly.

The embryo or germ is the principal part of the seed. All tissues of the germ consist of living cells that are very sensitive to heat. The endosperm that fills the whole inner part of the seed consists of thin-walled cells, filled with protoplasm and starch granules and serves as a kind of receptacle for reserve foodstuff for the developing embryo. The structures of a few important grains are shown in Figures 1.1 through 1.4.

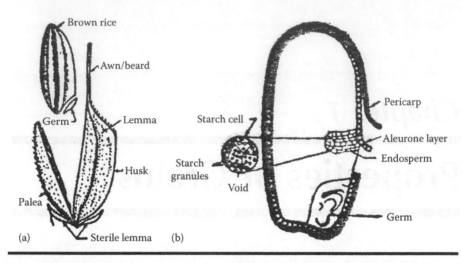

Figure 1.1 **(a) Different parts of paddy. (b) Structure of brown rice kernel (longitudinal section).**

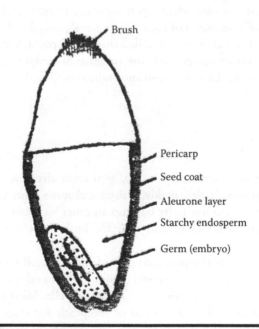

Figure 1.2 **Structure of wheat.**

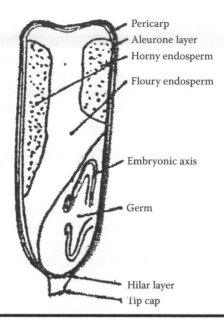

Pericarp
Aleurone layer
Horny endosperm
Floury endosperm
Embryonic axis
Germ
Hilar layer
Tip cap

Figure 1.3   Structure of shelled corn (longitudinal section).

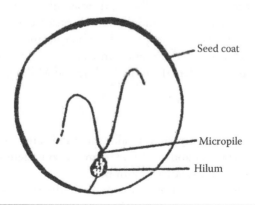

Seed coat
Micropile
Hilum

Figure 1.4   Whole arhar pulses (*Cajanus cajan*).

Chemical Composition

The grain is composed of both organic and inorganic substances, such as carbohydrates, proteins, vitamins, fats, ash, water, mineral salts, and enzymes. Paddy, corn, wheat, and buckwheat seeds are especially rich in carbohydrates, whereas legumes are rich in proteins and oilseeds in oils.

Generally, pericarp and husk contain cellulose, pentosan, and ash. The aleurone layer contains mainly albumin and fat. The endosperm contains the highest amount of carbohydrate in the form of starch, a small amount of reserve protein, and a very little amount of ash and cellulose, whereas the germ contains the highest amount of fat, protein, and a small amount of carbohydrate in the form of sugars and a large amount of enzyme.

Effects of Temperature on the Quality of Grain

Proteins

The proteins present in cereal grains and flour are hydrophilic colloids. The capacity of flour proteins to swell plays an important role in the preparation of dough. At temperatures above 50°C, denaturation and even coagulation of proteins take place. As a result, the water-absorbing capacity of the proteins and their capacity for swelling decrease.

Starch

Starch is insoluble in cold water but swells in hot water. Up to a temperature of 60°C, the quality of starch does not change appreciably. With a further increase in temperature, particularly above 70°C, and especially in the presence of high moisture in the grain, gelatinization and partial conversion of starch to dextrin take place. In addition, a partial caramelization of sugars with the formation of caramel may take place, which causes deterioration in color of the product. These effects will be discussed in detail in Part III on Parboiling and Milling.

Fats

Fats are insoluble in water. Compared to albumins and starch, fats are more heat-resistant. But at temperatures above 70°C, fats may also undergo a partial decomposition resulting in an increase of acid numbers.

In the range of temperatures from 40°C to 45°C, the rate of enzymatic activity on fats increases with the increase of moisture and temperature. With a further rise of temperature, the enzymatic activity begins to decrease, and at temperatures between 80°C and 100°C the enzymes are completely inactivated.

Vitamins

The heat-sensitive B vitamins present in the germ and aleurone layer are destroyed at high temperature.

The details of the structures and compositions of wheat, rice, corn, and pulses/legumes can be found in Pomeranz (1971), Potter (1986), and Kadam et al. (1982).

Physical Properties

The knowledge of important physical properties such as shape, size, volume, surface area, density, porosity, color, etc., of different grains is necessary for the design of various separating and handling, storing, and drying systems. The density and specific gravity values are also used for the calculation of thermal diffusivity and Reynolds number. A few important physical properties have been discussed here.

Sphericity

Sphericity is defined as the ratio of surface area of a sphere having same volume as that of the particle to the surface area of the particle. Sphericity is also defined as

$$\text{Sphericity} = \frac{d_i}{d_c}$$

where
  d_i is the diameter of largest inscribed circle
  d_c is the diameter of smallest circumscribed circle of the particle

The sphericity of different grains varies widely.

Bulk Density

The bulk density of a grain can be determined by weighing a known volume of grain filled uniformly in a measuring cylinder. Bulk densities can then be found at different moisture contents for various biomaterials. The following equation is used to calculate the bulk density of the material:

$$\rho_B = \frac{W}{V}$$

where
  ρ_B is the bulk density, g/cc or kg/m^3
  W is the weight of the material, kg or g
  V is the volume of the material, cc or m^3

True Density

The mass per unit volume of a material excluding the void space is termed as its true density.

The simplest technique of measuring true density is by liquid displacement method, where tube is commonly used. The expressions used for calculation of true volume are given as follows:

$$\text{Volume (cc)} = \frac{\text{Weight of displaced water, g}}{\text{Weight density of water, g/cc}} \quad \text{and}$$

$$\text{Volume (cc)} = \frac{(\text{Weight in air} - \text{weight in water}), g}{\text{Weight density of water, g/cc}}$$

However, the only limitation of this method is to use the materials impervious to the liquid used. Hence, the use of toluene has also been in practice for a long time. The expression used for calculating true density is

$$\text{True density, } \rho_t, \text{g/cc} = \left(-\frac{\text{Weight of the grain, g}}{\begin{array}{c}\text{Weight of toluene}\\\text{displaced by grain, g}\end{array}}\right) \times \left(\begin{array}{c}\text{Weight density}\\\text{of toluene, g/cc}\end{array}\right)$$

Air comparison pycnometer is an instrument that can be conveniently used to measure the true volume of a sample of any shape and size without wetting the sample. Thus, the true density is determined from the measurement of true volume of a sample of known weight with this instrument.

Porosity

It is defined as the percentage of volume of inter-grain space to the total volume of grain bulk. The percent void of different grains in bulk is often needed in drying, airflow, and heat flow studies of grains. Porosity depends on (a) shape, (b) dimensions, and (c) roughness of the grain surface.

Porosity of some crops is tabulated as follows:

Grain	Porosity, %
Corn	40–45
Wheat	50–55
Paddy	48–50
Oats	65–70

The grain porosity can be measured by using an air comparison pycnometer or by the mercury displacement method (Thompson and Issas, 1967).

Coefficient of Friction and Angle of Repose

Angle of repose and frictional properties of grains play an important role in selection of design features of hoppers, chutes, dryers, storage bins, and other equipment for grain flow.

The additional details on the method of determination of frictional coefficients are available in Chakraverty et al. (2003) and Dutta et al. (1988).

Coefficient of Friction

The coefficient of friction between granular materials is equal to the tangent of the angle of internal friction for the material. The frictional coefficient depends on (a) grain shape, (b) surface characteristics, and (c) moisture content.

Angle of Repose

The angle of repose of grain can be determined by the following method. Grain is poured slowly and uniformly onto a circular platform of 6.5 cm diameter to form a cone. The height of this cone is measured using a traveling microscope. The angle of repose of grain at different moisture contents is determined from the geometry of the cone formed (Dutta et al., 1988). It is the angle made by the surface of the cone with horizontal. It is calculated using the following equation:

$$\varphi_{AR} = \tan^{-1}\left\{\frac{2(H_c - H_p)}{D_p}\right\}$$

where
φ_{AR} is the angle of repose, degrees
H_c is the height of the cone, cm or m
D_p is the diameter of the platform, cm or m

Thermal Properties

Raw foods are subjected to various types of thermal treatment, namely, heating, cooling, drying, freezing, etc., for processing. The change of temperature depends on the thermal properties of the product. Therefore, knowledge of thermal properties, namely, specific heat, thermal conductivity, and thermal diffusivity, is essential to design different thermal equipment and solve various problems on heat transfer operation.

Specific Heat

The specific heat of a substance is defined as the amount of heat required to raise the temperature of unit mass through 1°C. The specific heat of wet grain may be considered as the sum of specific heat of bone dry grain and its moisture content. It can be expressed as follows:

$$c = \left(\frac{m}{100}\right)c_w + \left(\frac{100-m}{100}\right)c_d$$

$$\text{or} \quad c = \left(\frac{m}{100}\right) + \left(\frac{100-m}{100}\right)c_d, \text{ kcal/(kg °C) or kJ/(kg °C)}$$

where
c_d is the specific heat of the bone dry grain
c_w is the specific heat of water
m is the moisture content of the grain, percent (wet bulb temperature)

The specific heat of bone dry grain varies from 0.35 to 0.45 kcal/kg or 1.46 to 1.88 kJ/kg °C.

The aforementioned linear relationship between c and m exists above $m = 8\%$ moisture content only (Gerzhoi and Samochetov, 1958).

Specific Heat Measurement

The specific heat of grain can be determined by the method of mixture for which the experimental setup and the procedure are explained later.

A thermos flask of required capacity is used as a calorimeter. It is further insulated by centrally placing it in a thermocole container and filling the gap between the flask and the container with glass wool. A glass beaker of required capacity is also insulated all around by placing it in a thermocole box and used as an ice bath. A long precision mercury thermometer can be used to measure the temperature. Any balance with an accuracy of at least 0.1 mg can be used for weight measurements during the experiments.

The water equivalent of the calorimeter is first determined and it is calculated using the following heat balance equation:

$$W_e = W_{cw}\left\{\frac{t_e - t_c}{t_f - t_e}\right\}$$

where
W_e is the water equivalent of the flask calorimeter, g or kg
W_{cw} is the weight of cold distilled water, g or kg
t_f is the temperature of flask calorimeter (ambient), °C
t_c is the temperature of cold distilled water, °C
t_e is the equilibrium temperature of water, °C

The specific heat can be determined by taking about 15–25 g of grain in the calorimeter and then rapidly pouring 200 g of ice-cooled distilled water at a low temperature into it. It is then shaken thoroughly for 5 min and the equilibrium mixture temperature is recorded. The heat balance equation is used to calculate the specific heat of grain as

$$C_p = \frac{W_{cw}(t_e - t_c) - W_e(t_g - t_e)}{W_g(t_g - t_e)}$$

where
C_p is the specific heat of grain, cal/(g °C) or kcal/(kg °C) or kJ/(kg °C)
W_g is the weight of grain, g or kg
t_g is the temperature of grain and calorimeter, °C
t_c is the temperature of cold distilled water, °C
t_e is the equilibrium temperature of the mixture, °C
W_e is the water equivalent of the flask calorimeter, g or kg

The various other methods of specific heat measurement are discussed by Rahaman (1995).

Thermal Conductivity

The thermal conductivity is defined as the amount of heat flow through unit thickness of material over a unit area per unit time for unit temperature difference. The thermal conductivity of the single grain varies from 0.3 to 0.6 kcal/(m·h °C), whereas the thermal conductivity of grains in bulk is about 0.10 to 0.15 kcal/(m·h °C), which is due to the presence of air space in it. The thermal conductivity of air is 0.02 kcal/(m·h °C) only.

Thermal conductivity of the single grain is three to four times greater than that of the grain bulk. In the case of wheat bulk, the moisture content ranging from 10% to 20% (dry bulb temperature) can be expressed as follows (Gerzhoi and Samochetov, 1958):

$$K = 0.060 + 0.002M \text{ kcal/(m · h °C)}$$

where
K is the thermal conductivity
M is the moisture content (dry bulb temperature)

Thermal Conductivity Measurement

The thermal conductivity of a grain can be determined by the transient heat flow method using a thermal conductivity probe. The experimental setup and the procedure are explained next.

A thermal conductivity probe is placed in a sample holder equipped with a digital multimeter, a rheostat, a d.c. ammeter, and a 12 V storage battery. The thermal conductivity probe shown in Figure 1.5 consists of a 24 gauge iron–constantan wire of 300 mm length covered with teflon, which is used as a heater and housed in a hollow brass tube of 6.35 mm (1/4″) o.d. and a wall thickness of 1.59 mm (1/16″). The heater wire with a brass tube is centrally located in a cylindrical sample holder of 200 mm diameter and 300 mm height. The cylinder is made of 0.79 mm (1/32″) thick aluminum sheet. The ends of the sample holder are closed by 12.7 mm (1/2″) thick bakelite covers. The iron–constantan thermocouples are fixed; one at the middle point of the heater and the other

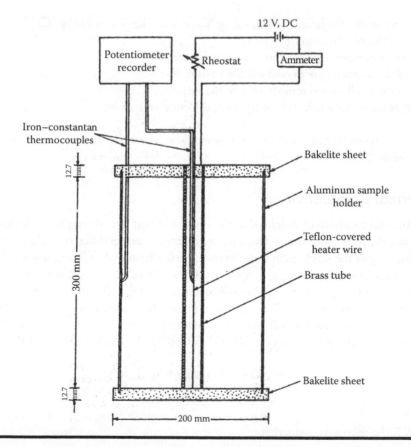

Figure 1.5   Schematic representation of a thermal conductivity probe.

to the inner wall of the sample holder to measure the temperatures at the said points. The heater is connected to the battery for the necessary power supply. The required strength of the current is adjusted with the help of a rheostat. An ammeter measures the current. The temperatures at the heater and the inside wall of the sample holder are recorded in terms of millivolts with a digital multimeter. The cold junction of the thermocouples is kept in an ice bath.

The moisture content of a grain sample is determined prior to the experiment by the standard oven drying method (105°C ± 2°C for 24 h). The resistance of the heating wire per unit length can be predetermined to be 2.0866 Ω/m (approximately). Similarly, a current of 1.25 A can be determined in preliminary trials to achieve a rise in temperature of the heater by 10°C–15°C in 10 min.

The sample holder is completely filled with the sample in an identical manner during each test to maintain the same bulk density.

A current of 1.25 A indicated by an ammeter is passed through the heater by adjusting the rheostat. The temperature of grain is recorded in terms of millivolts at every 30 s interval from the digital multimeter.

Thermal conductivity of a grain at a moisture content is calculated using the following equation:

$$K = \left\{ \frac{0.86\ I^2 R}{4\pi(t_2 - t_1)} \right\} \ln\left[\frac{\theta_2 - \theta_0}{\theta_1 - \theta_0} \right]$$

where
   K is the thermal conductivity, kcal/h m °C
   I is the current flow, A
   R is the resistance of the line heater, Ω/m
   θ_0 is the time correction factor, min
   t_2 and t_1 are temperatures at θ_2 and θ_1 times, °C
   θ_2 and θ_1 are times, min
   0.86 is conversion factor from W to kcal

The thermal conductivity probe and an analysis of the method are detailed in Rao and Rizvi (1986).

Aerodynamic Properties

For designing air and water conveying and separating systems (i.e., pneumatic or hydrodynamic systems), the knowledge of aerodynamic and hydrodynamic properties of the agricultural products is necessary. In this connection, the knowledge of terminal velocities of different crops in a fluid is necessary.

The air velocity at which an object remains suspended in a vertical pipe under the action of the air current is called terminal velocity of the object.

Thus, in free fall, the object attains a constant terminal velocity, V_t, when the gravitational accelerating force, F_g, becomes equal to the resisting upward drag force F_r.

Hence, $F_g = F_r$ when $V = V_t$

$$\text{or} \quad W\left[\frac{\rho_v - \rho_f}{\rho_v}\right] = \frac{1}{2}ca_v\rho_f V_t^2$$

$$V_t = \left[\frac{2W(\rho_v - \rho_f)}{\rho_v\rho_f a_v c}\right]^{1/2}$$

where

V_t is the terminal velocity, m/s

W is the weight of the particle, kg

ρ_v and ρ_f are mass densities of the particles and fluids, (kg s^2)/m^4

a_v is the projected area of the particle perpendicular to the direction of motion, m^2

c is the overall drag coefficient (dimensionless)

Grains	Terminal Velocity, m/s
Wheat	9–11.5
Barley	8.5–10.5
Small oats	19.3
Corn	34.9
Soybeans	44.3
Rye	8.5–10.0
Oats	8.0–9.0

Resistance of Grain Bed to Airflow

In the design of blowers for grain dryers, it is necessary to know the resistance exerted by the grain bed to the air current blown through it. The resistance is dependent upon (a) the bed thickness, (b) the air velocity, (c) the orientation of the grains, and (d) the type of grains.

Additional details of all physical properties of agricultural products biomaterials are available in Mohsenin (1980) and Rahaman (1995).

Symbols

a_v	Projected area, m^2
C, C_p	Specific heat, kcal/(kg °C), or cal/(g °C) or kJ/(kg °C)
c	Drag coefficient, dimensionless
K	Thermal conductivity, kcal/(m h °C) or kW/(m °C)
m	Moisture content, percent (wet bulb temperature)
M	Moisture content, percent (dry bulb temperature)
φ_{AR}	Angle of repose, degree
ρ_v and ρ_f	Mass densities of particles and fluid, (kg s^2)/m^4
ρ_B	Bulk density, kg/m^3 or g/cc
ρ_t	True density, kg/m^3 or g/cc
V_t	Terminal velocity, m/s
W_t	Weight of particle, kg or g

Chapter 2

Psychrometry

Ambient air is a mixture of dry air and water vapor. Moist air is necessary in many unit operations. To work out such problems, it is essential to have knowledge of the amount of water vapor present in air under various conditions, the thermal properties of such a mixture, and changes in the heat and moisture contents as it is brought in contact with water or wet solid. Particularly in grain drying, the natural or heated air is used as a drying medium. Although the proportion of water vapor in air is small, it has a profound effect on the drying process.

Problems in air–water vapor mixture including heating, cooling, humidification, dehumidification, and mixing can be solved with the help of mathematical formulae. As these calculations are time-consuming, special charts containing the most common physical and thermal properties of moist air have been prepared and are known as psychrometric charts. The psychrometric chart is, therefore, a graphical representation of the physical and thermal properties of atmospheric air.

The different terms used to express the physical as well as other thermodynamic properties of air–water vapor mixture are defined and discussed here.

Humidity

The absolute humidity, H, is defined as kilograms of water vapor present in 1 kg of dry air under a given set of conditions.

H depends upon partial pressure of water vapor, p_w, in air and total pressure, P.

Therefore, H can be expressed mathematically as follows:

$$H = \frac{18 p_w}{29(P - p_w)} \tag{2.1}$$

When $P = 1$ atm (for psychrometry),

$$H = \frac{18 p_w}{29(1 - p_w)} \text{ kg/kg} \tag{2.2}$$

As per p_w is small,

$$H = \frac{18 p_w}{29} \tag{2.3}$$

Again, from Equation 2.1

$$H = \frac{p_w}{\dfrac{29}{18}(P - p_w)} = \frac{p_w}{1.611(P - p_w)} \tag{2.4}$$

Rearranging Equation 2.4

$$p_w = \left(\frac{1.611 H}{1 + 1.611 H}\right) P \tag{2.5}$$

Saturated air is the air in which water vapor is in equilibrium with the liquid water at a given set of temperature and pressure.

Percentage Humidity

It is the ratio of the weight of water present in 1 kg of dry air at any temperature and pressure and the weight of water present in 1 kg of dry air, which is saturated with water vapor at the same temperature and pressure:

$$\text{Percentage humidity} = \left(\frac{H}{H_s}\right) \times 100 \tag{2.6}$$

Relative Humidity

Relative humidity (RH) is defined as the ratio of the partial pressure of water vapor in the air to the partial pressure of water vapor in saturated air at the same temperature:

$$\text{RH} = \left(\frac{p_w}{p_s}\right) \times 100$$

The relation between percentage humidity and RH

$$\text{Percentage humidity} = \text{RH}\left(\frac{1-p_s}{1-p_w}\right) \qquad (2.7)$$

Humid Heat

Humid heat is the number of kilocalories necessary to raise the temperature of 1 kg dry air and its accompanying water vapor through 1°C:

$$S = 0.24 + 0.45H, \text{ kcal}/(\text{kg}°\text{C})$$

$$= 1.005 + 1.88H, \text{ kJ}/(\text{kg}°\text{C}) \qquad (2.8)$$

Enthalpy

Enthalpy h' of an air and water vapor mixture is the total heat content of 1 kg of dry air plus its accompanying water vapor. If the datum temperature and pressure are 0°C and 1 atm, respectively, then the enthalpy at t°C for air and water vapor mixture is

$$h' = 0.24(t-0) + H[\lambda + 0.45(t-0)] = (0.24 + 0.45H)t + \lambda H \text{ kcal/kg}$$

$$= (1.005 + 1.88H)t + \lambda H \text{ kJ/kg} \qquad (2.9)$$

Humid Volume

Humid volume, v, is the total volume in cubic meter of 1 kg dry air and its accompanying water vapor:

$$v = \frac{22.4}{29}\left(\frac{t+273}{273}\right) + \frac{22.4}{18}H\left(\frac{t+273}{273}\right)$$

$$= (22.4/273)\,(t+273)\left[\frac{1}{29} + \frac{H}{18}\right]$$

$$= (0.00283 + 0.00456H)\,(t+273)\text{ m}^3/\text{kg} \qquad (2.10)$$

Saturated Volume

Saturated volume is the volume of 1 kg of dry air plus that of the water vapor necessary to saturate it.

Dew Point

Dew point is the temperature at which a mixture of air and water vapor has to be cooled (at constant humidity) to make it saturated.

Wet Bulb Temperature

Under adiabatic condition, if a stream of unsaturated air, at constant initial temperature and humidity, is passed over wetted surface (which is approximately at the same temperature as that of air), then the evaporation of water from the wetted surface tends to lower the temperature of the liquid water. When the water becomes cooler than the air, sensible heat will be transferred from the air to the water. Ultimately, a steady state will be reached at such a temperature that the loss of heat from the water by evaporation is exactly balanced by the sensible heat passing from the air into the water. Under such conditions, the temperature of the water will remain constant and this constant temperature is called wet bulb temperature.

Wet Bulb Theory

By definition of wet bulb temperature,

$$q = (h_G + h_r)A\,(t_G - t_w) = \lambda_w 18\,K_G A(p_w - p_G) \tag{2.11}$$

where

q is the sensible heat flowing from air to the wetted surface
h_G is the heat transfer coefficient by convection from the air to the wetted surface, kcal/(h m² °C) or kW/(m² °C)
h_r is the heat transfer coefficient corresponding to radiation from the surroundings, kcal/(h m² °C) or kW/(m² °C)
t_G and t_w are the temperatures of air and interface, °C
p_G and p_w are the partial pressures of water vapor in air and interface, atm
A is the area of the wetted surface, m²
K_G is the mass transfer coefficient, kgmol/(h m² atm)
λ_w is the latent heat of water vapor diffusing from the wetted surface to the air, kcal/kg or kJ/kg

Therefore,

$$p_w - p_G = \frac{h_G + h_r}{18\lambda_w K_G}(t_G - t_w) \tag{2.12}$$

$$\text{If } h_r = 0, \quad p_w = \frac{29H_w}{18} \quad \text{and} \quad p_G = \frac{29H_G}{18}$$

Then

$$H_w - H_G = \frac{h_G}{29\lambda_w K_G}(t_G - t_w) \tag{2.13}$$

The ratio h_G/K_G may be considered as constant. If the ratio h_G/K_G is constant, then Equation 2.13 can be used to determine the composition of the air–water vapor mixture from the observed values of t_G, the dry bulb temperature and t_w, the wet bulb temperature.

It is apparent from Equation 2.13 that the wet bulb temperature depends only upon the temperature and humidity of the air, provided h_r is negligible and h_G/K_G is constant.

It may be noted that the equation for the adiabatic cooling line is (Figure 2.2b)

$$H_s - H_G = \frac{s}{\lambda_s}(t_G - t_s) \tag{2.14}$$

where
  t_s is the temperature of water
  H_s is the saturated humidity
  λ_s is the latent heat of evaporation at t_s
  s is the humid heat

If $h_G/(29\,K_G) = s$, Equations 2.13 and 2.14 become identical. Fortuitously, for air–water vapor, $h_G/(29\,K_G) = s = 0.26$ at a humidity of 0.047. Therefore, under ordinary conditions the adiabatic cooling line can be used for wet bulb problems.

Introduction of Psychrometric Chart

Usually a psychrometric chart is prepared for 1 atm pressure, where humidities are plotted as ordinates against temperatures as abscissa. Any point on this chart represents the humidity and temperature of a given sample of air. The psychrometric chart is bound by an extreme left-hand curve representing humidities of saturated air (100% RH) and the horizontal x-axis giving various dry bulb temperatures (0% RH). The family of curved lines below the 100% RH line represents various percent RH as shown in Figure 2.1. Values of H for the saturation curve can be calculated by putting saturated pressure values from a steam table for different temperatures in Equation 2.4. The vapor pressure of water in air for different humidities is calculated by Equation 2.5 and added to the plot in the position shown in Figure 2.1. The oblique isovolume straight lines (humid volume lines) are plotted in the chart with steeper slopes than those of wet bulb lines. They are not exactly parallel. The humid volume at any temperature and humidity can be found from these lines. The humid volumes corresponding to these lines can be computed by Equation 2.10. The humid heat sometimes plotted

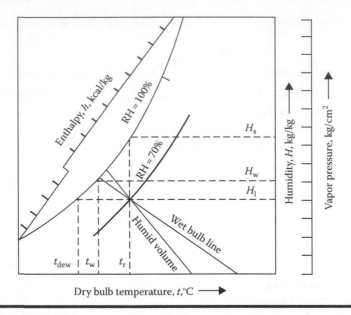

Figure 2.1 **Introduction of psychrometric chart (1 atm) pressure.**

against humidity can be calculated by Equation 2.8. The values of the enthalpy lines are usually indicated on a scale on the upper left-hand side of the chart. The wet bulb lines presented in the chart for different temperatures and humidities are actually adiabatic cooling lines. The straight wet bulb lines are inclined at angles of slightly unequal magnitudes.

Use of Psychrometric Chart

The psychrometric chart can be used to find out the following:

1. Dry bulb temperature
2. Wet bulb temperature
3. Dew point temperature
4. Absolute humidity
5. Relative humidity
6. Humid volume
7. Enthalpy

Any one of the aforementioned physical properties of air and water vapor mixture can be obtained from the psychrometric chart, provided two other values are known. Figure 2.1 shows that the meeting point of any two property lines represent the state point from which all other values can be obtained.

The following points may be noted from the psychrometric chart:

1. The t_G, t_w, and dew point temperatures are equal when RH is 100%.
2. The pressure of water vapor nearly doubles for each 10°C rise in temperature.
3. The rate of heat transfer from air to water (grain moisture) is proportional to $(t_G - t_w)$.

Psychrometric representation of several operations, namely, heating and cooling, drying, mixing, cooling, and dehumidification of moist air, is given in Figure 2.2a–d.

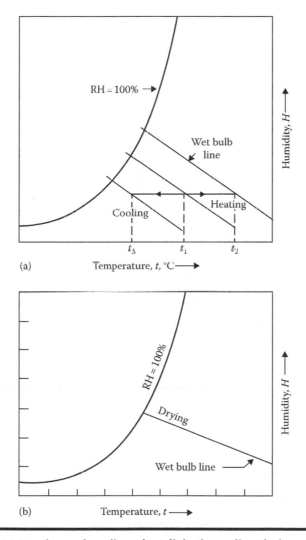

Figure 2.2 (a) Heating and cooling. (b) Adiabatic cooling/drying.

(continued)

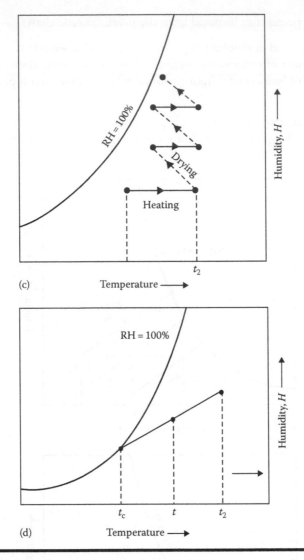

Figure 2.2 (continued) **(c) Heating, drying, reheating, and recycling. (d) Cooling and dehumidifying.**

Problems on Psychrometry

Solved Problems

1. Moist air at 25°C dry bulb and 45% RH is heated to 80°C. Calculate the humid volume, percentage humidity, and humid heat at the initial condition and check the results from chart. Find also the final condition of the air.

Data given

Initial condition: Dry bulb temperature = 25°C RH = 45%
   Final condition = 80°C

From the psychrometric chart

Humid heat = 0.244 kcal/kg °C
Humid volume = 0.856 m³/kg
Humidity = 0.009 kg/kg
Saturated humidity = 0.02 kg/kg

Percent humidity = $\dfrac{0.009}{0.02} \times 100 = 45\%$

Enthalpy = 11.5 kcal/kg = 11.5 × 4.184 = 48.12 kJ/kg

Final condition

Humid volume = 1.015 m³/kg
Relative humidity = 3%
Enthalpy = 25.5 kcal/kg = 25.5 × 4.184

By calculation

$$\text{Humid heat} = 0.24 + 0.45H$$

$$= 0.24 + 0.45 \times 0.009 = 0.24405 \text{ kcal / kg}$$

$$\text{Humid volume} = (1.005 + 1.88 \times 0.009) = 0.022 \text{ kJ / kg}$$

$$= \frac{22.4}{273 \times 29}(t + 273) + \frac{22.44H}{273 \times 18}(t + 273)$$

$$= (0.00283 + 0.00455H)(t + 273)$$

When t = 25°C and H = 0.009 kg/kg

$$\text{Humid volume} = (0.00283 + 0.00455 \times 0.009) \times 298$$

$$= (0.00283 + 0.000041) \times 298$$

$$= 0.856 \text{ m}^3 / \text{kg}$$

$$\text{Enthalpy} = (0.24 + 0.45H)t + \lambda H = (1.005 + 1.88H)t + \lambda H$$

$$= 0.24405 \times 25 + 598 \times 0.009 = 1.02 \times 25 + 2501.4 \times 0.009$$

[therefore, λ = 598 kcal/kg = 2501.4 kJ/kg]

Therefore, h = 11.48 kcal/kg = 48.06 kJ/kg.

2. In a grain dryer, one stream of air at 50 m³/min at 25°C and 23°C wet bulb temperature is mixed with another air stream at 50 m³/min at 60°C and 52°C wet bulb temperatures (Figure 2.3).

    Determine the dry bulb and wet bulb temperatures of the mixture.

Solution

Stream = 1
    Rate of flow = 50 m³/min
    Dry bulb temperature = 25°C
    Wet bulb temperature = 23°C

From the psychrometric chart

Enthalpy = 16.5 kcal/kg = 69.036 kJ/kg
Humid volume = 0.866 m³/kg

$$\text{Therefore, } m_1 = \frac{50}{0.866} = 57.7 \text{ kg/ min}$$

Stream = 2
    Rate of flow = 50 m³/min
    Dry bulb temperature = 60°C
    Wet bulb temperature = 52°C

From the chart

Enthalpy = 72.5 kcal/kg = 303.34 kJ/kg
Humid volume = 1.084 m³/kg

$$\text{Therefore, } m_1 = \frac{50}{1.084} = 46.12 \text{ kg/min}$$

We know that

$$\frac{m_1}{m_2} = \frac{h_2 - h_3}{h_3 - h_1}$$

$$\frac{57.7}{46.12} = \frac{72.5 - h_3}{h_3 - 16.5} = \frac{303.34 - h_3}{h_3 - 69.04}$$

Therefore, enthalpy of the final mixture state

$$h_3 = 41.39 \text{ kcal/kg} = 173.17 \text{ kg/kg}$$

From the psychrometric chart

Dry bulb temperature = 41°C
Wet bulb temperature = 40.75°C

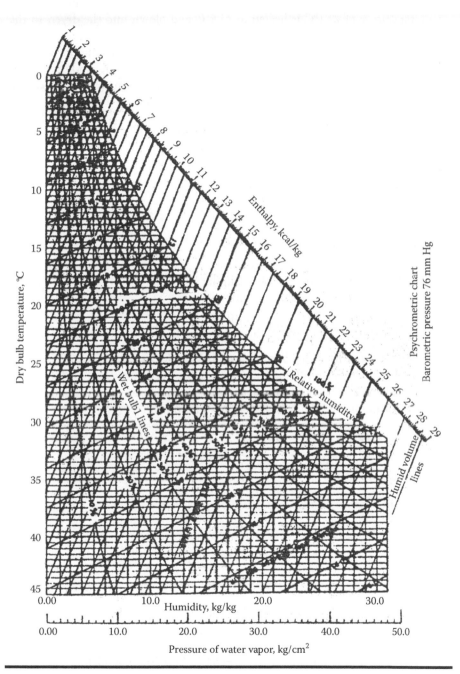

Figure 2.3   Humidity–temperature diagram (psychrometric chart).

3. The air to use in a dryer at a dry bulb temperature of 26.66°C and wet bulb temperature of 21.1°C is heated to 71.1°C and blown into the dryer. In the dryer, it cools along an adiabatic cooling line and leaves the dryer fully saturated. Find the dew point temperature at the initial condition, absolute humidity of initial air, percentage humidity of initial air, amount of heat needed to heat 2.8 m³/min of entering air, and temperature of the air leaving the dryer.

Solution

Data supplied

Dry bulb temperature = 26.66°C
Wet bulb temperature = 21.10°C
Heated to 71.10°C
Volume of air entering = 2.8 m³/min

From the psychrometric chart

1. Dew point temperature = 18°C
2. Humidity ratio, $H = 0.0132$ kg/kg
3. Saturated humidity, $H_s = 0.022$ kg/kg
4. Percent $H = \dfrac{H}{H_s} \times 100 = \dfrac{0.0132}{0.022} \times 100 = 60\%$
5. Relative humidity at initial condition = RH = 64%
6. Initial humid volume, $v = 0.867$ m³/kg dry air (at $H = 0.0132$)
7. Temperature of air leaving dryer = 31.8°C
8. Amount of heat needed to heat = 2.8 m³/min

$$\frac{\text{Volume of air}}{\text{Humid volume}} \times \Delta t \times \text{Humid heat}$$

Humid heat

$$s = 0.24 + 0.45H$$

$$= 0.24 + 0.45 \times 0.0132$$

$$= 0.24 + 0.0058$$

$$= 0.246 \text{ kcal/(kg °C)}$$

$$\text{Hence, heat required} = \frac{2.8}{0.867} \times (71.11 - 26.66) \times 0.246$$

$$= 35.3 \text{ kcal/min}$$

Exercises

1. The air supply for a dryer has a dry bulb temperature of 27°C and wet bulb temperature of 21°C. It is heated to 94°C and blown through the dryer. In the dryer, it cools along adiabatic line and leaves the dryer saturated. Find the following:
   a. Dew point temperature of the initial air
   b. Absolute humidity of the initial air
   c. Percentage humidity
   d. Amount of the heat needed to heat 100 m³/min entering air
   e. Temperature of the air leaving the dryer
2. Temperature and dew point of the air entering a dryer are 70°C and 26°C. What additional data can be obtained from psychrometric chart?
3. Air is heated by a heating system from 30°C, 80% RH to 60°C. Find out the relative humidity wet bulb temperature, dew point temperature of the heated air. Determine the quantity of heat added per kilogram of dry air.
4. The dry bulb and wet bulb temperatures of an air supply are 60°C and 40°C, respectively. Calculate the RH, humid volume, dew point temperature, and enthalpy at 60°C and check the answers by the psychrometric chart.
5. A grain dryer requires 300 m³/min of heated air at 45°C. The atmospheric air is at 24°C and RH is 80%. Calculate the amount of heat required per hour to raise the air temperature from 24°C to 45°C. Check the answer with the help of the psychrometric chart.

Symbols

A	Area, m²
H, H_G, H_w, H_s	Humidity, humidity at dry bulb temperature, humidity at wet bulb temperature, and saturation humidity, respectively, kg/kg
h_G	Heat transfer coefficient, kcal/(h m² °C) or kW/m² °C
h_r	Heat transfer coefficient equivalent to radiation, kcal/(h m² °C)
h_t	Enthalpy, kcal/kg or kJ/kg
K_G	Mass transfer coefficient, kgmol/(h m² atm)
P	Total pressure, atm
p_a	Partial pressure of water vapor in air, atm
p_w	Partial pressure of water vapor in air at the interface, atm
q	Rate of heat transfer, kcal/h or kJ/h
RH	Relative humidity, percent
S	Humid heat kcal/(kg °C) or kJ/(kg °C)
t_G, t_w, t_d	Dry bulb, wet bulb, and dew point temperatures, respectively, °C
t_s	Temperature of water, °C
v	Humid volume, m³/kg
λ	Latent heat of evaporation of water, kcal/kg or kJ/kg

Chapter 3

Theory of Grain Drying

Generally, the term "drying" refers to the removal of relatively small amount of moisture from a solid or nearly solid material by evaporation. Therefore, drying involves both heat and mass transfer operations simultaneously. In convective drying, the heat required for evaporating moisture from the drying product is supplied by the external drying medium, usually air. Owing to the basic differences in drying characteristics of grains in thin layer and deep bed, the whole grain drying process is divided into thin layer drying and deep bed drying.

Thin Layer Drying

Thin layer drying refers to the grain drying process in which all grains are fully exposed to the drying air under constant drying conditions, i.e., at constant air temperature and humidity. Generally, up to 20 cm thickness of grain bed (with a recommended air–grain ratio) is taken as thin layer.

All commercial flow dryers are designed on thin layer drying principles.

The process of drying should be approached from two points of view: the equilibrium relationship and the drying rate relationship.

For convenience, a few terms used in describing the drying process are defined and discussed.

Moisture Content

Usually, the moisture content of a substance is expressed in percentage by weight on wet basis. But the moisture content on dry basis (d.b.) is more simple to use in calculation as the quantity of moisture present at any time is directly proportional to the moisture content on d.b.

The moisture content, m, percent, wet basis is

$$m = \frac{W_m}{W_m + W_d} \times 100 \tag{3.1}$$

where
W_m is the weight of moisture
W_d is the weight of bone dry material

The moisture content, M, d.b., percent is

$$M = \frac{W_m}{W_d} \times 100 = \frac{m}{100 - m} \times 100 \tag{3.2}$$

The moisture content, X, d.b, is sometimes expressed in decimal also as follows:

$$X = \frac{M}{100} \tag{3.3}$$

Two additional useful equations for moisture content are given later for the following calculation:

$$\frac{W_m'}{W_1} = \frac{m_1 - m_2}{100 - m_2} = \frac{M_1 - M_2}{100 + M_1} \tag{3.3a}$$

$$\frac{W_m'}{W_2} = \frac{m_1 - m_2}{100 - m_1} = \frac{M_1 - M_2}{100 + M_2} \tag{3.3b}$$

where
W_1 is the initial weight of wet material = $(W_m + W_d)$, kg
W_2 is the final weight of dried product, kg
W_m' is the weight of moisture evaporated, kg
m_1 and m_2 are initial and final moisture contents, respectively, percent, wet basis
M_1 and M_2 are initial and final moisture contents, respectively, d.b., percent

Moisture Measurement

Moisture content can be determined by direct and indirect methods. Direct methods include air-oven drying method (130°C ± 2°C) and distillation method. Direct methods are simple and accurate but time-consuming, whereas indirect methods are convenient and quick but less accurate.

Direct Methods

The air-oven drying method can be accomplished in a single stage or double stage in accordance with the grain samples containing either less or more than 13% moisture content (Hall, 1957).

Single Stage Method

Single stage method consists of the following steps:

1. Grind 2–3 g sample.
2. Keep the sample in the oven for about 1 h at 130°C ± 2°C.
3. Place the sample in a desiccator and then weigh it after cooling.

Double Stage Method

1. In this method, keep 25–30 g whole grain sample in the air oven at 130°C ± 2°C for 14–16 h so that its moisture content is reduced to about 13%.
2. Then follow the same procedure as in single stage method.

Other Methods

Place the whole grain sample in the air oven at 100°C ± 2°C for 24–36 h depending on the type of grain and then weigh it.

The vacuum oven drying method is also used for determining the moisture content.

However, moisture determination should be made according to the standard procedure for each grain which is laid down by the Government or by the Association of Agricultural Chemists.

Brown–Duvel Distillation Method

The distillation method directly measures the volume of moisture, in cc condensed in a measuring cylinder by heating a mixture of 100 g grain and 150 cc oil in a flask at 200°C for 30–40 min.

Moisture content can be measured by the toluene distillation method also.

Indirect Methods

Indirect methods are based on the measurement of a property of the grain that depends upon moisture content.

Two indirect methods are described as follows.

Electrical Resistance Method

Resistance type moisture meter measures the electrical resistance of a measured amount of grain sample at a given compaction (bulk density) and temperature. The electrical resistance varies with moisture, temperature, and degree or compaction.

The universal moisture meter (the United States), Tag–Happenstall moisture meter (the United States), and Kett moisture meter (Japan) are some of the resistance-type moisture meters. They take only 30 s for the moisture measurement.

Dielectric Method

The dielectric properties of grain depend on its moisture content. In this type of moisture meter, 200 g grain sample is placed between the condenser plates and the capacitance is measured. The measured capacitance varies with moisture, temperature, and degree of compaction.

The Motomco moisture meter (the United States) and Burrows moisture recorder (the United States) are some of the capacitance-type of moisture meters. They take about 1 min to measure the moisture. These are also known as safe crop moisture testers as they do not damage the grain sample.

Equilibrium Moisture Content

When a solid is exposed to a continual supply of air at constant temperature and humidity, having a fixed partial pressure of the vapor, p the solid will either lose moisture by evaporation or gain moisture from the air until the vapor pressure of the moisture of the solid equals p. The solid and the gas are then in equilibrium and the moisture content of the solid in equilibrium with the surrounding conditions is known as equilibrium moisture content (EMC) (Figure 3.1). The EMC is useful to determine whether a product will gain or lose moisture under a given set of temperature and relative humidity conditions. Thus, EMC is directly related to drying and storage. Different materials have different EMCs. The EMC is dependent upon the temperature and relative humidity of the environment and on the variety and maturity of the grain. The EMC of different grains at different temperatures and humidities are given in Table A.1. A plot of the equilibrium relative humidity and moisture content of a particular material at a particular temperature (usually 25°C) is known as equilibrium moisture curve or isotherm. Grain isotherms are generally S-shaped and attributed to multimolecular adsorption.

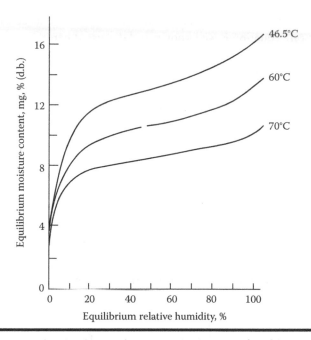

Figure 3.1   Desorption isotherms for *Patnai-23* variety of paddy. (From Kachrew, R.P. et al., *Bull. Grain Technol.*, 3(9), 186, 1971.)

Determination of EMC

Generally, EMC is determined by two methods: (1) the static method and (2) the dynamic method. In the static method, the grain is allowed to come to equilibrium with the surrounding still air without any agitation, whereas in the dynamic method, the air is generally mechanically moved. As the static method is time-consuming, at high relative humidities mold growth in the grain may take place before equilibrium is reached. The dynamic method is faster and is thus preferred. The EMC is to be determined under constant relative humidity and temperature conditions of air. Generally, a thermostat is used to control the temperature and aqueous acid or salt solutions of different concentrations are used to control the relative humidity of air.

EMC Models

A number of EMC equations, namely, BET equation (1938), Harkin and Jura's equation (1944), Henderson's equation (1952), Chung and Fost's equation (1967), etc., have been developed for different ranges of relative humidities. A few purely empirical EMC equations, namely, Haynes' equation (1961), etc., have also been proposed for different ranges of relative humidities for different cereal grains. Of them, Henderson's equation is well known and discussed here.

Henderson (1952) developed the following equation to express the equilibrium moisture curve mathematically as

$$1 - RH = \exp[-cTM_e^n] \qquad (3.4)$$

where
  RH is the equilibrium relative humidity, decimal
  M_e is EMC, d.b., percent
  T is the temperature, °K
  c and n are product constants, varying with materials

Values of c and n for some grains are given in Table A.2.

But Henderson's equation has been found to be inadequate for many cereal grains. A few useful modified forms of Henderson's equations for different cereal grains are given as follows.

$$1 - \left(\frac{p'_g}{p'_s}\right) = \exp\left[-J'M_e^{k'}\right] \qquad (3.5)$$

where
  $J' = 4.1606 \times 10^{-9}(t + 17.78)^{3.3718}$
  $k' = 11.6300(t + 17.78)^{-0.41733}$

Thompson (1965) proposed the following EMC equation for corn:

$$1 - \left(\frac{p'_g}{p'_s}\right) = \exp[-3.8195 \times 10^{-5}(1.8t + 82)M_e^2]$$

$$= \exp[-6.875 \times 10^{-5}(t + 45.55)M_e^2] \qquad (3.6)$$

Modified Henderson's equation is

$$1 - RH = [-A(t + C)M_e^B]$$

Modified Chung–Pfost's equation is

$$RH = \exp\left[\frac{-A}{t + C}\exp\left(\frac{-BM_e}{100}\right)\right]$$

Modified Halsey's equation is

$$RH = \exp[-\exp(A + Bt)M_e^{-c}]$$

Modified Smith's equation is

$$M_e = (A + Bt) - [(C + Dt)\ln(1 - RH)]$$

Modified Oswin's equation is

$$M_e = (A + Bt)\left(\frac{RH}{1 - RH}\right)^c$$

Guggenheim–Anderson–de Boer (GAB)'s equation is

$$M_e = \frac{ABC \times RH}{(1 - B \times RH)(1 - B \times RH + B \times C \times RH)} \tag{3.7}$$

$$B = B_1 \exp\left(\frac{h_1}{RT}\right); \quad C = C_1 \exp\left(\frac{h_2}{RT}\right)$$

Chung–Pfost's equation is

$$\ln[-\ln(RH)] = \ln A - BM_e$$

Halsey's equation is

$$\ln[-\ln(RH)] = A - B \ln(M_e)$$

Oswin's equation is

$$\ln(M_e) = \ln A + B\left[\ln\frac{RH}{1 - RH}\right]$$

where
  M_e is EMC, % (d.b.)
  RH is the equilibrium relative humidity decimal
  A, B, and C are constants
  t is the temperature, °C
  B_1, C_1, h_1, and h_2 are the coefficients
  T is the absolute temperature
  R is the universal gas constant

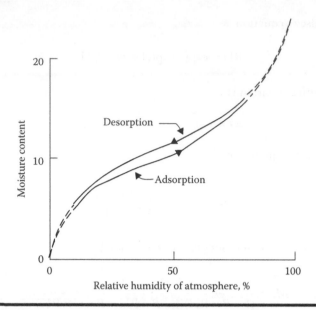

Figure 3.2   Relation between equilibrium moisture content of paddy and relative humidity showing hysteresis.

Hysteresis

Many solid materials, including cereal grains, exhibit different equilibrium moisture characteristics depending upon whether the equilibrium is reached by adsorption/ sorption or desorption of the moisture. This phenomenon is known as hysteresis, which is shown in Figure 3.2.

Bound Moisture

This refers to the moisture contained by a substance that exerts less equilibrium vapor pressure than that of the pure liquid at the same temperature (Figure 3.3). The bound moisture may be contained inside the cell walls of the plant structure, in loose chemical combination with the cellulosic material, held in small capillaries and crevasses throughout the solid.

Unbound Moisture

This refers to the moisture contained by a substance that exerts equilibrium vapor pressure equal to that of the pure liquid at the same temperature (Figure 3.3).

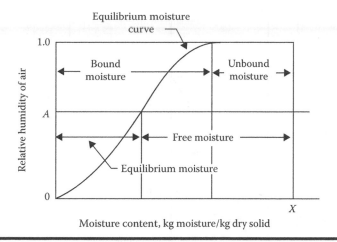

Figure 3.3 Types of moisture.

Free Moisture

Free moisture is the moisture contained by a substance in excess of the equilibrium moisture, $X - X_E$ (Figure 3.3). Only free moisture can be evaporated and the free water content of a solid depends upon the vapor concentration in the air.

The aforementioned relations are shown in Figure 3.3 for a solid of moisture content X exposed to air of relative humidity RH.

A typical drying curve is shown in Figure 3.4. This figure clearly shows that there are two major periods of drying, namely, the constant rate period and the falling rate period.

The plots of moisture content versus drying time or drying rate versus drying time or drying rate versus moisture content are known as drying curves (Figures 3.4 through 3.8).

Constant Rate Period

Some crops including cereal grains at high moisture content are dried under constant rate period at the initial period of drying. Falling rate period follows subsequently. For example, wheat is dried under constant rate period when its moisture content exceeds 72%.

In the constant rate period, the rate of evaporation under any given set of air conditions is independent of the solid and is essentially the same as the rate of evaporation from a free liquid surface under the same condition. The rate of drying during this period is dependent upon (a) difference between the temperature of air and temperature of the wetted surface at constant air velocity and relative humidity; (b) difference in humidity between air stream and wet surface at

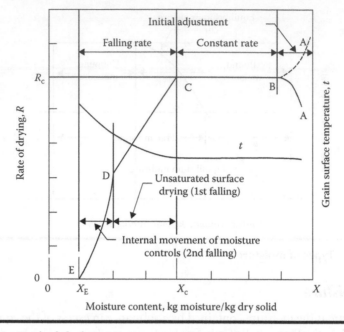

Figure 3.4 Typical drying rate curve, constant drying condition.

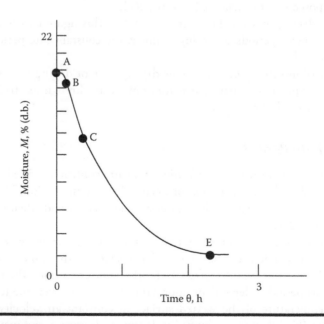

Figure 3.5 Moisture content versus drying time.

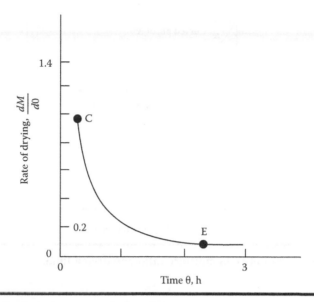

Figure 3.6   Drying rate versus drying time.

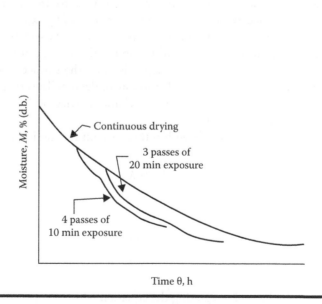

Figure 3.7   Effects of tempering on intermittent drying.

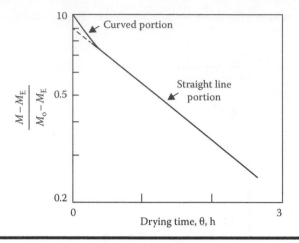

Figure 3.8 Relation between moisture ratio and drying time.

constant air velocity and temperature; and (c) air velocity at constant air temperature and humidity.

Under adiabatic and controlled drying air conditions, the temperature of the wetted surface attains the wet bulb temperature. In the constant rate period, drying takes place by surface evaporation and moisture moves by vapor pressure difference. The moisture content at which the drying rate ceases to be constant is known as the critical moisture content of the solid. The average critical moisture content, X_c, for a given type of material depends upon the surface moisture concentration, bed thickness of the material, and rate of drying. The critical moisture content of a product also depends upon the characteristics of the solid such as shape, size, and drying conditions.

If the drying takes place entirely within the constant rate period so that

$$X_1, X_2 > X_{c'}$$

then, by definition,

$$R = -\frac{W_d}{A}\frac{dX}{d\theta}$$

Separating the variables and integrating the equation within proper limits, we get

$$\text{Time of drying, } \theta_c = -\frac{W_d}{A}\left(\frac{X_1 - X_2}{R_c}\right)$$

where

W_d is the weight of dry solid, kg

A is the wet surface, m^2

X_1 is the initial moisture content, kg moisture/kg dry solid

X_2 is the final moisture content, kg/kg

X_c is the critical moisture content, kg/kg

R_c is the rate of drying in the constant rate period, kg moisture evaporated/(h m^2)

θ_c is the drying time, h

The constant drying rate of a crop can also be expressed as follows by use of wet bulb temperature theory:

$$\frac{dw}{d\theta} = 18 \ KA \ (p_s - p_a) = \frac{h_{fa} A(t_a - t_s)}{\lambda_s} \tag{3.8}$$

where

$dw/d\theta$ is the constant rate of drying, kg/h

p_a and p_s are water vapor pressures at t_a and t_s, respectively, atm

t_a and t_s are air and water temperatures, respectively, °C

h_{fa} is the film heat transfer coefficient of air at air–water interface, kcal/(h m^2 °C) or kW/(h m^2 °C)

λ_s is the latent heat of water at t_s, kcal/kg or kJ/kG

A is the water surface area, m^2

K is the water vapor transfer coefficient at the water–air interface, kg mol/(h m^2 atm)

Falling Rate Period

Cereal grains are usually dried entirely under falling rate period.

The falling rate period enters after the constant drying rate period and corresponds to the drying cycle where all surface is no longer wetted and the wetted surface continually decreases, until at the end of this period, the surface is dry. The cause of falling off in the rate of drying is due to the inability of the moisture to be conveyed from the center of the body to the surface at a rate comparable with the moisture evaporation from its surface to the surroundings.

The falling rate period is characterized by increasing temperatures both at the surface and within the solid. Furthermore, changes in air velocity have a much smaller effect than during the constant rate period. The falling rate period of drying is controlled largely by the product and is dependent upon the movement of moisture within the material from the center to the surface by liquid diffusion and the removal of moisture from the surface of the product.

The falling rate period of drying often can be divided into two stages: (1) unsaturated surface drying and (2) drying where the rate of water diffusion within the product is slow and also the controlling factor. Practically, all cereal grains are dried under falling rate period if their moisture contents are not very high.

Many theories have been proposed to describe the moisture movement phenomena in cereal grains. Of them, the following are most popular:

1. Liquid movement due to moisture concentration differences (liquid diffusion surface)
2. Liquid movement due to surface forces (capillary flow)
3. Liquid movement due to moisture diffusion in the pores (surface diffusion)
4. Vapor movement due to differences in vapor pressures (vapor diffusion)
5. Vapor movement due to temperature differences (thermal diffusion)
6. Liquid and vapor movement due to total pressure differences (hydrodynamic flow)

Drying Equations

1. On the basis of the earlier mechanisms, mathematical models have been developed (Luikov, 1966) to describe the drying of capillary porous products as follows:

$$\frac{\delta M}{\delta \theta} = \nabla^2 K_{11}M + \nabla^2 K_{12}t + \nabla^2 K_{13}P \tag{3.9}$$

$$\frac{\delta t}{\delta \theta} = \nabla^2 K_{21}M + \nabla^2 K_{22}t + \nabla^2 K_{23}P \tag{3.10}$$

$$\frac{\delta M}{\delta \theta} = \nabla^2 K_{31}M + \nabla^2 K_{32}t + \nabla^2 K_{33}P \tag{3.11}$$

where K_{11}, K_{22}, and K_{33} are all phenomenological coefficients and the other K terms are coupling coefficients that resulted from the combined effects of moisture, temperature, and pressure.

In grain drying analyses, the effects of total pressure and temperature gradients need not be considered. Therefore, the final simplified Luikov's equation will be of the form:

$$\frac{\delta M}{\delta \theta} = \nabla^2 K_{11}M \tag{3.12}$$

If it is accepted that the movement of moisture takes place by liquid or vapor diffusion, then the transfer coefficient K_{11} may be replaced by the diffusion coefficient D_v.

If D_v is taken as constant, then Equation 3.12 can be written as

$$\frac{\delta M}{\delta \theta} = D_V \left[\frac{\delta^2 M}{\delta r'^2} + \frac{c'}{r'} \frac{\delta M}{\delta r'} \right] \tag{3.13}$$

where
   $c' = 0$ for planar symmetry
   $c' = 1$ for cylindrical body
   $c' = 2$ for sphere

Under the following boundary conditions,

$$M(r',0) = M_o(\text{IMC}) \quad \text{and} \quad M(r'_o, \theta) = M_e(\text{EMC})$$

The solutions of Equation 3.13 are as follows:

$$\frac{M - M_e}{M_o - M_e} = \frac{8}{\pi^2} \sum_{n'=0}^{\infty} \frac{1}{(2n'+1)^2} \exp\left[\frac{(2n'+1)^2 \pi^2}{4} X'^2 \right] \quad \text{for infinite plane}$$

$$\tag{3.14}$$

$$\frac{M - M_e}{M_o - M_e} = \frac{6}{\pi^2} \sum_{n'=1}^{\infty} \frac{1}{n'^2} \exp\left[\frac{n'^2 \pi^2}{9} X'^2 \right] \quad \text{for sphere} \tag{3.15}$$

$$\frac{M - M_e}{M_o - M_e} = \sum_{n'=1}^{\infty} \frac{4}{\lambda_{n'}^2} \exp\left[-\frac{\lambda_{n'}}{4} X'^2 \right] \quad \text{for infinite cylinder} \tag{3.16}$$

where $\lambda_{n'}$ are the roots of the Bessel function of zero order

$$X' = \frac{A}{V} (D_V \theta)^{1/2}$$

where
   A is the surface area
   V is the volume of the body

$$\frac{A}{V} = \frac{1}{\text{half thickness}} \text{ for plane}$$

$$\frac{A}{V} = \frac{3}{\text{radius}} \text{ for sphere}$$

$$\frac{A}{V} = \frac{2}{\text{radius}} \text{ for cylinder}$$

2. If the moisture contents X_1 and X_2 are both less than X_c so that drying occurs under conditions changing R (i.e., under falling rate period), the drying time in this period may be expressed as follows.

   The rate of drying is by definition

$$R = -\frac{W_d}{A} \int \frac{dx}{d\theta}$$

Rearranging and integrating over the time interval while the moisture content changes from its initial value X_1 to its final value X_2:

$$\theta_f = \int_0^{\theta_f} d\theta \frac{W_d}{A} = \int_{x_2}^{x_1} \frac{dx}{R} \tag{3.17}$$

General case: For any shape of falling rate curve, Equation 3.17 may be integrated graphically by determining the area under a curve of $1/R$ as ordinate, X as abscissa, the data for which may be obtained from the rate of drying curve.

3. Based on Newton's equation for heating or cooling of solids, a simple drying equation can be derived as follows:

$$\text{Newton's equation } = \frac{dt}{d\theta} = -K(t - t_e)$$

If the temperature term t is replaced by the moisture term M, then

$$\frac{dM}{d\theta} = -K(M - M_e) \tag{3.18}$$

where
   M is the moisture content (d.b.), %
   θ is the time, h
   M_e is EMC (d.b.), %
   K is the drying constant, 1/h

Rearranging Equation 3.18, we get

$$\frac{dM}{M - M_e} = -Kd\theta$$

Integrating the aforementioned equation within proper limits, we get

$$\frac{M - M_e}{M_o - M_e} = \exp[-K\theta] \tag{3.19}$$

or

$$\theta = \frac{1}{K} \ln \frac{M_o - M_e}{M - M_e} \tag{3.20}$$

$\dfrac{M_o - M_e}{M - M_e}$ is known as the moisture ratio (MR)

4. The similar form of Equations 3.14 and 3.15 can be derived assuming concentration difference as the driving force in diffusion of liquid through solid:

$$\frac{\delta C}{\delta\theta} = D_V \frac{\delta^2 C}{\delta x^2} \tag{3.21}$$

where
  D_V is the diffusivity, m²/h
  C is the concentration
  θ is the time, h
  x is the distance from the center of the material, m

If the concentration term be replaced by the moisture content term, M, the aforesaid equation will be of the form:

$$\frac{\delta M}{\delta\theta} = D_V \frac{\delta^2 M}{\delta x^2} \tag{3.22}$$

The solutions of the equation are as follows:

$$\frac{M - M_e}{M_o - M_e} = \frac{8}{\pi^2} \left[\exp\left(-D_V\theta \frac{\pi^2}{4a^2}\right) + \frac{1}{9} \exp\left(-9D_V\theta \frac{\pi^2}{4a^2}\right) \right.$$

$$\left. + \frac{1}{25} \exp\left(-25D_V\theta \frac{\pi^2}{4a^2}\right) + \ldots \right] \tag{3.23}$$

for a slab of infinite length.

Where a is equal to half of the thickness of the slab,

$$\frac{M - M_e}{M_o - M_e} = \frac{6}{\pi^2}\left[\exp\left(-D_V\theta\frac{\pi^2}{r^2}\right) + \frac{1}{4}\exp\left(-4D_V\theta\frac{\pi^2}{r^2}\right) + \ldots\right] \quad \text{for a sphere}$$

(3.24)

where r is the radius of the sphere.

If MR is plotted against θ on a semilog graph paper, a curve of the type shown in Figure 3.8 is obtained. The curvature portion of Figure 3.8 results from the effects of second, third, and following terms in the series.

The equation of the straight line portion of the curve can be expressed as follows.

As θ increases, the terms other than first approach zero.

Neglecting higher terms of the equation,

$$\frac{M - M_e}{M_o - M_e} = \frac{6}{\pi^2}\left[\exp\left(-D_V\theta\frac{\pi^2}{r^2}\right)\right] = \frac{6}{\pi^2}e^{-K\theta}$$

(3.25)

i.e., $\dfrac{M - M_e}{M_o - M_e} = B\exp[-K\theta]$

where

$K = D_V\dfrac{\pi^2}{r^2}$ for a sphere

$B = 6/\pi^2$, B is the shape factor

Determination of Drying Constant

1. *Graphical method*: For straight portion of the curve shown in Figure 3.8, the drying constant, K, can be worked out easily by finding out the slope of the straight line. This method is used in example 3.

2. *Half-life period method*: If the time of one-half response in a drying process be defined as the number of hours necessary to obtain a moisture content ratio of one-half, then the drying equation

$$\frac{M - M_e}{M_o - M_e} = \exp[-K\theta] \text{ can be written as}$$

$$\frac{1}{2} = \exp[-K\theta_{1/2}] \quad \text{or} \quad \theta_{1/2}\frac{\ln 2}{K}$$

$$\text{and} \quad \frac{1}{4} = \exp[-K\theta_{1/4}] \quad \text{or} \quad \theta_{1/4}\frac{\ln 4}{K}$$

Therefore, by knowing the values of $\theta_{1/2}$ or $\theta_{1/4}$ K can be calculated.

Remarks on Thin-Layer Drying Equations

None of the theoretical equations presented in this chapter represents the drying characteristics of grains accurately over a wide range of moisture and temperature, on account of the following limitations:

1. The theoretical drying equations are based on the concept that all grains in thin layer are fully exposed to the drying air under constant drying conditions (at constant drying air temperature and humidity) and dried uniformly. Therefore, there is no gradient in a thin layer of grain that is not true for finite mass depths.
2. The grain-drying equations developed from diffusion equation are based on the incorrect assumptions that D_V and K are independent of moisture and temperature.
3. It is not possible to choose accurate boundary conditions and shape factors for drying of biological materials.
4. Drying equation developed from Newton's equation for heating or cooling does not take into account the shape of the material.

Therefore, the uses of the theoretical drying equations are limited. However, if accurate results are not desired and the values of D_V and K are known, then the theoretical drying equations can be used and give fairly good results within a limited range of moisture.

Many empirical drying equations for different cereal grains are found to be useful and frequently used as they give more accurate results in predicting drying characteristics of a particular grain for a certain range of moisture, temperature, airflow rate, and relative humidity. A few empirical drying equations are presented later.

The following equations are for wheat:

$$MR = 1 - 8.78(D_V\theta)^{1/2} + 13.22(D_V\theta) \quad \text{for } (D_V\theta)^{1/2} < 0.0104 \quad (3.26)$$

$$MR = 0.509 \times \exp[-58.4\, D_V\theta] \quad \text{for } (D_V\theta)^{1/2} \geq 0.0104 \quad (3.27)$$

where $D_V = 7.135e^{-19944/T}$.

$$(3.28)$$

$$D_V = m^2/h, \quad \theta = h, \quad \text{and} \quad T = {}^\circ K$$

Based on drying equation for planar symmetry, the following expression is developed for diffusivity for thin layer drying of corn:

$$D_{Vcom} = 5.853 \times 10^{-10} \exp[-12,502/T] \quad (3.29)$$

On the basis of drying equation for sphere, the following expression for drying constant, K_{com} has been developed:

$$K_{com} = 5.4 \times 10^{-1} \exp[-9041/T] \tag{3.30}$$

where
  $K = 1/s$
  $T = °K$

Drying Models

Generally moisture content, M, data at any thin layer convection drying time, t, for different drying air temperature, T and initial moisture content, M_o, conditions are converted to MR, in decimal and fitted against the drying time. Some useful convection drying models are presented here for convenience.

Model	Model Name	Reference
$MR = \exp(-Kt)$	Newton	Liu and Bakker-Arkema (1997)
$MR = \exp(-Kt^n)$	Page	Doymaz (2004)
$MR = a \exp(-Kt)$	Henderson and Pabis	Pal and Chakraverty (1997)
$MR = a \exp(-Kt) + c$	Logarithmic	
$MR = a \exp(-Kt) + b \exp(-k_1t)$	Two-term	Turkan et al. (2007)
$MR = 1 + at + bt^2$	Wang and Singh	Wang and Singh (1978)
$MR = a \exp(-K_1t) + b \exp(-gt) + c \exp(-ht)$	Modified Henderson and Pabis	Togrul and Pehlivan (2003)

$MR = (M - M/M_o - M_e)$, decimal; M_o is initial moisture content, kg water/kg dry, matter, t is the drying time; M is moisture content at any during time, kg water/ kg dry matter.
k, k_1, g, and h are drying constants, 1/min or 1/h.
a, b, c, and n are coefficients; t is the drying time, min or h, and T is the drying air temperature, °C.

Effects of Different Factors on the Drying Process

The drying rate is dependent upon many factors, namely, air, temperature, airflow rate, relative humidity, exposure time, types, variety and size of the grain, initial moisture content grain depth, etc. Of them, the first four factors are important drying process variables that have been discussed later. The effects of some of the factors are shown in Figures 3.9 through 3.11.

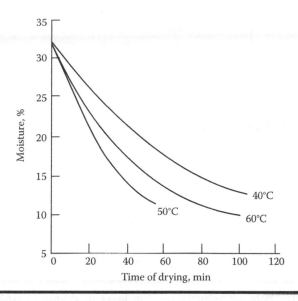

Figure 3.9 **Effects of air temperatures on drying characteristics of parboiled paddy. (From Bhattacharya, K.R. and Indudharaswamy, Y.M.,** *Cereal Chem.,* **44(6), 592, 1967.)**

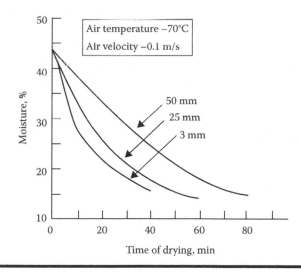

Figure 3.10 **Effects of grain thickness on thin layer drying of wheat. (From Gerzhoi, A.P. and Samochetov, V.F.,** *Grain Drying and Grain Dryers,* **3rd review and enl. edn., Ginzburg, A.S. (Ed.), Khieboizdat, Moscow, Russia, 1958.)**

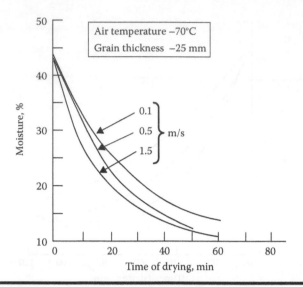

Figure 3.11 **Effects of air velocity on thin layer of drying of wet corn. (From Gerzhoi, A.P. and Samochetov, V.F., *Grain Drying and Grain Dryers*, 3rd review and enl. edn., Ginzburg, A.S. (Ed.), Khieboizdat, Moscow, Russia, 1958.)**

Effects of Air Temperature

Simmonds et al. (1953a) showed that the rate of drying of wheat was sharply dependent upon the temperature of air varying from 21°C to 77°C. The rate of drying increases with the rise of air temperature. But the EMC falls as air temperature increases. These observations are true for other cereal grains also.

The effects of air temperature on the quality of grains have been discussed in Chapter 1.

Effects of Air Velocity

It is generally assumed that the internal resistance to moisture movement of agricultural materials is so great when compared to the surface mass transfer resistance that the air rate past the particles has no significant effect on the time of drying or on the drying coefficient. Henderson and Pabis (1961, 1962) found that air rate had no observable effect on thin layer drying of wheat when airflow was turbulent. According to them, airflow rate varying from 10 to 68 cm^3/s/cm^2 had no significant effect on the drying rate of wheat. But in cases of paddy and corn, it has been found that air rate has some effect on rate of drying.

However, the recommended airflow rates per unit mass of different grains are given in Table A.7.

Effects of Air Humidity

When the humidity of the air increases, the rate of drying decreases slightly. The effect, however, is much smaller in comparison to the effect of temperature changes.

Effects of Air Exposure Time

In the case of intermittent drying, drying rate of grain depends on its exposure time to the drying air in each pass. Total drying time, which is the sum of all exposure times, is dependent upon exposure time. Total drying time reduces as exposure time decreases (Chakraverty, 1975).

Deep Bed Drying

In deep bed drying, all the grains in the dryer are not fully exposed to the same condition of drying air. The condition of drying air at any point in the grain mass changes with time and at any time it also changes with the depth of the grain bed. Over and above, the rate of airflow per unit mass of grain is small compared to the thin layer drying of grain. All on-farm static bed batch dryers are designed on deep bed drying principle. The condition of drying in deep bed is shown in Figure 3.12.

The drying of grain in a deep bin can be taken as the sum of several thin layers. The humidity and temperature of air entering and leaving each layer vary with time depending upon the stage of drying and moisture removed from the dry layer until the EMC is reached. Little moisture is removed; rather a small amount may be added to the wet zone until the drying zone reaches it. The volume of drying zone varies

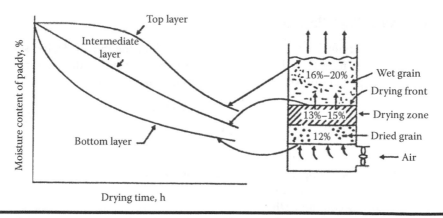

Figure 3.12   Deep bed drying characteristics at different depths.

with the temperature and humidity of entering air, the moisture content of grain, and the velocity of air movement. Drying will cease as soon as the product comes in equilibrium with the air.

Time of Advance of Drying Front

The time period taken by the drying front to reach the top of the bin is called the maximum drying rate period.

The time taken by the drying front to reach the top of the bed can be calculated by the following equation:

$$\frac{W_d(M_1 - M_X)}{100} = AG(H_s - H_1)\,\theta_1$$

or

$$\theta_1 = \frac{W_d(M_1 - M_X)}{AG\,(H_s - H_1)\times 100} \tag{3.31}$$

where

M_1 is the initial moisture content of grain (d.b.), %

M_X is the average moisture content (d.b.), % at the end of drying front advance at the top

θ_1 is the time of advance, h

A is the cross-sectional area of the dryer, through which air passes, m^2

G is the mass flow rate of dry air, kg/h m^2

H_s is the humidity of the saturated air leaving the dryer, kg/kg

H_1 is the humidity of the air entering into the dryer, kg/kg

W_d is the weight of dry grain in the bin, kg

Decreasing Rate Period

As soon as the drying front reaches the top of the bin, the rate of drying starts decreasing and is termed as decreasing rate period. The time of drying for this decreasing rate period can be expressed by the following equation:

$$\theta_2 = \left(\frac{1}{K}\right)\ln\left(\frac{M_X - M_e}{M - M_e}\right) \tag{3.32}$$

where

θ_2 is the time of drying during decreasing rate period, h

M_e is EMC of the grain (d.b.)

K is the drying constant, 1/h

M is the average moisture content (d.b.) at the end of decreasing rate period

M_X is the average initial moisture content (d.b.) at the beginning of decreasing rate period

The total drying time for grains in the bin is the sum of the time required for the maximum drying rate and decreasing rate periods:

$$\text{Total drying time, } \theta = \theta_1 + \theta_2$$

Deep bed drying problems can be solved by Hukill's analysis also (Hall, 1957). The effects of air temperature, air velocity, bed depth, and initial moisture content of grain on deep bed drying characteristics of grain are shown in Figures 3.13 through 3.16.

From these figures, the following general observations can be made in regard to deep bed drying:

1. The rate of drying of a grain bed increases with an increase in drying air temperature (Figure 3.13)
2. The bottom layer of a static grain bed is dried more than the top layer. The moisture gradient across the grain bed increases as the inlet drying air temperature increases (Figure 3.13).
3. The rate of drying of a grain bed is increased with an increase in airflow rate (Figure 3.14).
4. The drying zone passes more quickly through a grain bed if the initial moisture content of grain is low (Figure 3.15).
5. Moisture gradient across the grain bed increases with an increase in bed depth (Figure 3.16).

Various deep bed drying models for grain have been discussed in detail by Brooker et al. (1992) and Hall (1957, 1980).

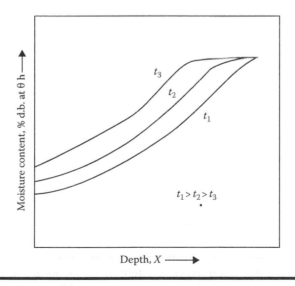

Figure 3.13   Effect of drying air temperature *t* on the moisture content distribution within a fixed bed of grain after a drying period θ.

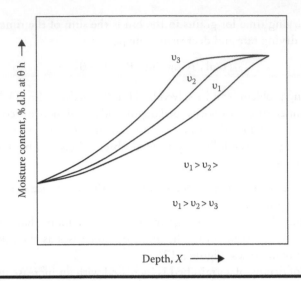

Figure 3.14   Effects of airflow rate υ on the moisture content distribution within a fixed bed of grain after a drying period θ.

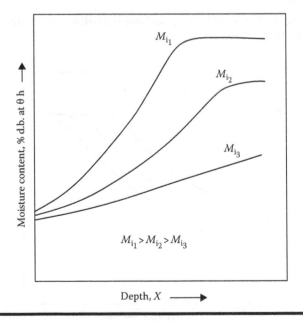

Figure 3.15   Effect of initial moisture content M_i on the moisture content distribution within a fixed bed of grain after a drying period θ.

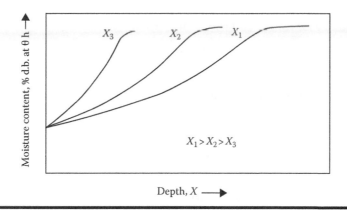

Figure 3.16 **Effect of bed depth X on the moisture content distribution within a fixed bed of grain after a drying period θ.**

Remarks on Deep Bed Drying

1. If drying air is at high relative humidity and relatively low temperature is used, then the total drying time will be very long due to slow rate of drying, which may cause spoilage of grains.
2. The correct choice of airflow rate is very important.
3. Drying air at high temperature cannot be used due to development of moisture gradients within the grain bed. It leads to nonuniform drying of grain. In general, an air temperature of 40°C (15°C rise) is recommended for deep bed drying.

Mass and Heat Balance in Grain Drying

Mass Balance

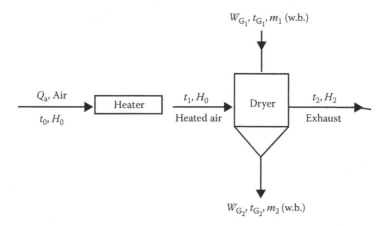

$$\text{Let } W = W_{G_1} - W_{G_2}$$
$$\text{or} \quad W_{G_1} = W_{G_2} + W \tag{3.33}$$

Amount of dry material entering into the dryer

$$= \frac{100 - m_1}{100} W_{G_1}, \text{ kg/h}$$

Amount of dry material leaving the dryer

$$= \frac{100 - m_2}{100} W_{G_2}, \text{ kg/h}$$

$$\text{But } \frac{100 - m_1}{100} W_{G_1} = \frac{100 - m_2}{100} \times W_{G_2} \tag{3.34}$$

$$W_{G_2} = W_{G_1} \times \frac{100 - m_1}{100 - m_2}$$

But inserting Equation 3.34 in Equation 3.33

$$W = W_{G_1} = W_{G_1} \times \frac{100 - m_1}{100 - m_2} = W_{G_1} \left(\frac{m_1 - m_2}{100 - m_2} \right)$$

Similarly,

$$W = W_{G_1} \times \frac{m_1 - m_2}{100 - m_2} = W_{G_1} \left(\frac{m_1 - m_2}{100 - m_2} \right) = W_{G_2} \times \frac{m_1 - m_2}{100 - m_1} \tag{3.35}$$

If f_d, the dryer factor be defined as follows:

$$f_d = \frac{W}{W_{G_1}} \times 100 = \frac{W_{G_1} - W_{G_2}}{W_{G_1}} \times 100 = \left(\frac{m_1 - m_2}{100 - m_2} \right) \times 100 \tag{3.36}$$

Heat Balance

$$W_d \times CP_G \times (t_{G_2} - t_{G_1}) + W_d \times (X_1) \times Cp_w(t_{G_2} - t_{G_1}) + W_d \times (X_1 - X_2)\lambda$$

Sensible heat   Latent heat

$$= Q_a \times (0.24 + 0.45H_o) \times (t_1 - t_2) \times \theta = Q_a(1.005 + 1.88H) \times (t_1 - t_2) \times \theta \quad (3.37)$$

where
  W_d is d.b. materials, kg
  Q_a is the air rate, kg/h
  θ is the time, h
  X_1 and X_2 the moisture contents (d.b.), decimal

Dryer Performance

Dryer performance can be expressed in terms of various efficiency factors, which are given next.

Thermal Efficiency

Thermal efficiency can be defined as the ratio of the latent heat of evaporation credited to the heat energy of the fuel charged. Thermal efficiency can be expressed mathematically as follows:

$$\frac{(dM/d\theta)W_d\lambda}{q} \quad (3.38)$$

where

$$q = \frac{60VA}{\upsilon}(h_1 - h_o) \quad (3.39)$$

$$\frac{dM}{d\theta} = \text{drying rate, kg/(h kg)}$$

  W_d is the weight of dry material, kg
  λ is the latent heat of evaporation, kcal/kg, or kJ/kg
  q is the rate of heat flow, kcal/h, or kJ/h
  V is the air rate, m³/(min m²)
  A is the area, m²
  υ is the humid volume of air (at the point of rate measurement), m³/kg
  h_1 and h_o are enthalpies of drying and ambient air, kcal/kg, or kJ/kg

Heat Utilization Factor

Heat utilization factor (HUF) may be defined as the ratio of temperature decrease due to cooling of the air during drying and the temperature increase due to heating of air.

$$HUF = \frac{\text{Air temperature decrease during drying}}{\text{Air temperature increase during heating}}$$

$$HUF = \frac{\text{Heat utilizer}}{\text{Heat supplied}} = \frac{t_1 - t_2}{t_1 - t_0} \qquad (3.40)$$

HUF may be more than unity under certain drying conditions.

Coefficient of Performance

The coefficient of performance (COP) of a grain dryer is expressed mathematically as follows:

$$COP = \frac{t_2 - t_0}{t_1 - t_0} \qquad (3.41)$$

where
t_2 is the dry bulb temperature of exhaust air, °C
t_0 is the dry bulb temperature of ambient air, °C
t_1 is the dry bulb temperature of drying air, °C

Relationship between HUE and COP

$$HUE = 1 - COP \qquad (3.42)$$

Effective Heat Efficiency

The effective heat efficiency (EHE) is mathematically defined as follows:

$$EHE = \frac{t_1 - t_2}{t_1 - t_{w_1}} \qquad (3.43)$$

where t_{w_1} is wet bulb temperature of drying air, °C.

EHE considers the sensible heat in drying air as being the effective heat for drying.

Problems on Moisture Content and Drying

Solved Problems on Moisture Content

1. Two tons of paddy with 22% moisture content on wet basis is to be dried to 13% moisture content on d.b. Calculate the weight of bone dry products and water evaporated.

   Solution

   Weight of bone dry sample

   $$= 2000 - \frac{2000 \times 22}{100}$$

   $$= 1560 \text{ kg}$$

   Moisture content on d.b. for 22% moisture on wet basis

   $$= \frac{22}{100 - 22} \times 100$$

   $$= 28.2\% \text{ (d.b.)}$$

   Therefore, water evaporated

   $$= 1560 \times (0.282 - 0.13)$$

   $$= 237.2 \text{ kg}$$

   Amount of dried product

   $$= 2000 - 237.2$$

   $$= 1762.8 \text{ kg}$$

2. Determine the quantity of parboiled paddy with 40% moisture content on wet basis required to produce 1 ton of product with 12% moisture content on wet basis. Work out the problem on wet basis and check the answer using d.b.

   Solution

   On wet basis: Weight of paddy with 12% moisture on wet basis = 1 ton.
   Weight of bone dry paddy

   $$= 1 - \frac{12 \times 1}{100}$$

   $$= 0.88 \text{ ton}$$

Let x be the amount of water present in the paddy with 40% moisture content.

Therefore,

$$\frac{x}{0.88 + x} \times 100 = 40$$

$$x = \frac{40 \times 0.88}{60} = 0.587 \text{ ton}$$

Therefore, quantity of paddy with 40% moisture content on wet basis

$$= 0.587 + 0.88$$

$$= 1.467 \text{ ton}$$

On dry basis: 40% m.c. (w.b.) = 66.66% (d.b.)
Similarly, 12% m.c. (w.b.) = 13.65% (d.b.)
Amount of moisture evaporated

$$= 0.88 \left(\frac{66.66 - 13.65}{100} \right)$$

$$= 0.467 \text{ ton}$$

Total weight of paddy should be 1 + 0.467 = 1.467 ton.

3. Determine the values of c and n from the Henderson's equation for the following data obtained from thin layer paddy drying studies:

$$\text{(i) RH} = 30\%, \quad t = 50°C, \quad M_e = 10.5\%$$

$$\text{(ii) RH} = 55\%, \quad t = 50°C, \quad M_e = 15.5\%$$

Solution

Henderson's equation is expressed as

$$1 - \text{RH} = \exp[-cTM_e^R]$$

Putting condition (i) in Henderson's equation, we get

$$1 - 0.3 = \exp[-c(50 + 273)\,(10.5)^n]$$

$$0.7 = \exp[-c \times 323 \times (10.5)^n]$$

$$e^{-0.357} = \exp[-c \times 323 \times (10.5)^n] \qquad (1)$$

$$0.357 = c \times 323 \times (10.5)^n$$

Substituting condition (ii) in Henderson's equation, we get

$$1 - 0.55 = \exp[-c(50 + 273)\,(15.5)^n]$$

$$\text{or} \quad 0.45 = \exp[-c \times 323 \times (15.5)^n]$$

$$\text{or} \quad e^{-0.796} = \exp[-c \times 323 \times (15.5)^n]$$

$$\text{or} \quad 0.796 = c \times 323 \times (15.5)^n$$

(2)

Dividing Equation 2 by Equation 1,

$$\frac{0.796}{0.357} = \frac{c \times 323 \times (15.5)^n}{c \times 323 \times (10.5)^n}$$

$$\text{or} \quad 2.23 = \left(\frac{15.5}{10.5}\right)^n$$

$$\text{or} \quad (1.475)^n = 2.23$$

Therefore, $n = 2.07$.

Substituting the value of n in Equation 1,

$$0.357 = c \times 323\,(10.5)^{2.07}$$

$$\text{or} \quad c = \frac{0.357}{323 \times 130}$$

Therefore, $c = 8.5 \times 10^{-6}$.

Exercises

1. Calculate the amount of moisture evaporated from 100 kg of grain for drying it from an initial moisture content of 25% to a final moisture content of 13% on wet basis.
2. Draw a graph showing moisture content of grain on wet basis versus moisture content on d.b. Take moisture content of grain on wet basis from 10% to 60% at equal intervals of 5%.
3. One thousand kilograms of parboiled paddy is to be dried from 32% to 13% moisture content (w.b.). Calculate the amount of moisture to be evaporated.
4. Determine the EMC of sorghum at RH = 10% and t = 60°C using Henderson's equation, where $c = 6.12 \times 10^{-6}$ and $n = 2.31$.
5. Determine the values of c and n from Henderson's equation for the following data:
   (a) RH = 40%, t = 60°C, M_e = 8.65%
   (b) RH = 80%, t = 60°C, M_e = 14.62%

Solved Problems on Drying

1. In an experiment on drying of raw paddy at an air temperature of 55°C, the following data were obtained. Initial weight of the sample = 1000 g.
   Initial moisture content = 30.8% (d.b.)

S. No	Drying Time (min)	Moisture Removed (g)
1	0	0.0
2	10	22.9
3	20	38.2
4	40	57.4
5	60	68.8
6	80	78.0
7	100	84.0
8	140	101.0
9	180	112.5
10	220	121.0
11	260	128.4
12	300	131.8

Prepare a drying rate curve for the experiment.

Solution

IMC = 30.8% (d.b.)

Therefore, the amount of bone dry material in the sample

$$= \frac{100}{130.8} \times 1000 = 764.5 \text{ g}$$

The amount of moisture present in the sample = 235.5 g.
   After 10 min, 22.9 g of water was removed.
   Therefore, moisture content of the sample after 10 min

$$= \frac{235.5 - 229}{764.5} \times 100$$

$$= 27.8\% \text{ (d.b.)}$$

Drying rate, R in gram of water per minute per 100 g of bone dry material is expressed as follows:

$$R = \frac{\text{Amount of moisture removed}}{\text{Time taken} \times \left(\dfrac{\text{Total bone dry weight of sample in gram}}{100} \right)}$$

For example, for the second reading,

$$R = \frac{229}{10 \times \left(\dfrac{764.5}{100} \right)}$$

$$= 0.300 \frac{\text{g of water}}{\text{minimum 100 g of b.d. material}}$$

Similar calculations are made for all readings and the following table is prepared.

S. No.	Drying Time (min)	Moisture Removed (g)	Moisture Present in the Sample (g)	Moisture Content (d.b.) (%)	Average Moisture Content (d.b.) (%)	Drying Rate R (Gram of Water/min) 100 g of d.b. Materials)
1	0	0.0	235.5	30.8		
2	10	22.9	212.6	27.8	29.30	0.299
3	20	38.2	1973	25.80	26.80	0.249
4	40	57.8	178.10	23.29	24.545	0.189
5	60	68.8	166.70	21.80	22.545	0.150
6	80	78.0	157.5	20.60	11.20	0.128
7	100	84.0	151.5	19.81	20.205	0.110
8	140	101.0	134.5	17.59	18.70	0.944
9	180	112.5	123.0	16.08	16.835	0.081
10	220	121.0	114.5	14.97	15.525s	0.072
11	260	128.4	107.1	14.0	14.485	0.065
12	300	131.8	103.7	13.58	13.79	0.054

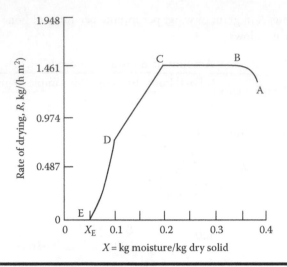

Figure 3.17   Drying curve (for Problem 2).

The drying rate curve is obtained by plotting the rate of drying against the average moisture content.

2. The drying curve for a batch of solid dried from 25% to 6% moisture (w.b.) is shown in Figure 3.17. The initial weight of solid is 159 kg and drying surface is 1 m²/39 dry material. Determine the time for drying.

Solution

The time required for drying up to "C" or time for constant rate period,

$$\theta_0 = \frac{W_d}{A}\left(\frac{X_1 - X_0}{R_0}\right)$$

$$\frac{W_d}{A} = \frac{1}{1/39}\,\text{kg/m}^2$$

$$= 39\,\text{kg/m}^2$$

$$X_1 = \frac{25}{100 - 25} = \frac{25}{75} = 0.333\,\text{kg/kg dry material}$$

$X_0 = 0.2$ (from Figure 3.17)

$$R_0 = 1.461 \text{ kg/(m}^2) \text{ (h) (from Figure 3.17)}$$

$$\theta_0 = 39 \times \frac{0.333 - 0.2}{1.461} = 3.55 \text{ h}$$

Therefore, $\theta_0 = 3.55$ h

for falling rate period
Time required for drying can be calculated by two methods:

1. Approximate method by assuming straight line
2. Graphical method

1. By assuming a straight line

$$\theta_f = \frac{W_d}{A} \times \frac{X_0 - X_2}{R_m}$$

where

$$R_m = \frac{R_0 - R_2}{\ln(R_0/R_2)}$$

$R_0 = 1.461$ kg/(m^2) (h) (from Figure 3.17)

$$X_2 = \frac{6}{94} = 0.0637 \text{ kg/kg of dry material}$$

$R_2 = 0.075$ kg/m^2 h (from Figure 3.17)

$$R_m = \frac{1.461 - 0.075}{2.3 \log \dfrac{1.461}{0.075}} = \frac{1.386}{23 \log 19.5} = \frac{1.386}{23 \times 1.29}$$

$$= 0.466 \text{ kg/h m}^2$$

$$\theta_f = 39 \frac{0.200 - 0.0637}{0.466}$$

$$= 39 \times \frac{0.1363}{0.466} = 11.4 \text{ h}$$

$$\theta_f = 11.4 \text{ h}$$

2. In a graphical solution, divide the falling rate curve in two parts:
   (i) Unsaturated surface drying
   (ii) Drying under control of internal movement of moisture
   In unsaturated surface drying, the falling rate curve is a straight line; therefore, the aforesaid method can be used up to the point "D"

$$\theta_{f_1} = \frac{W_d}{A} \frac{X_c - X_d}{R_{md}}$$

where

$$R_{md} = \frac{R_c - R_d}{\ln \dfrac{R_c}{R_d}}$$

From the curve,

$X_d = 0.1$ kg/kg
$X_c = 0.2$ kg/kg
$R_c = 1.461$ kg/h m^2
$R_d = 0.73$ kg/h m^2

$$R_{md} = \frac{1.461 - 0.73}{2.3 \log_{10} \dfrac{1.461}{0.73}} = \frac{0.731}{2.3 \log_{10} 2} = \frac{0.731}{2.3 \times 0.3010}$$

$$= 1.056 \text{ kg/h m}^2$$

$$\theta_{f_1} = 39 \times \frac{0.2 - 0.1}{1.056} = 3.69 \text{ h}$$

θ_{f_1} is the time for falling rate period for unsaturated surface drying = 3.69 h.
   For the second part in which internal movement of moisture controls, a curve is plotted relating $1/R$ and X.

$$\text{The area under that curve} = \int \frac{dX}{R}$$

X	0.1000	0.0950	0.0870	0.0785	0.0700	0.0637
R	0.73	0.568	0.460	0.290	0.180	0.075
$\dfrac{1}{R}$	1.37	1.76	2.195	3.450	5.55	13.35

$$\text{Area under the curve} = 1 \times 60 + \frac{1}{100} \times 1371$$

$$= 73.71$$

Area of each square = 0.001667 m² h/kg

$$\text{Total area,} \int \frac{dX}{R} = 0.001667 \times 73.71$$

$$= 0.1228 \text{ m}^2 \text{ h/kg}$$

Therefore, time required for drying the second part

$$\theta_{f_2} = \frac{W_d}{A} \int \frac{dX}{R} = 39 \times 0.1288$$

$$= 4.78 \text{ h}$$

Total time required for drying in the falling rate period

$$\theta_{f_1} = \theta_{f_1} + \theta_{f_2} = 3.69 + 4.78 = 8.47 \text{ h}$$

Therefore, total time required for the entire period of drying = time for constant rate period + time for falling rate period

$$\theta = \theta_c + \theta_f = 3.55 + 8.47$$

$$= 12.02 \text{ h}$$

3. In an experiment on thin layer drying of parboiled wheat at a drying air temperature of 75°C, the following drying data were obtained.

Drying Time (min)	0	10	20	30	40	50	70	90	110	130	150	180	210
MC (d.b.)	88.0	59.66	45.02	33.59	23.35	19.86	12.54	8.42	6.59	5.45	4.996	4.99	4.86

The EMC of parboiled wheat at 70°C was found to be 4.75% (d.b.). Find out the drying constant, K, by graphical method.

Solution

Using the aforementioned drying data, the following table is prepared.

Drying Time (min)	0	10	20	30	40	50	70	90	110
$(M - M_e)/(M_o - M_e)$	1.0	0.702	0.515	0.369	0.263	0.193	0.0996	0.046	0.024

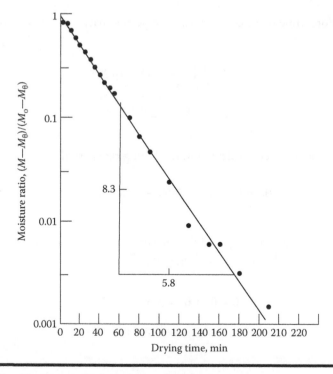

Figure 3.18 Graph to calculate drying constant.

Drying Time (min)	130	150	180	210
$(M - M_e)/(M_o - M_e)$	0.00895	0.00575	0.00807	0.0014

The MR $(M - M_e)/(M_o - M_e)$ is plotted against drying time on a semilog graph paper (Figure 3.18). From the graph, the slope of straight line, K, is calculated as follows:

$$K = -\left(\frac{8.3}{5.8} \times \frac{1}{0.432 \times 100}\right) = -0.03313 \text{ min}^{-1}$$

Exercises

1. The following data were obtained on the tray drying of sand. Obtain the drying rate curve for the test:
   (i)   Area of tray = 0.20 m^2
   (ii) Thickness of bed = 2.54 cm
   (iii) Weight of dry sand = 10.0 kg

S. No.	Drying Time (h)	Total Moisture Present in the Sample (kg)
1	0.00	0.050
2	0.50	1.835
3	1.50	1.410
4	2.00	1.215
5	2.50	1.015
6	3.00	0.811
7	4.00	0.430
8	4.50	0.272
9	5.00	0.163
10	7.00	0.009
11	7.50	0.000

2. In an experiment on thin layer drying of parboiled wheat with drying air at 85°C, the following data were obtained.

   The initial and EMCs were found to be 83% (d.b.) and 4.50% (d.b.), respectively.

Time (min)	Weight of the Sample during Drying (g)
0	200.0
15	159.0
30	138.0
45	128.0
60	122.0
90	117.0
130	114.5
210	114.25

Determine the drying constant, K, by graphical method.

Symbols

A	area, m²
a	half of the thickness of the slab, cm
B	shape factor, dimensionless
C	concentration
c	a constant, dimensionless
c'	0 or 1 or 2
Cp_g	specific heat of grain, kcal/(kg °C) or kJ/(kg k)
Cp_w	specific heat of water, kcal/(kg °C) or kJ/(kg k)
d.b.	dry basis
d.b.	dry bulb temperature, °C
d_c	diameter, mm
D_V	diffusivity, m²/h
G	mass flow rate of base dry air, kg/(h m²)
H	humidity, kg/kg
H_0	humidity of atmospheric air, kg/kg
H_2	humidity of drying air, kg/kg
H_2	humidity of exhaust air, kg/kg
H_a	humidity at t_a, °C
H_s	humidity of saturated air, kg/kg
h_{fa}	film heat transfer coefficient, kcal/(m²h °C) or kW/(m² °C)

h	enthalpy, kcal/kg, or kJ/kg
h_o	enthalpy of atmospheric air, kcal/kg, or kJ/kg
h_1	enthalpy of drying air, kcal/kg, or kJ/kg
J'	a constant in modified Henderson's equation
K	drying constant, 1/h
K'	a constant in modified Henderson's equation
K	water vapor transfer coefficient, kg/(h m² atm)
K_{11}, K_{12}, and K_{13}	phenomenological coefficients in Luikov's equation
K_{12}, K_{13}, K_{21}, K_{23}, K_{31}, and K_{32}	coupling coefficients in Luikov's equation
M	moisture content, d.b., percent
M_1	initial moisture content of grain, IMC, d.b., percent
M_2	final moisture content of grain, FMC, d.b., percent
M_E	equilibrium moisture content, EMC, d.b., percent
MR	moisture ratio $(M - M_e)/(M_o - M_e)$
M_X	average moisture content at the end of maximum period of drying
n	a constant dimensionless
n'	whole numbers
p_a	partial pressure of water vapor at t_a, atm
p_s	partial pressure of water vapor at t_s, atm
p_g'	water vapor pressure of the grain, kg/cm²
p_s'	saturated water vapor pressure at equilibrium temperature of the system, kg/cm²
Q_a	rate of air supply, kg/h
q	heat flow rate, kcal/h
R	rate of drying kg/(h m²)
R_c	constant rate period drying, kg/(h m²)
r	radius of the sphere, m
r_0'	radius of a body, m
r''	coordinate of a body, m
RH	relative humidity, decimal
T	temperature, °K
t_0	temperature of the inlet or ambient air, °C
t_1	temperature of the drying air (heated air), °C
t_2	temperature of the exhaust air, °C
t_a	temperature of the air, °C
t_s	temperature of the saturated air, °C
t_{w0}	wet bulb temperature of the ambient air, °C

t_{w1}	wet bulb temperature of the drying air, °C
V	volumetric airflow rate, $m^3/(min\ m^2)$
υ	humid volume, m^3/kg
W	grain flow rate at the inlet condition, kg/h
W_d	weight of bone dry material, kg
W_m	weight of moisture, kg
W_{G1}	grain flow rate at the outlet condition, kg/h
w	moisture removed, kg
w.b.	wet bulb temperature, °C
X	moisture content, d.b., decimal
X_1	initial moisture content, d.b., decimal
X_2	final moisture content, d.b., decimal
X_C	critical moisture content, d.b., decimal
X_E	equilibrium moisture content, d.b., decimal
X'	dimensionless quantity
x	distance from the center, m
θ	time, h
θ_1	drying time for maximum rate period in deep bed drying, h
θ_2	drying time for decreasing rate period in deep bed drying, h
θ_c	drying time for constant rate period, h
θ_f	drying time for falling rate period, h
λ	latent heat of vaporization, kcal/kg, or kJ/kg
$\lambda_{n'}$	roots of Bessel function
λ_s	latent heat at t_s, kcal/kg, or kJ/kg

Chapter 4

Methods of Grain Drying

So far, drying systems have not been classified systematically. However, drying methods can be broadly classified on the basis of either the mode of heat transfer to the wet solid or the handling characteristics and physical properties of the wet material. The first method of classification reveals differences in dryer design and operation, while the second method is most useful in the selection of a group of dryers for preliminary consideration in a given drying problem.

According to the mode of heat transfer, drying methods can be divided into (a) conduction drying; (b) convection drying; and (c) radiation drying. There are other methods of drying also, namely, dielectric drying, chemical or sorption drying, vacuum drying, freeze-drying, etc.

Of them, convection drying is commonly used for drying of all types of grain and conduction drying can be employed for drying of parboiled grain.

Conduction Drying

When the heat for drying is transferred to the wet solid mainly by conduction through a solid surface (usually metallic), the phenomenon is known as conduction or contact drying. In this method, conduction is the principal mode of heat transfer and the vaporized moisture is removed independently of the heating media. Conduction drying is characterized by the following:

1. Heat transfer to the wet solid takes place by conduction through a solid surface, usually metallic. The source of heat may be hot water, steam, flue gases, hot oil, etc.
2. Surface temperatures may vary widely.

3. Contact dryers can be operated under low pressure and in inert atmosphere.
4. Dust and dusty materials can be removed very effectively.
5. When agitated, a more uniform dried product along with increased drying rate is achieved by using conduction drying. Conduction drying can be carried out either continuously or batch-wise. Cylinder dryers, drum dryers, and steam tube rotary dryers are some of the continuous conduction *dryers*. Vacuum tray dryers, freeze-dryers, and agitated pan dryers are examples of batch conduction dryers.

Convection Drying

In convection drying, the drying agent (hot gases) in contact with the wet solid is used to supply heat and carry away the vaporized moisture, and the heat is transferred to the wet solid mainly by convection. The characteristics of convection drying are as follows:

1. Drying is dependent upon the heat transfer from the drying agent to the wet material, the former being the carrier of vaporized moisture.
2. Steam heated air, direct flue gases of agricultural waste, etc., can be used as drying agents.
3. Drying temperature varies widely.
4. At gas temperatures below the boiling point, the vapor content of the gas affects the drying rate and the final moisture content of the solid.
5. If the atmospheric humidities are high, natural air drying needs dehumidification.
6. Fuel consumption per kilogram of moisture evaporated is always higher than that of conduction drying.

Convection drying is most popular in grain drying. It can be carried out either continuously or batch-wise. Continuous tray dryers, continuous sheeting dryers, pneumatic conveying dryers, rotary dryers, and tunnel dryers come under the continuous system, whereas tray and compartment dryers and batch through circulation dryers are batch dryers.

Convection drying can be further classified as follows:

Pneumatic of fluidized bed drying: When the hot gas (drying agent) is supplied at a velocity higher than the terminal velocity of the wet solid, the drying of the wet solid occurs in a suspended or fluidized state. This phenomenon is known as *fluidized bed drying*.

Drying may be carried out in a semi-suspended state or *spouted bed condition* also.

Generally, the convection drying is conducted under ordinary state, i.e., drying agent is supplied at a velocity much lower than the terminal velocity of the wet material.

In *natural air drying*, the unheated air as supplied by nature is utilized. In *drying with supplemental heat*, just sufficient amount of heat (temperature rise within

5°C–10°C) only is supplied to the drying air to reduce its relative humidity so that drying can take place.

In *heated air drying*, the drying air is heated to a considerable extent.

The natural air drying and drying with supplemental heat methods, which may require 1–4 weeks or even more to reduce the grain moisture content to safe levels, are generally used to dry grain for short-term storage in the farm. Heated air drying is most useful when a large quantity of grain is to be dried within a short time and marketed at once. It is used for both short- and long-term storage.

Comparative advantages and disadvantages of the three convective drying methods are given as follows.

Natural Air Drying

Advantages

- Lowest initial investment and maintenance cost
- No fuel cost
- No fire hazard
- Least supervision
- Least mold growth compared to supplemental heat

Disadvantages

- Very slow drying rate, drying period may be extended to several weeks
- Weather-dependent
- More drying space necessary in comparison to heated air drying
- Useful particularly for short-term storage in the farm
- Not useful for humid tropics

Supplemental Heat Drying

Advantages

- Lower cost of equipment and maintenance
- Independent of weather
- Requires less supervision
- Most efficient use of bin capacity

Disadvantages

- Fire hazard to a certain extent
- Danger of accelerated mold growth
- Rate of drying is still low
- Useful, particularly for short-term storage in the farm

Heated Air Drying

Advantages

- Independent of weather
- Fast drying
- High drying capacity per fan horsepower
- Used for both long- and short-term storage of grains

Disadvantages

- Higher initial investment and maintenance cost
- Considerable fuel expenditure
- Danger of fire hazard
- Requires skilled manpower for control of drying condition
- By direct firing with liquid fuel, products contaminated with flue gases

The fluidized bed and spouted bed drying systems are detailed in Das and Chakraverty (2003).

Radiation Drying

Radiation drying is based on the absorption of radiant energy of the sun and its transformation into heat energy by the grain. Sun drying is an example of radiation drying. Radiation drying can also be accomplished with the acid of special infrared radiation generators, namely, infrared lamps. Moisture movement and evaporation is caused by the difference in temperature and partial pressure of water vapor between grain and surrounding air. The effectiveness of sun drying depends upon temperature and relative humidity of the atmospheric air, speed of the wind, type and condition of the grain, etc.

Sun Drying

Sun drying is the most popular traditional method of drying. A major quantity of grain is still dried by the sun in most of the developing countries.

Advantages

- No fuel or mechanical energy is required
- Operation is very simple
- Viability, germination, baking qualities are fully preserved
- Microbial activity and insect/pest infestation are reduced

- Labor-oriented
- No pollution

Disadvantages

- Completely dependent on weather
- Not possible round the clock or round the year
- Excessive losses occur due to shattering, birds, rodents, etc.
- Requires specially constructed large floor area, restricting the capacity of mill to a certain limit
- The entire process is unhygienic
- Unsuitable for handling of large quantity of grain within a short period of harvest

Infrared Drying

Infrared rays can penetrate into the irradiated body to a certain depth and transform into heat energy. Special infrared lamps, or metallic and ceramic surfaces heated to a specified temperature by an open flame, may be used as generators of infrared radiation.

Advantages

- Small thermal inertia
- Simplicity and safety in operation of lamp radiation dryers

Disadvantages

- High expenditure of electric power
- Low utilization factor

Radiation dryers have been used in many countries for drying the painted surfaces of machinery, and in timber processing, textile industry, and cereal grain and other food industries.

Solar Drying

A solar drying system can be effectively utilized in tropical and subtropical countries where solar energy is abundantly available and solar insolation is also appreciably high. Various designs and capacities of solar dryers with the variation in efficiency are available.

One of the major components of a solar dryer is an absorber that receives and absorbs solar spectrum of radiation. This absorber transforms the radiation energy to thermal energy for heating air flowing over it, which can be utilized as a drying medium. Solar air dryers can be classified into natural convection and forced convection dryers. These can be operated in either direct mode or indirect mode.

Natural convection solar dryers do not require any fan or blower for air circulation in the dryers. Hence, these dryers are cheap and easy to operate. But the drying operation would be slow and take longer drying time. The simple solar cabinet dryers come under this category.

In the category of forced convection solar dryers, blowers circulate adequate quantity of air through the drying material. These dryers cart dry a comparatively large quantity of material with a reduced drying time. Bin type–forced convection solar dryers are quite common.

Integrated Hybrid-Type Solar Grain Dryer

Solar energy is not available round the clock and the year round. Hence, an auxiliary air-heating system is necessary for a continuous drying operation. An improved solar energy storage system may not be economical. In view of these common problems, a solar-cum-husk-fired flue gas grain drying system of 1 ton/day capacity has been patented by Chakraverty, Das, and IIT, Kharagpur, India, for commercial exploitation.

It comprises an inclined roof cum solar flat plate collector for heating air, a husk-fired furnace, and a grain dryer. It is suitable for drying grain for both seed and food purposes.

Its principle, structure, and operations are detailed in Chakraverty et al. (1987).

Dielectric and Microwave Drying

In dielectric drying, heat is generated within the solid by placing it in a fixed high frequency current. In this method, the substance is heated at the expense of the dielectric loss factor. The molecules of the substance, placed in a field of high frequency current, are polarized and begin to oscillate in accordance with the frequency. The oscillations are accompanied by friction and thus, a part of the electrical energy is transformed into heat. The main advantage of this method is that the substance is heated with extraordinary rapidity.

The amount of microwave energy dissipated as heat in a certain volume of material can be expressed by the following equation (Das and Chakraverty, 2003):

$$\left(\frac{P}{V}\right) = 2\pi\gamma \in_1 E_{loc}^2$$

where
P is the microwave power dissipated as heat, W
V is the volume of material, m^3
γ is the frequency of microwave field, Hz
$\in_1$ is the loss factor of material
E_{loc} is the local electrical field strength within the material, V/m

The dielectric drying has now been in use in different industries such as timber, plastics, and cereal grain processing.

Chemical Drying

Various chemicals such as sodium chloride, calcium propionate, copper sulfate, ferrous sulfate, urea, etc., have been tried for preservation of wet paddy. Of these, common salt has been proved to be effective and convenient for arresting deteriorative changes during storage. When wet paddy is treated with common salt, water is removed from the rice kernel by osmosis. The common salt absorbs moisture from paddy but it cannot penetrate into the endosperm through the husk layer. This is a unique property of the paddy, which has rendered the application of common salt preservation possible.

Advantages

- It not only dries paddy but also reduces the damage due to fungal, microbial, and enzymatic activities and heat of respiration.
- It does not affect the viability of the grain.
- The milling quality of paddy is satisfactory.
- Loss of dry matter is negligible.
- It does not affect the quality of rice bran.

Disadvantages

- The moisture may be retained on the husk due to the presence of sodium chloride.
- The useful life of gunny will be shortened.
- The color of husk changes to dark yellow.
- The common salt–treated paddy requires an additional drying subsequently.
- Economy of the process has yet to be established.

Sack Prying

This method is particularly suitable for drying of small quantities of seed to prevent mixing of varieties and conserve strain purity and viability.

The grain bags are laid flat over holes cut on the floor of a tunnel system so that heated air can be forced up through the grain from an air chamber underneath.

Usually an air temperature of 45°C with an airflow rate of 4 m³/min at 3–4 cm static pressure per bag of 60 kg is used for fastest drying rate. The sacks are turned once during the drying operation. The sack drying process involves higher labor cost.

Chapter 5

Grain Dryers

Grain dryers can be divided into two broad categories: unheated air dryers and heated air dryers. Different types of grain dryers of both groups have been discussed in this chapter.

Unheated Air Dryers

Unheated or natural air drying is usually performed in the grain storage bin. That is why unheated air drying is also known as in-bin or in-storage drying.

Natural air drying is commonly used for on-farm drying for a relatively small volume of grain. Either a full-bin or a layer drying system is employed in natural air drying. The period of drying for either system may be as long as several weeks, depending on the weather. In layer drying, die is filled with a layer of grain at a time and drying is begun. After the layer is partially dried, other layers of grain are added periodically, perhaps daily, with the continuation of drying until the bin is full and the whole grain mass is dried. In full-bin drying, a full bin of grain is dried as a single batch. Then the drying bin is used for storage purposes. The airflow rate provided is relatively low. Though natural air is supposed to be used, an air heating system should be kept so that supplemental heat may be supplied to the natural air during rainy seasons and periods of high humidity weather for highly moist grains. Natural air drying cannot be used if the ambient relative humidity exceeds 70%. Similarly, grains containing moisture higher than 20% should not be dried with natural air.

Various types of unheated air dryers with different constructions, shapes, grain feeding, and discharging mechanisms and aeration systems are available. Some of the common types of dryers are described here.

As in natural air drying, the grain is aerated (for drying) and stored in the same unit; the complete installation simply consists of a storage unit equipped with ducts for air distribution and devices for air exhaustion and a blower.

Storage Unit

Any shape of grain holding bin such as semicircular, circular, square, or rectangular and of any material like metal, wood, concrete, asbestos, or mineral agglomeration can be used, provided the bin is made moisture proof. Different types of units are shown in Figures 5.1 through 5.7. Of the many types of bins used in grain drying, some of the common types are described as follows.

Round Metal Bin

A container with false perforated floor, having 4.5 m diameter and 3 m height, can hold about 25 tons of paddy. The bin is fitted with a cover at the top in such a way that only the exhaust air can escape through it but rain cannot enter into the bin. In some cases, exhaust air is allowed to escape through the side walls of the dryer also (Figure 5.3). The round bins can also be made of concrete of ferrocement. They are usually constructed of several rings sealed together.

Rectangular or square bins fitted with a false perforated floor or main duct and laterals are also in use (Figure 5.4).

Screen Tunnel Quonset-Type Storage Unit

The unit is fitted with a central horizontal screen-type duct and a special air outlet system near the top of each vertical wall (Figure 5.7).

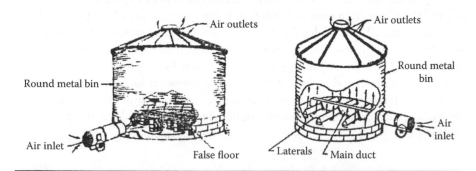

Figure 5.1   Types of air distribution systems used in bin drying.

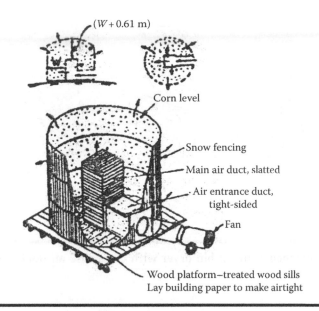

Figure 5.2   An inexpensive, easily built crib for the mechanical drying of ear corn.

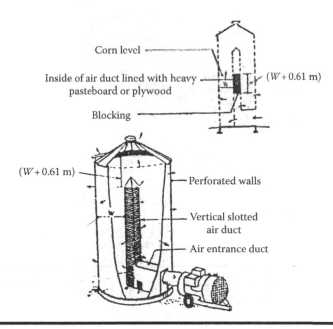

Figure 5.3   High round crib with perforated walls (permanent structure) for drying of ear corn.

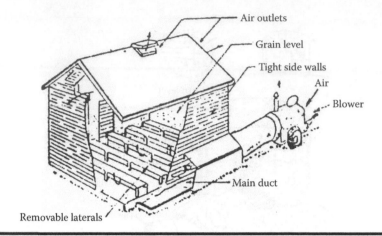

Figure 5.4 Rectangular metal bin dryer with cross-wise air ducts—permanent construction.

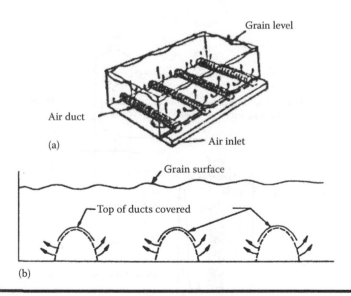

Figure 5.5 (a) Rectangular metal bin dryer with cross-wise air ducts—temporary construction. (b) Most desirable ducting system.

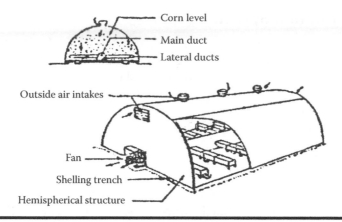

Figure 5.6   General purpose building for drying and storing of grain (permanent structure).

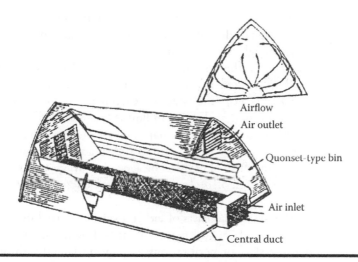

Figure 5.7   Types of air distribution systems used in bin drying.

The bins are generally made circular to ensure uniform distribution of air and avoid stagnant pockets. The quonset type has the same advantages in this respect as the cylindrical bin.

Aeration System

Both propeller and centrifugal types of blowers are used for aeration. Centrifugal blowers may have either forward-curved or backward-curved blades.

The airflow and static pressure requirements for different types and depths of grains are given in Table A.7.

Air Distribution System

Sufficient care should be taken in selecting and designing the air distribution system so that air is uniformly distributed throughout the grain bulk and void pockets are avoided. There are four major systems of air distribution:

1. Perforated floor
2. Central horizontal duct
3. Main duct and laterals
4. Vertical slatted duct

1. *Perforated floor*: The circular storage bin (Figure 5.1) can be fitted with the perforated false floor through which unheated air is blown. Though the system is suitable for small- and medium-sized round bins and for small depths of grain, it is used for large rectangular bins and for higher grain depths.
2. *Central horizontal duct*: This system is used in the quonset-type units (Figure 5.7). This type of duct with openings in the wall can distribute air more uniformly through the grain bulk.
3. *Main duct and laterals*: The system of main duct and laterals is most commonly used and is adopted in round, square, and even rectangular bins (Figures 5.1b, 5.4, and 5.6). The laterals are open at the bottom and raised off the floor of the bin so that the air can flow through the mass. The laterals are inverted V or U or rectangular in shape and are made of wood or steel or concrete or ferrocement. The laterals are spaced in accordance with the size of the storage unit, quantity of grain to be aerated or dried, and depth of the grain (Figures 5.8 through 5.10). In round bins, the ducts can also be placed in the form of a ring on the bin floor.
4. *Vertical ducts*: This system consists of either a vertical slatted duct (Figures 5.2 and 5.3) or a central vertical perforated tube (Figure 5.11a). The air is blown through the slots or perforations and spread laterally through the grain mass.

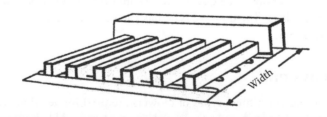

Figure 5.8 Main side duct and laterals.

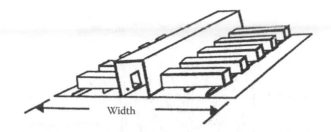

Width

Figure 5.9   Main central duct and laterals.

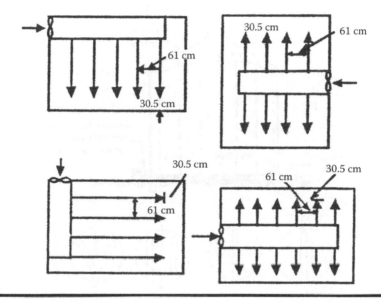

Figure 5.10   Four common floor layouts for the main duct and laterals in bins.

Heated Air Dryers

Flat Bed–Type Batch Dryer

This is a static, deep bed, batch dryer. This type of batch dryer is very simple in design and most popular for on-farm drying in many countries.

Construction

The rectangular box–type batch dryers are shown in Figures 5.12 through 5.14. The size of the dryer depends on the area of the supporting perforated screen on which the grain is placed. The holding capacity of these dryers ranges from 0.25 to 1 ton

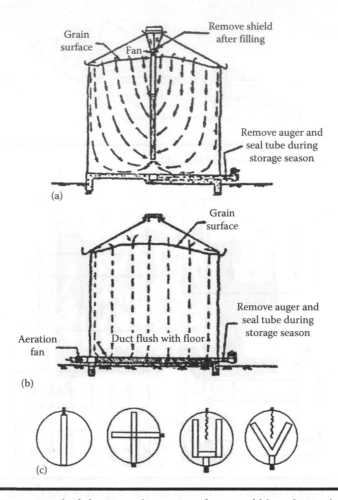

Figure 5.11 **(a) Vertical duct aeration system for round bins. (b) Horizontal duct aeration system for round bins. (c) Different duct patterns for aeration.**

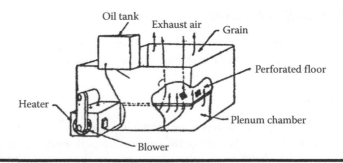

Figure 5.12 **Rectangular flat bed–type batch dryer (Japan).**

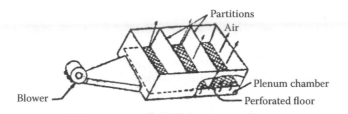

Figure 5.13 Rectangular flat bed–type batch dryer with partitions.

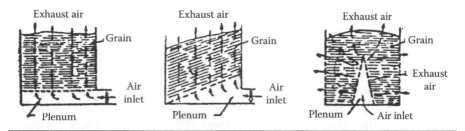

Figure 5.14 Some aeration systems for deep bed batch dryers.

per batch. The horsepower of the motor for the blower ranges from 1/4 to 1. For convenience, an oil burner can be used but for economy a husk-fired furnace should be used for the heat supply.

Operation

The grain is placed on supporting screen and the heated air is forced through the deep bed of grain. After drying the grains to the desired moisture level, it is discharged manually.

The temperature of the heated air should be limited to 45°C. The drying rate varies with the drying temperature.

Airflow rate varies from 20 to 40 m³/min per 1000 kg of raw paddy depending on the initial moisture content.

Advantages

- It is of fairly reasonable price.
- Intermittent drying can also be used.
- Operation is very simple.
- It can be manufactured locally using various types of materials like steel sheet, wood piece, etc.
- It can be used for seed drying and also for storage purpose after drying.

Disadvantages

- Rate of drying is slow.
- Uneven drying results in higher percentage of broken grains.
- Holding capacity is small compared to flow dryers.

Recirculatory Batch Dryer (PHTC Type)

This is a continuous flow non-mixing type of grain dryer.

Construction

The dryer consists of two concentric circular cylinders made of a perforated (2 mm diameter) mild steel sheet of 20 gauge. The two cylinders are set 15–20 cm apart. These two cylinders are supported on four channel sections. The whole frame can be supported by a suitable foundation or may be bolted to a frame made of channel section. A bucket elevator of suitable capacity is used to feed and recirculate the grain into the dryer. A centrifugal blower blows the hot air into the inner cylinder, which acts as a plenum. The hot air from the plenum passes through the grain moving downward by gravity and comes out of the outer perforated cylinder. A torch burner is employed to supply the necessary heat with kerosene oil as fuel. The designs of PHTC dryer for 1/2, 1, and 2 tons holding capacity are available. The PHTC dryer of 2 tons holding capacity developed at PHTC, IIT, Kharagpur, India, is shown in Figure 5.15.

Operation

The grain is fed to the top of the inside cylinder. While descending through the annular space from the feed end to the discharge end by gravity, the grain comes in contact with a cross flow of hot air. The exhaust air comes out through the perforations of the outer cylinder and the grain is discharged through the outlet of the hopper. The feed rate of grain is controlled by closing or opening the gate provided with the outlet pipe of the discharge hopper. The grain is recirculated till it is dried to the desired moisture level (Figures 5.16 through 5.18).

Advantages

- Price is reasonable.
- It is a simplest design amongst all flow-type dryers.
- It is easy to operate.
- It can be used on the farm and rice mill as well.
- Operating cost is low with husk-fired furnace.

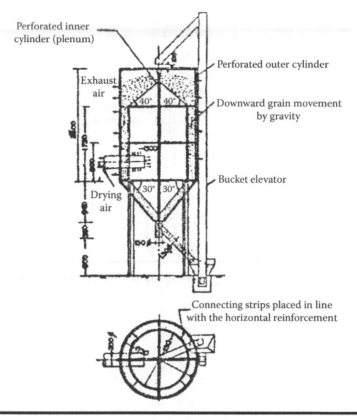

Figure 5.15 PHTC recirculating batch *dryer* (holding capacity—2 tons).

Disadvantages

- Drying is not so uniform as compared to mixing type.
- Perforations of the cylinders may be clogged with the parboiled paddy after using it for a long time.

Louisiana State University Dryer

This is a continuous flow-mixing type of grain dryer, which is popular in India and the United States (Figure 5.19).

Construction

It consists of (1) a rectangular drying chamber fitted with air ports and the holding bin, (2) an air blower with duct, (3) grain discharging mechanism with a hopper bottom, and (4) an air heating system.

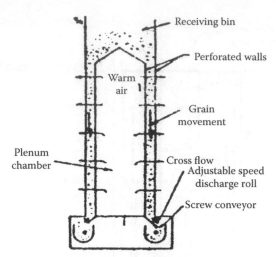

Figure 5.16   Continuous flow type non-mixing double columnar dryer.

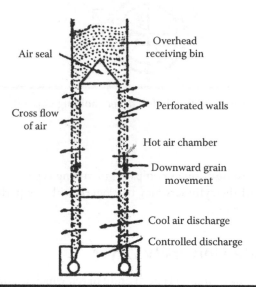

Figure 5.17   Continuous flow type non-mixing double screen columnar dryer with grain cooling chamber.

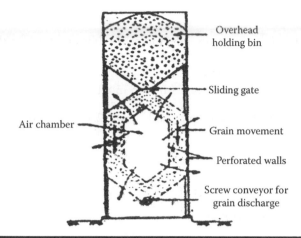

Figure 5.18 Columnar dryer with overhead tempering bin.

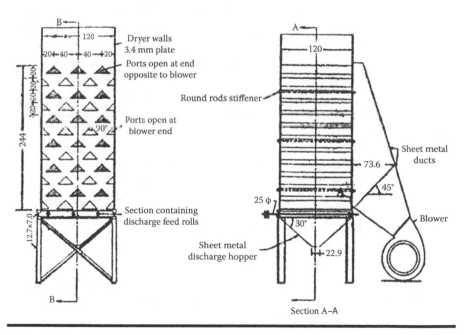

Figure 5.19 LSU-type dryer details.

Rectangular Bin

Usually, the following top square sections of the bin are used for the design of Louisiana State University (LSU) dryers:

(i) 1.2 m × 1.2 m, (ii) 1.5 m × 1.5 m,

(iii) 1.8 m × 1.8 m, and (iv) 2.1 m × 2.1 m.

The rectangular bin can be divided into two sections, namely, top holding bin and bottom drying chamber.

Air Distribution System

Layers of inverted V-shaped channels (called inverted V-ports) are installed in the drying chamber. Heated air is introduced at many points through the descending grain bulk by way of these channels. One end of each air channel has an opening and the other end is sealed. Alternate layers have air inlet and air outlet channels. In the inlet layers, the channel openings face the air inlet plenum chamber but they are sealed at the opposite wall, whereas in the outlet layers, the channel openings face the exhaust but are sealed on the other side. The inlet and outlet ports are arranged one below the other in an offset pattern. Thus, air is forced through the descending grain while moving from the feed end to the discharge end. The inlet ports consist of a few full-size ports and two half-size ports at two sides. All these same-sized ports are arranged in equal spacing between them. The number of ports containing a dryer varies widely depending on the size of the dryer.

Each layer is offset so that the top of the inverted V-ports helps in splitting the stream of grain and flowing the grains between these ports, taking a zigzag path.

In most models, the heated air is supplied by a blower.

Grain Discharging Mechanism

Three or more ribbed rollers are provided at the bottom of the drying chamber, which can be rotated at different low speeds for different discharge rates of grains. The grain is discharged through a hopper fixed at the bottom of the drying chamber.

Causing some mixing of grain and air, the discharge system at the base of the dryer also regulates the rate of fall of the grain.

Air Heating System

The air is heated by burning either gaseous fuels such as natural gas, butane gas, etc., or liquid fuels such as kerosene, furnace oil, fuel oil, etc., or solid fuels such as coal, husk, etc. Heat can be supplied directly through a gas burner or oil burner or husk-fired furnace and indirectly by the use of heat exchangers. Indirect heating is always less efficient than direct firing system. However, oil-fired burners or gas burners should be immediately replaced by husk-fired furnaces for economy of grain drying.

The heated air is introduced at many points in the dryer so as to be distributed uniformly through the inlet ports and the descending grain bulk. It escapes through the outlet ports.

This type of dryer is sometimes equipped with a special fan to blow ambient air from the bottom cooling section in which the dried or partially dried warm grain comes in contact with the ambient air.

In general, the capacity of the dryer varies from 2 to 12 tons of grain, but sometimes dryers of higher capacities are also installed. Accordingly, power requirement varies widely.

Recommended airflow rate is 60–70 m³/min/ton for parboiled paddy and optimum air temperatures are 60°C and 85°C for raw and parboiled paddy, respectively. A series of dryers can also be installed.

Advantages

- ■ Uniformly dried product can be obtained if the dryer is designed properly.
- ■ The dryer can be used for different types of grains.

Disadvantages

- ■ It requires high capital investment.
- ■ Cost of drying is very high if oil is used as fuel.

Baffle Dryer

This is a continuous flow-mixing type of grain dryer (Figures 5.20 and 5.21).

Construction

The baffle dryer consists of (1) grain receiving bin, (2) drying chamber fitted with baffles, (3) plenum fitted with hot air inlet, (4) grain discharge control device, and (5) hopper bottom. A number of baffles are fitted with the drying chamber to divert the flow and effect certain degree of mixing of grain. The two baffle plates with the outer and inner sides are set 20 cm apart for the passage of the grain in the drying chamber. The dryer is made of mild steel sheet.

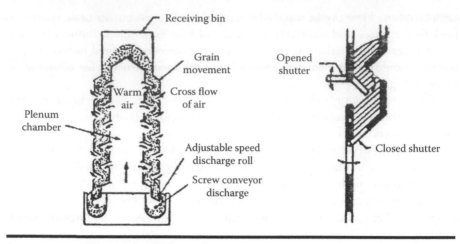

Figure 5.20 Mixing-type baffle dryer.

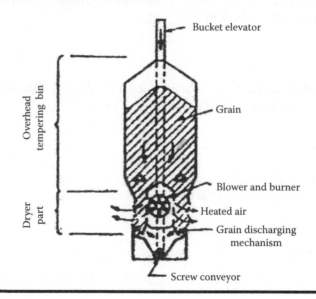

Figure 5.21 Baffle-type tempering dryer.

Operation

Grain is fed at the top of the receiving bin and allowed to move downward in a zigzag path through the drying chamber where it encounters a cross flow of hot air. On account of the zigzag movement, a certain degree of mixing of grain takes place. The partially dried grain discharged from the hopper bottom is recirculated by a bucket elevator until it is dried to the desired moisture level.

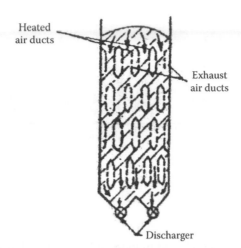

Figure 5.22 Multiple air ducts-type dryer (self-explanatory).

Some of the dryers are fitted with a large overhead bin at the top, which acts as an overhead tempering bin. This type of tempering dryer is shown in Figure 5.21. Multiple air ducts-type dryer is shown in Figure 5.22.

Advantage

■ Uniformly *dried* product is obtained.

Disadvantages

■ Ratio of the volume of plenum to the total volume of the dryer is relatively high.
■ Grains on the baffle plates move slowly than that of other sections.

Other advantages and disadvantages are the same as described in the LSU dryer.

Rotary Dryer

This is a continuous dryer (Figure 5.23) as it produces the final dried product continuously. Horizontal rotary dryers of various designs have been developed by different countries for the drying of parboiled paddy. Some of them are also fitted with external steam jackets and internal steam tubes. As parboiled paddy can stand high temperature without significant increase of cracks in grains, these dryers can be employed for rapid drying of parboiled paddy using temperatures as high as 100°C–110°C. In India, Jadavpur University, Calcutta, introduced a rotary dryer of 1 ton/h capacity for the drying of parboiled paddy. The construction and operation of the same dryer are described as follows.

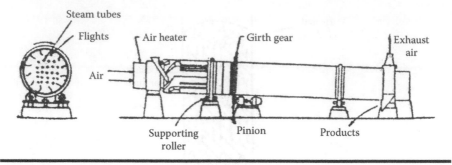

Figure 5.23   Steam tube rotary dryer.

Construction

It consists of a cylindrical shell 9.15 m long and 1.22 m in diameter, with 48 pairs of 5 and 3.75 cm size steam pipes in two concentric rows inside the shell in combination with common steam inlet and condensate outlet fittings. The shell is equipped with six longitudinal flights 9.15 m long and 15.24 cm wide for the lifting and forward movement of the parboiled paddy toward the discharge end while it is being dried. Over the feed end breeching box, there are feed hopper and screw conveyor with an adjustable sliding gate. The dryer is equipped with an air blower and a small steam tube heat exchanger for supplying heated air at the entrance of the feed end breeching box. The cylindrical shell of the dryer is rotated at 2–6 rpm by a motor through speed reduction gear, pulley, and belt drive system.

Operation

The soaked and steamed paddy is fed to the dryer by the screw feeder. Heated air at about 80°C is blown (from the feed end) through the dryer in the same direction as the paddy moves and exhausted through the exhaust pipe. Heated air acts here mainly as a carrier of moisture from the dryer. While traveling from the feed end to the discharge end of the dryer, the parboiled paddy comes in contact with the steam-heated pipes for a very short time in each rotation and is gradually dried to about 16% moisture content in a single pass. Therefore, drying is accomplished mainly by the conduction of heat from the steam pipe to the grain. The traveling time of the grain in the dryer is adjusted to 30–45 min by adjusting the inclination and rpm of the dryer. The hot paddy discharged from the dryer is then aerated by passing it through a cup and cone-type cooler.

Advantages

- Fast rate of drying
- Uniform drying of all grains
- Milling quality of parboiled paddy is high if it is dried in two passes under optimum drying conditions

Disadvantages

- Complicated design
- Needs careful attention
- Higher capital investment
- Higher power requirement
- Operating cost may be high due to higher consumption of electricity and steam
- The dryer being horizontal, larger floor space is required
- Generally, only 30% of the dryer volume is utilized
- It cannot be used for all types of freshly harvested grains

Chapter 6

Selection, Design, Specifications, and Testing of Grain Dryers

Selection of Dryers

Many factors have to be considered before the final selection of the most suitable type of dryer for a given application. Such a selection is further complicated by the availability of a large variety of dryers. Commercial dryers are usually not flexible enough to compensate for design inaccuracies or for problems associated with the handling of different types of food materials that have not been previously considered. For this reason, it is particularly important that all pertinent points be considered and that drying tests be conducted prior to the final selection of a dryer for a given problem. The following procedure is recommended for the selection of the most suitable dryer to produce the desirable product, economically.

Preliminary Dryer Selection

The important factors to be considered in the preliminary selection of a crop dryer are as follows:

1. Physicochemical properties of the crop being handled
2. Drying characteristics of the crop
   (a) Type of moisture
   (b) Initial, final, and equilibrium moisture contents

(c) Permissible drying temperature

(d) Drying curves and drying times for different crops with different dryers

3. Flow of the crop to and from the dryer

(a) Quantity to be handled per hour

(b) Continuous or batch operation

(c) Processes during drying and subsequent to drying

4. Product qualities

(a) Color

(b) Flavor

(c) Shrinkage

(d) Contamination

(e) Uniformity of drying

(f) Decomposition or conversion of product constituents

(g) Overdrying

(h) State of subdivision

(i) Product temperature

(j) Bulk density

(k) Case hardening

(l) Cracking and other desirable qualities of the end products

5. Dust recovery problems

6. Facilities available at the site of proposed installation

a. Space

b. Temperature, humidity, and cleanliness of air

c. Available fuels

d. Available electric power

e. Permissible noise, vibration, dust, or heat losses

f. Source of wet feed

g. Exhaust gas outlets

Comparison of Dryers

The dryers so selected are to be evaluated on the basis of drying performance and cost data.

Drying Test

Drying tests described in this chapter for a given crop have to be carried out with the dryers under consideration to determine the drying performance, operating conditions, and product characteristics. An approximate cost analysis is also useful for evaluation of the dryers.

Final Selection of Dryer

From the results of the drying tests and cost analyses, the final selection of the most suitable dryer can be made.

For successful introduction of any grain dryer at the farm level, a few additional points are to be borne in mind in the selection and design of grain drying system. They are as follows:

1. The dryer should be of proper size matching with the demand of a farmer, miller, or any organization.
2. The price of the dryer should be reasonable.
3. The dryer should be simple in design and should be made of different, cheap, and locally available materials so that it can be manufactured locally.
4. It should be easy to operate.
5. It should be possible to make the dryer portable, if necessary.
6. The operating cost should be minimum. Solar-cum-furnace (fired with agricultural waste like husk, shells, etc.) air heating system should be introduced in grain drying to minimize the cost of drying.
7. The repair and maintenance requirements should be minimum.
8. It should be possible to use the dryer for different grains and to be used as a storage bin later for its maximum utilization.

Design of Grain Dryers

As indicated earlier, heated air grain dryers can be divided into three major groups:

1. Static deep bed batch dryers
2. Continuous flow batch dryers (either mixing or non-mixing type)
3. Continuous dryers. Grain dryers mainly consist of (a) drying chamber; (b) air distribution system; (c) direct or indirect air heating system; (d) blower; (e) control system (if any); and (f) grain conveying system (for flow dryers)

The following important factors are taken into consideration in the design of heated air grain dryers:

1. Dryer factors
   a. Size, shape, and type of dryer
   b. Grain feeding rate
   c. Total drying time
   d. Airflow pattern and air distribution system

     e. Depth of grain bed in the dryer
     f. System of cooling grain (if any)
2. Air factors
     a. Velocity and airflow rate per unit mass of the grain
     b. Temperatures and relative humidities of the heated air and exhaust air
     c. Static pressure of the air at which it is blown
     d. Average ambient conditions
3. Grain factors
     a. Type, variety, and condition of grain
     b. Initial and final moisture contents of grain
     c. The *usage* of dried grain
     d. Latent heat of vaporization of grain moisture
4. Heating system
     a. Type of fuel and rate of fuel supply
     b. Type of burner (for liquid fuel) or type of furnace (for solid fuel)
     c. Type of heat exchanger (for indirect heating system)

Some of the important design factors have been discussed very briefly in subsequent paragraphs.

Size, Shape, and Type of Dryer

The size or capacity of a dryer is decided by the amount and variety of grain to be dried per day or for the whole season.

Sizes of the dryer are expressed either in terms of holding capacity or amount of grain to be dried per unit time or the amount of grain passing through the dryer per unit time (throughput capacity).

Thickness of grain layer exposed to the heated air is generally restricted to 20 cm for continuous flow dryers. The designs of the continuous flow dryers are based on thin layer drying principles whereas static batch dryers are designed on deep bed drying principles. Airflow requirements for different depths and for different grains are given in Table A.7. Total drying time required for various air temperatures for different grains is obtained from the drying curves given in Chapter 3.

Choice of a grain dryer largely depends on the situation. Continuous flow dryers are normally used for commercial purpose, whereas static deep bed batch dryers are used for on-farm drying. Constructional features of different types of heated air dryers have been described in Chapter 5. Farm-level batch dryers can be made of locally available materials, namely, wood, bamboo, etc., if necessary. But commercial big dryers are made of mild steel sheet, angle iron, and channel section supports.

Calculation of Air, Heat, and Fuel Requirements for Heated Air Dryers

The airflow rate required for heated air drying systems can be calculated as follows.

The rate of airflow required for drying may be calculated by making heat balance. The heated air drying system is represented by

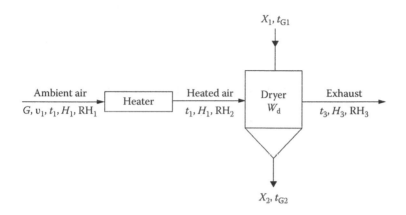

where
    G is the airflow rate, m³/min
    H_1 and H_2 are humidities of ambient and heated air, kg/kg
    H_3 is the humidity of exhaust air, kg/kg
    RH_1, RH_2, and RH_3 are relative humidities of ambient, heated, and exhaust air, respectively, %
    t_1, t_2, and t_3 are dry bulb temperatures of ambient, heated, and exhaust air, respectively, °C
    W_d is the total weight of bone dry grain in the dryer, kg
    X_1 and X_2 are initial and final moisture contents of grain (d.b.), kg/kg
    t_{G_1} and t_{G_2} are initial and final grain temperatures, °C
    v_1 is the initial humid volume, m³/kg

Heat supplied by drying air, q_a, kcal:

$$q_a = (0.24 + 0.45\,H_1)G' \times (t_2 - t_3)\theta$$

where
    G' is the rate of air supply, kg/min
    θ is the total drying time, min

Amount of heat required

Heat required for evaporation of moisture from the grain, q_1, kcal

$$q_1 = W_d(X_1 - X_2)\lambda$$

where λ is the average value of latent heat of vaporization of moisture from the grain, kcal/kg.

Sensible heat required to raise the temperature of the grain and its moisture, q, kcal:

$$q = W_d C_g(t_{G2} - t_{G1}) + W_d C_x(t_{G2} - t_{G1})X_1$$

where C_g and C_w are specific heats of grain and water, respectively, kcal/kg °C.

Therefore,

$$q_a = q_1 + q$$

or

$$G' = (0.24 + 0.45H_1)(t_2 - t_3)\theta$$

$$= W_d\left[(X_1 - X_2)\lambda + C_g(t_{G2} - t_{G1}) + C_w(t_{G2} - t_{G1})X_1\right]$$

$$\text{or} \quad G' = \frac{W_d\left[(X_1 - X_2)\lambda + C_g(t_{G2} - t_{G1}) + C_w(t_{G2} - t_{G1})X_1\right]}{(0.24 + 0.45H_1)(t_2 - t_3)\theta}$$

$$\therefore G = G' \times v_1$$

where v_1 is the humid volume.

Fuel consumption

The rate of fuel consumption can be calculated as follows:

$$f = \frac{q_a'}{\eta \times \eta_b \eta_{ex} \times C_n}$$

where

f is the fuel rate, kg/h

q_a' is the total heat required to heat the drying air, kcal/h

C_n is the calorific value of fuel, kcal/kg of fuel

η is the efficiency of the heating system

η_{ex} is the efficiency of the heat exchanger

η_b is the efficiency of the boiler, if any

Drying Air Temperature

Correct choice of drying air temperature for a given type of grain is very important as it has effects on the quality of the dried product. The highest allowable air temperature of drying of grain depends on the type and condition of grain and the usage of dried grain. The upper limits of drying air temperatures for different grains to be used for food, feed, and seed purposes are different and are given in Table A.7.

Grain Parameters

The grain factors that affect the rate of drying are as follows:

1. Type, variety, and condition of grain
2. Initial/harvest moisture content, final moisture, and equilibrium moisture contents of the grain
3. Structure and chemical composition of the kernel, seed coat, husk, etc.
4. Foreign materials present in the grain

The earlier stated factors are, therefore, to be considered in the design of grain dryers. Over and above, data on physical properties such as bulk density, angle of repose, porosity, angle of internal friction, flow properties of grain, aerodynamic properties, and thermal properties (specific heat, thermal conductivity, etc.) are required in the design of a grain dryer and are thus taken into consideration. Some of these properties are tabulated in Tables A.3 through A.7.

Airflow Pattern and Air Distribution

Any one of the three systems of airflow, namely, cross flow, counter flow, and cocurrent flow, can be adopted in flow-type grain dryers. Generally, cross flow of air is preferred. Double screen (e.g., PHTC type) and baffle types of columnar dryers have a plenum chamber and LSU dryer has inverted V-shaped air channels for uniform distribution of air throughout the drying chamber. The deep bed batch dryer has the plenums at the bottom of the grain drying chamber. These systems had been shown in Chapter 5.

Conveying and Handling System

Suitable conveying equipment for loading, discharging, recirculating, and shifting of grain before, during, and after drying of grain are necessary for the grain drying system. Bucket elevators, vertical screw lifts for feeding, and hopper bottom with proper inclination for grain discharging are commonly used. A forced discharge mechanism with slowly rotating fluted rolls is used for better control of the feed rate and drying rate. Dried grain from the dryer is usually conveyed to different places by belt conveyor or screw conveyor and bucket elevator.

Air Heating System

Generally, direct firing systems are used for gaseous and liquid fuels and indirect heating system using heat exchangers is employed for solid fuels. But direct flue gas from the husk-fired furnace can also be efficiently used for grain drying. In view of the present energy crisis, the liquid or gaseous fuel burning system should be immediately replaced by the agricultural waste (husk, shell, bagasse, etc.) fired furnace for the supply of heated air economically. The drying cost can be further reduced by introducing a solar-cum-husk-fired grain drying system.

Testing of Grain Dryers

No generalized test procedure can be adopted for all types of grain dryers. The testing method for static deep bed batch dryer cannot be the same as that of continuous flow thin layer dryers. It is always preferable that test procedure for each type of dryer be designed separately.

However for convenience, the dryer testing method can be broadly divided into two major heads: simple method and rigorous method. Either of these two methods can be adopted in accordance with the objectives of the test.

Simple Method

A simple test procedure is so designed as to determine the approximate performance of the grain dryer.

Simple Test Procedure for Continuous Flow Dryer

Besides the test items tabulated in the "Specifications" section, the following items are to be taken into consideration for continuous flow dryers:

1. Moisture content after each pass (%)
2. Residence time in the dryer for each circulation (h)
3. Number of passes
4. Tempering time (h)
5. Average rate of moisture reduction or rate of moisture evaporation in each circulation (kg/h)

6. Rate of grain recirculation (tons/h)
7. Drying air temperature at each pass (°C)
8. Weight of remaining grain in the dryer, elevator, etc. (kg)

Rigorous Method

Rigorous test procedures for some batch and continuous flow dryers are given as follows. The whole test procedure can be grouped into the following major heads:

1. Checking construction
2. Drying performance test
3. Fan/blower performance test
4. Control system performance test
5. Handling equipment performance test
6. Checking of different dryer parts after disassembling (after the drying tests)

Checking of Construction

The purpose of this test is to ascertain the major dimensions, material of construction, and other necessary specifications of the dryer and its accessories.

Investigation items: Specifications of (a) dryer as a whole; (b) drying chamber with air distribution system; (c) blower; (d) heating system; and (e) conveying units such as bucket elevator, grain distributor, screw conveyor, belt conveyor, etc.

The specifications of the earlier items have been discussed earlier.

Drying Performance Test

The objectives of this test are to determine the drying performance of a dryer on the basis of rate of drying, rate of consumption of fuel and electricity, heat utilization, quality of the dried grain, and other operating conditions.

The investigation items have already been tabulated.

Blower Performance Test

The objective of this test is to determine the performance of the fan/blower attached with the dryer.

Investigation items: (a) Power input, kW; (b) airflow rate, m^3/min; (c) static and total pressure, mm water; (d) static pressure efficiency; and (e) vibration, noise, and other working conditions of the blower.

Performance of the Control System

The objective of this test is to find out the accuracy of (a) control of drying air temperature and the temperature of heating unit, (b) control of airflow rate, etc., and (c) other working conditions of the whole control system.

Investigation items: (a) Accuracy of the temperature control with the heating unit, (b) accuracy of the temperature control of the drying air, (c) variation of heated air temperature at different points, (d) airflow rate control and any other control system, and (e) any mechanical trouble with the system.

Performance of the Handling Equipment

The rated and actual capacities and other working conditions of the conveying and handling equipment are to be found out. This has been discussed earlier.

Investigation after Disassembling

This is necessary to investigate the conditions of different parts of various units after completion of the drying test.

Specifications

Specifications of Dryer and Its Accessories

Name of the Unit/Part	S. No.	Items
	1.	Name of dryer
	2.	Type of dryer
	3.	Model of dryer
	4.	Manufacturer's name and address
	5.	Holding capacity of dryer (tons)
	6.	Dimensions of dryer (cm)
Dryer	7.	Total height of dryer, and ground clearance (cm)
	8.	Total weight of dryer (kg)
	9.	Safety measures of dryer (if any)
	10.	Hp required for feed rolls

(continued) Specifications of Dryer and Its Accessories

Name of the Unit/Part	S. No.	Items
	11.	Total hp requirement
	12.	Total power requirement (kW)
	13.	Price of dryer (Rs.)
	1.	Type of drying section (screen, baffle, etc.)
	2.	Dimensions (cm)
	3.	Grain holding capacity (kg)
	4.	Dimensions of the plenum (for double screen and baffle type)
	5.	Size and number of air ports (for LSU type)
Drying chamber	6.	Ratio of plenum volume to the whole drying chamber volume
	7.	Thickness of grain layer (grain depth) in drying section
	8.	Mesh no. of the screen (for double screen)
	9.	Dimensions and position of hopper
	10.	Grain distribution mechanism
	11.	Grain discharging mechanism
	1.	Blower type
	2.	Dimensions (cm)
	3.	Rated capacity of blower (m³/min)
	4.	Static pressure (mm water)
	5.	Diameter of discharge outlet (cm)
Blower	6.	Diameter of inlet port (cm)
	7.	Rated hp of the blower motor
	8.	Recommended speed of blower (rpm)

(continued)

(continued) Specifications of Dryer and Its Accessories

Name of the Unit/Part	S. No.	Items
	9.	No. of rotary blades/impeller vanes
	10.	Diameter of rotary blades/impeller (cm)
	1.	Drying air temperature range and control
	2.	Airflow rate control
Control system	3.	Grain flow rate control
	4.	Any other control *Handling equipment*
	1.	Capacity (tons/h)
	2.	Total height (cm)
	3.	Height from the ground level (cm)
Bucket elevator	4.	Shape and size of bucket
	5.	Number of buckets
	6.	Position of grain inlet and outlet
	7.	Rated speed of main shaft (rpm)
	8.	Material of construction
	9.	Power requirement
	1.	Capacity (tons/h)
	2.	Speed of shaft (rpm)
	3.	Outer diameter of screw (cm)
	4.	Outer diameter of shaft (cm)
	5.	Pitch (cm)
Screw conveyor	6.	Length of the screw (cm)
	7.	Clearance between the screw tip and housing troughs (mm)
	8.	Material and thickness of flight (cm)
	9.	Power requirement

(continued) Specifications of Dryer and Its Accessories

Name of the Unit/Part	S. No.	Items
		Air heating system
		Type of fuel used and its calorific value (kcal/kg)
		Direct or indirect firing
	1.	Dimensions (cm)
	2.	Built-in type or separate
	3.	Type of burner
	4.	Nozzle diameter (cm)
Oil burner	5.	Burner rating (kg/min)
	6.	Fuel and air ratio (kg/kg)
	7.	Required pressure for atomizing (kg/cm^2)
	8.	Fuel feeding method and control
	9.	Material of construction
	10.	Capacity of the fuel storage tank (L)
	11.	Method of temperature control
	12.	Method of ignition and extinguishing
	13.	Safety device

The simple test procedure for a batch dryer is tabulated as follows.

Test Procedures for the Performance of the Static Deep Bed Batch Dryer

Type and Model No. of Dryer		
Type of Grain and Variety		
	1.	Initial weight of wet grain (kg)
	2.	Final weight of dried grain (kg)
	3.	Initial moisture content (%)
Grain	4.	Final moisture content (%)

(continued)

(continued)   Test Procedures for the Performance of the Static Deep Bed Batch Dryer

Type and Model No. of Dryer		
Type of Grain and Variety		
	5.	Dryer loading time (h)
	6.	Dryer unloading time (h)
	7.	Average drying grain temperature (°C)
	1.	Airflow rate (maximum) (m³/min)
	2.	Airflow rate (minimum) (m³/min)
	3.	Maximum static pressure (mm water)
	4.	Minimum static pressure (mm water)
Air	5.	Average ambient d.b. temperature (°C)
	6.	Average ambient w.b. temperature (°C)
	7.	Average heated air d.b. temperature (°C)
	8.	Average heated air w.b. temperature (°C)
	9.	Average exhaust air d.b. temperature (°C)
	10.	Average exhaust air w.b. temperature (°C)
	1.	Total drying time (h)
	2.	Cooling time (if any) (h)
	3.	Total moisture evaporation (kg/h)
Drying capacity	4.	Rate of moisture evaporation (kg/h)
	5.	Rate of dried grain productions (tons/h)
	1.	Air heating method (oil-fired burner/husk-fired furnace/steam heat exchanger)
	2.	Type of air heating (direct/indirect)
Heater and fuel	3.	When oil-fired burner/husk-fired furnace is used (a) Type of fuel and cal. value (b) Total fuel consumption (kg) (c) Rate of fuel consumption (kg/h)

(continued)   Test Procedures for the Performance of the Static Deep Bed Batch Dryer

Type and Model No. of Dryer:		
Type of Grain and Variety		
	4.	When stem heat exchanger is used
		(a) Incoming steam pressure (kg/cm²)
		(b) Rate of condensate outflow (kg/h)
		(c) Temperature of condensate (°C)
	1.	Power consumption for blowing air to burner (kW)
	2.	Power consumption for pumping oil to burner (kW)
Power	3.	Power consumption for blowing heated air (kW)
	4.	Power consumption for loading and unloading grain (kW)
	5.	Power consumption for running feed rolls (kW)
	1.	Germination of grain before drying (%)
	2.	Germination of grain after drying (%)
Quality of dried grain	3.	Head yield before drying (%—for paddy)
	4.	Total yield before drying (%)
	5.	Head yield after drying (%)
	6.	Total yield after drying (%)
	7.	Other quality factors

Problems on Dryer Design

Solved Problems

1. Design a PHTC-recirculating batch dryer (Figure 6.1) having holding capacity of 2 tons of paddy with 15% m.c. (w.b.).

   Assume the following data:

   Ambient air temperature = 30°C
   Relative humidity of ambient air = 70%
   Initial m.c. of paddy = 30% (w.b.)

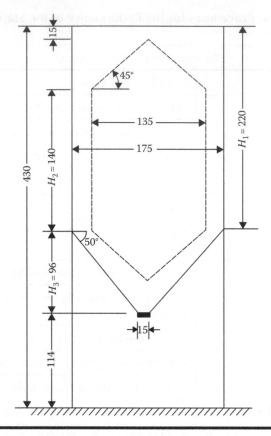

Figure 6.1 Specifications of PHTC dryer as per solution of Problem 1.

Final m.c. of paddy = 15% (w.b.)
Grain inlet temperature = 30°C
Grain outlet temperature = 70°C
Heated air temperature = 85°C
Exhaust air temperature = 40°C
Latent heat of water vapor = 600 kcal/kg
Angle of repose = 45°
Thickness of grain bed to be dried = 20 cm
Bulk density of paddy grain at 15% m.c. = 575 kg/m³
Diameter of plenum chamber = 135 cm
Diameter of dryer = 175 cm
Hopper angle = 50°
Drying time = 3 h

Solution

Assumptions

1. Distance between top of dryer and top of plenum chamber = 15 cm
2. Angle of conical portion of plenum chamber = 45°
3. Diameter of grain outlet = 15 cm
4. Specific heat of grain = 0.4 kcal/kg °C

Height of the dryer

$$\text{Height, } H = H_1 + H_3$$

$$H_1 = H_2 + \frac{135}{2} \tan 45° + 15 = H_2 + 82.5$$

$$H_3 = \frac{175 - 15}{2} \tan 50° = 95.34 = 96 \text{ cm}$$

Volume of plenum chamber

$$V_1 = \text{volume of inner cylinder} + \text{volume of inner cones}$$

$$V_1 = \frac{\pi}{4}(135)^2 H_2 + 2 \times \frac{1}{3} \times \pi \left(\frac{135}{2}\right)^2 \left(\frac{135}{2} \tan 45°\right) \tag{6.1}$$

or $V_1 = 14,313.88 H_1 - 536,770.53$

Suppose V_2 = Volume of outer cylinder and hopper bottom

$$\text{Therefore, } V_2 = \frac{\pi}{4}(175)^2 H_1 + \frac{1}{3} \times \pi \times \tan 50° \left[\left(\frac{175}{2}\right)^3 - \left(\frac{15}{2}\right)^3\right]$$

$$= 24,055.9 H_1 + 831,557 \text{ cm}^3 \tag{6.2}$$

Volume of the drying chamber $= V_2 - V_1$

$$= 24,055.9 H_1 + 831,557 - 14,313.88 H_1 + 536,770$$

$$= 9742.02 H_1 + 1,368,327 \tag{6.3}$$

$$\text{Volume of the drying chamber} = \frac{2000}{575} = 3.478 \text{ m}^3$$

$$= 3{,}478{,}000 \text{ cm}^3 \qquad\qquad (6.4)$$

Hence, from Equations 6.3 and 6.4

$$9742.02 H_1 + 1{,}368{,}327 = 3{,}478{,}000$$

$$H_1 = 216 \text{ cm}$$

$$H_1 = 220 \text{ cm}$$

$$H_2 = 220 - 82.5 = 137.5 \text{ cm}$$

$$\text{Therefore, } H_2 \approx 140 \text{ cm}$$

$$H_3 = 90 \text{ cm}$$

Air requirement
 Bone dry paddy = 2000 (1 − 0.15) = 1700 kg
 Initial moisture content = 30% (w.b.) = 42.857% (d.b.)
 Final moisture content = 15% (w.b.) = 17.647% (d.b.)

Weight of moisture evaporated

$$= \text{Weight of bone dry paddy} \times (X_1 - X_2)$$

$$= 1700(0.42857 - 0.17647)$$

$$= 428.57 \text{ kg}$$

From psychrometric chart
 Absolute humidity of ambient air = 0.019 kg/kg
 Humid heat of ambient air

$$S = 0.24 + 0.45 H$$

$$= 0.24 + 0.45 \times 0.019$$

$$= 0.24855 \text{ kcal/kg °C}$$

Let G be the rate of air supply in kg/min.

Heat supplied by the air in 180 min

$$= G \times S \times (t_2 - t_1) \times \theta$$
$$= G(0.24855)(875 - 40) \times 3 \times 60$$
$$= 2013.255G \tag{6.5}$$

Heat utilized

1. As sensible heat of grain

$$= \text{B.D. grain} \times \text{specific heat of grain} \times \text{temperature rise}$$
$$= 1700 \times 0.4 \times (70 - 30)$$
$$= 27,200 \text{ kcal} \tag{6.6}$$

2. As sensible heat of water

$$= \text{Total weight of water} \times \text{specific heat of water} \times \text{temperature rise}$$
$$= 1700 \times 0.42857 \times 1.0 \times (70 - 30)$$
$$= 29,140 \text{ kcal} \tag{6.7}$$

3. As latent heat of water vapor

$$= \text{Water evaporated} \times \text{latent heat of water}$$
$$= 428.57 \times 600$$
$$= 257,100 \text{ kcal} \tag{6.8}$$

$$\text{Total heat utilized} = \text{sum of 6.6 through 6.8}$$
$$= 313,440 \text{ kcal} \tag{6.9}$$

Suppose, heat loss = 10%

$$\text{Net heat required} = \frac{313,440}{0.9} = 348,266.6 \text{ kcal}$$

Hence, 2013.255 G = 348,266.6

$$G = 172.987 \text{ kg/min}$$

From the psychrometric chart, humid volume of the ambient air = 0.88 m³/kg

$$\text{So, air required} = 172.987 \times 0.884$$

$$= 152.92 \text{ m}^3/\text{min}$$

$$\text{Air requirement} = 155 \text{ m}^3/\text{min}$$

Static pressure drop

Surface area of plenum chamber

$$= \pi \times 135 \times 140 + \left[\frac{1}{2} \pi \times 135 \times \left(\frac{135}{2} \sec 45° \right) \right] \times 2$$

$$= 59,383.8 + 28,627.76$$

$$= 88,011.56 \text{ cm}^2$$

Since maximum 50% of the area is perforated, area through which air passes = 44,005.78 cm²

$$\text{Air requirement per m}^2 = \frac{155}{44,005.78}$$

$$= 35.1 \text{ m}^3/\text{min/m}^2$$

From Shedd's curve (*Agri. Engg. Handbook*), static pressure drop for 32.12 m³/min/m² = 8.13 cm of water per 30.48 cm grain depth.

$$\text{Depth of grain} = 20 \text{ cm}$$

$$\text{So, pressure drop} = \frac{8.13}{30.48} \times 20 = 5.42 \text{ cm of water}$$

$$\text{Density of air} = 1.13/\text{kg/m}^3$$

$$\text{Pressure drop in terms of air column} = \frac{5.42}{100} \times \frac{1000}{1.13}$$

$$= 47.95 \text{ m}$$

Hp required

$$= \frac{\text{Height of air column (m)} \times \text{air flow rate (kg/min)}}{4500}$$

$$= \frac{47.95}{4500} \times 172.987$$

$$= 1.868 \text{ hp}$$

$$\approx 2 \text{ hp}$$

2. Design a LSU dryer of 2 tons holding capacity with paddy at 15% m.c. (w.b.). Data for grain and air parameters are the same as in the previous problem. Additional data are given as follows.

Cross section of the dryer = 1.2 × 1.2 m²
Air velocity in the air ports = 5 m/s
Pitch of the air ports = 40 cm
Row to row spacing = 20 cm
Grain residence time = 30 min

Solution

The heat and mass balance for the dryer are the same as those in the previous problem.

Hence, air required = 155 m³/min.

Since velocity of air inside the air port or duct is 5 m/s, total cross-sectional area of ducts required

$$= \frac{155}{5 \times 60} = 0.5167 \text{ m}^2$$

Let the height of drying chamber be *h* cm.

Therefore, volume of drying chamber

$$V = \text{Volume of the drying chamber} - \text{volume of ducts}$$

$$= 1.2 \times 1.2 \times h - 0.5167 \times 1.2 \tag{1}$$

$$V = (1.44/h - 0.62)\ \text{m}^3$$

The holding capacity of the dryer is given as 2 tons of paddy at 15% m.c. (w.b.) and the bulk density of paddy at 15% m.c. is 575 kg/m³.

$$\text{Hence,} \quad \text{volume } V = \frac{2000}{575} = 3.478\ \text{m}^3 \tag{2}$$

Substituting for V in Equation (1)

$$1.44\ h - 0.620 = 3.478$$

$$1.44\ h = 4.098$$

$$h = 2846\ \text{m}$$

Height of the grain holding bin = 35 cm (assumed).
Height of the hopper bottom = 60 cm (approximately).
According to spacing, number of ducts in a row = 3 and number of rows in 285 m = 14.
Total number of ducts = 39 (leaving 1 row for discharge rolls).

$$\text{Cross-sectional area of each duct} = \frac{0.5167}{39} \times 10^4\ \text{cm}^2$$

$$= 132.5\ \text{cm}^2$$

$$\text{Cross-sectional area of each duct} = 1/2b \times b/2$$

$$\text{or} \quad \frac{b^2}{4} = 132.5$$

$$\text{or} \quad b = 23.02 = 23\ \text{cm}$$

Let there be three discharge rolls having shaft diameter 2.5 cm and flute diameter 7.5 cm.

Volume discharged by each roll in one revolution

$$= \frac{\pi}{4}(D_0^2 - D_i^2) \times L$$

$$= \frac{\pi}{4}(7.5^2 - 2.5^2) \times 120$$

$$= 4712.39 \text{ cm}^3$$

Therefore, volume discharged by 3 rolls = 0.014137 m^3/revolution

Weight of paddy discharged = $0.014137 \times 575 = 8.129$ kg/revolution

Since the grain retention time in the dryer is 30 min,

$$\text{Grain discharge rate} = \frac{2000}{30} = 66.67 \text{ kg/min}$$

Therefore,

$$\text{roller speed} = \frac{66.67}{8.129} = 8.2 \text{ rpm} \approx 8 \text{ rpm}$$

Specifications of the LSU dryer (Figure 6.2) of 2 tons grain holding capacity:
   Total height of the dryer = 3.8 m
   Height of drying chamber = 2.85
   Height of the holding bin = 0.35 m
   Height of the hopper bottom = 0.60 m
   Number of feed rolls = 3
   Grain outlet diameter = 20 cm
   Space between inlet and outlet duct = 20 cm
   Pitch of ducts in a row = 40 cm
   Blower capacity = 155 m^3/min with static pressure 5 cm (W.G.)
   Blower motor hp = 2
   Duct dimensions: height = 11.5 cm, width = 23 cm
   Speed of the discharge roll = 8.0 rpm

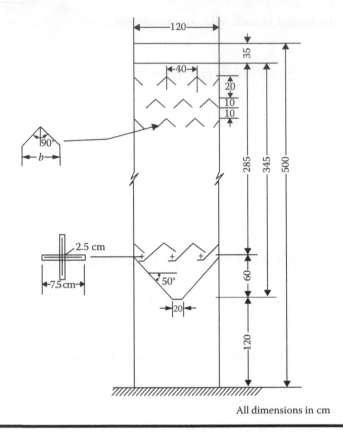

All dimensions in cm

Figure 6.2 Specifications of LSU dryer as per solution of Problem 2.

Exercises

1. From the following data, determine the rate of air supply in a grain dryer:
    (a) Holding capacity of the dryer = 6 tons
    (b) Initial m.c. of the parboiled paddy = 30% (w.b.)
    (c) Final m.c. of the parboiled paddy = 15% (w.b.)
    (d) Ambient air temperature = 30°C
    (e) Heated air temperature = 85°C
    (f) Exhaust air temperature = 40°C
    (g) Grain temperature (at inlet) = 30°C
    (h) Grain temperature (at outlet) = 70°C
    (i) Relative humidity of air before heating = 70%
    (j) Total drying time = 3 h
    (k) Loss of heat to the surroundings = 20%

2. Design a recirculating PHTC dryer of 1.25 tons holding capacity. Diameter of the plenum chamber (inner cylinder) is 90 cm. Assuming other necessary data as in Example No. 1, determine (i) height of the inner and outer cylinders, (ii) total height of the dryer, (iii) airflow rate, and (iv) hp requirement of the blower.

3. Design an LSU dryer of 2 tons/h capacity. The square cross section of the rectangular chamber is 2.02 × 2.02 m². Assume other necessary data as in Example No. 2. Calculate (i) total height of the dryer, (ii) number of air ports required, (iii) speed of the discharge roll, and (iv) hp of the motor for the discharge rolls.

4. Assuming all necessary data, design a baffle-type grain dryer of 1.25 tons holding capacity.

5. A static deep bed rectangular batch dryer of 1 ton holding capacity has to be designed. The temperatures of the ambient and drying air are 25°C and 40°C, respectively. Assuming other necessary data, determine (i) the dimensions of the drying chamber and plenum chamber and (ii) capacity of the blower.

Symbols

C_g	specific heat of grain, kcal/kg °C
C_w	specific heat of water, kcal/kg °C
C_n	calorific value of fuel, kcal/kg
f	rate of fuel consumption, kg/min
G	volumetric airflow rate, m³/min
G'	airflow rate, kg/min
H_1	humidity of ambient air, kg/kg
H_2	humidity of heated air, kg/kg
H_3	humidity of exhausted air, kg/kg
X_1	initial moisture content of grain (d.b.), kg/kg
X_2	final moisture content of grain (d.b.), kg/kg
q	heat required for heating the grain and moisture, kcal
q_1	latent heat required for evaporating grain moisture, kcal
q_a	total heat required for drying, kcal
q_a'	heat required to heat the drying air, kcal
RH_1	relative humidity of ambient air, %
RH_2	relative humidity of drying air, %
RH_3	relative humidity of exhaust air, %
t_1	temperature of the ambient air, °C
t_2	temperature of the drying air, °C
t_3	temperature of the exhaust air, °C
v_1	humid volume of the ambient air, m³/kg
W_d	weight of bone dry grain, kg

GRAIN STORAGE

Chapter 7

Food Grain Storage

The challenge of feeding an ever growing human population cannot be met with the increase of food production alone. During the last two decades, production of food has been increased significantly. In order to make further progress, it requires inputs of improved seeds, fertilizers, pesticides, and advanced postharvest technologies of preservation and utilization, apart from improved production practices and irrigation. The losses during growing crops and postharvest handling, processing, storage, and distribution systems vary between 20% and 60% in some of the countries of the world (Food Industries, 1979). It is estimated that 6%–11% food grains are lost during postharvest operations (Mc Farlance, 1988). If these losses are minimized, the shortage of food in many countries can probably be eliminated. In this context, the improvement grain storage system is inevitably the first step toward this direction. Reviews on various aspects of grain storage have been discussed earlier in Jayas et al. (1995). It may be mentioned that food grains are stored for later use as seed, reserved food, or buffer stock. Insect microorganisms and rodents not only consume the edible and inedible parts of the growing crops and stored grains but also lead to the deteriorations in quality. Hence, integrated approach for the control of pests is essential for maintaining quality during growth, production, as well as at the postharvest handling and storage periods.

Grain Storage Principles

General

Food grains are living organisms. Hence, the grain should be stored as a living seed. A grain is physiologically quite stable after harvesting and this stability as well as its viability should be preserved in a good storage method. Under natural conditions, however, stored grain undergoes chemical changes within itself. Its further deterioration is caused by external living organisms, such as insects, microorganisms, molds, fungi, mites, and rodents. The stored grain in bulk is a system in which deterioration results from interactions among physical, chemical, physiological, and biological variables. Some of the variables are temperature; moisture; oxygen; storage structure; physical, chemical, and biological properties of grain bulks; microorganisms; and insects, mites, rodents, and birds. In fact, they seldom act alone or all at once. They interact with the grain in groups, among themselves. Initially, the rate of deterioration is slow, but as the favorable combinations of variables are set and the storage period is prolonged, a very high loss in grain quality and quantity occurs (Figure 7.1).

The major variables that cause various changes and deteriorations in food grains during storage are given as follows:

1. Physical temperature and humidity
2. Chemical moisture and oxygen (or O_2:CO_2)
3. Physiological respiration and heating
4. Biological insects, fungi, molds, mites, microorganisms, and rodents

The deterioration of food grains may be either qualitative or quantitative, or both. Insects, microorganisms, fungi, molds, mites, and rodents may cause both these damages. Of the aforementioned variables, moisture and temperature are the most crucial ones as far as storage of food grains is concerned. The soundness of grain is also an important factor from the standpoint of viability (Sinha, 1973).

Moisture

The moisture of food grain is the first factor to be paid serious attention as it limits the development of bacteria, fungi, mites, and insects that cause spoilage of stored grains. The important points to be kept in mind while planning safe storage are that (i) moisture contents of grains below 13% arrest the growth of most of the microorganisms and mites; (ii) moisture contents below 10% limit development of most of the stored grain insect pests; and (iii) moisture contents within a grain bulk are seldom uniformly distributed and are changeable. The need for regular routine measurement of moisture within a grain bulk cannot be overemphasized.

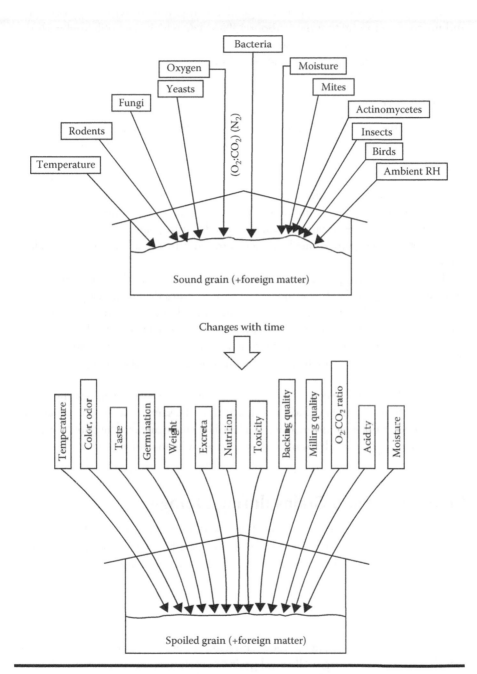

Figure 7.1 **Deteriorations of food grains during storage.**

The limit of moisture for safe storage of cereal grains, in regard to insect pest and microorganism infestation, is about 13%–14%, which is equilibrated with the atmospheric relative humidity of around 70%–75%. If a cereal grain is to be stored for a long period, its moisture should be below 12%.

Temperature

The temperature is another important basic factor to be considered along with the grain moisture and equilibrium relative humidity. The relative humidity, which is in equilibrium with the grain moisture, varies with the temperature. For the storage of brown rice, low temperature is generally used in Japan as the rates of chemical and biochemical reactions are always slow at the low temperatures. The metabolic heat produced exclusively by dry grain is about 1×10^{-7} cal/scc and by wet grain is $\sim 1.3 \times 10^5$ cal/scc. The amount of heat produced by fungi, insects, and other organisms infesting the grain is much higher compared to the aforementioned values (Sinha, 1973).

In considering temperature for safe grain storage system, the following important points are to be kept in mind: (i) generally, mites do not develop below 5°C nor insect below 15°C; (ii) most of the storage fungi do not develop below 0°C; and (iii) the effect of temperature on an organism can be correlated with the amount of grain moisture. The rate of respiration of grain, the growth of microorganisms, and the chemical and enzymatic reactions during storage also accelerate up to a certain temperature (Sinha, 1973).

When the grain temperature rises to around 20°C, it starts getting easily infested with insects and microorganisms, and, at the same time, its rate of respiration becomes rapid with the expense of chemical constituents. The grain temperature is always to be considered in conjunction with its moisture content.

Changes in Food Grains during Storage

The changes taking place in a food grain during storage are the chemical changes within itself and the deteriorations caused by various living organisms. However, these changes and deteriorations occur almost simultaneously during the storage.

Chemical Changes

Oxidation, enzymatic reactions, and respiration influence the chemical changes in cereal grains during storage. All cereal grains contain certain enzymes that decompose their constituents such as starch, proteins, and lipids. These enzymatic activities are enhanced with the rise in grain moisture and temperature. During storage, the lipase that is inherently present in rice bran hydrolyzes its fats into free fatty

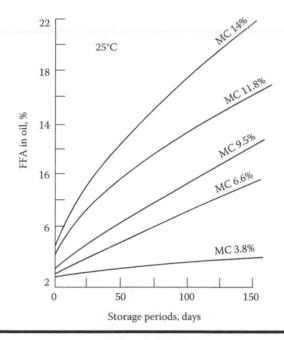

Figure 7.2 Development of free fatty acid (FFA) in brown rice at different moisture contents (MC) at 25°C. (From Chikubu, S., *Training Manual for Training in Storage and Preservation of Food Grains*, APO Project, Japan, 1970.)

acids and glycerol. Free fatty acids increase rapidly when both grain moisture and temperature are high (Figure 7.2). Moreover, with the growth of mold, decomposition of fat is further accelerated by the action of its enzymes. It is known that *n*-capraldehyde and *n*-valeraldehyde produced by the autooxidation of lipid are related to the generation of stale flavor in old rice. Starch, the main constituent of rice, is converted into dextrins and maltose by the action of amylase in the rice grain. It results in an increase in reducing sugars. But in grain this development is not pronounced when the moisture content of rice is around 14% and the temperature is also at a lower level (Chikubu, 1970).

When moisture content of a rice kernel is considerably at a higher level, the carbohydrates are fermented. As a result, alcohol and acetic acid are produced with the formation of acid odor. During storage of paddy, reducing sugars and acidity increase whereas nonreducing sugars decrease. The percentage of germination, nonreducing sugars, and acidity are the three most sensitive characteristics of grain storage (Figure 7.3). The changes in protein during storage are comparatively small and slow.

Besides enzymatic reactions, oxidation by the surrounding air causes changes in color and flavor. Regarding other constituents of cereal grains, vitamins are gradually diminished under ordinary storage conditions.

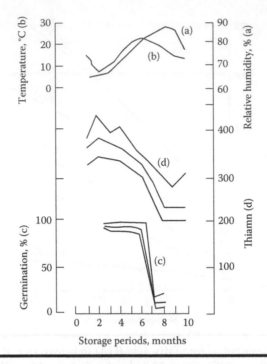

Figure 7.3 Changes in germination and thiamin content of brown rice during storage. (From Chikubu, S., *Training Manual for Training in Storage and Preservation of Food Grains***, APO Project, Japan, 1970.)**

Physiological Changes

Respiration

The life of a food grain is manifested by respiration. It should be noted carefully that, in aerobic respiration, a complete oxidation of the hexose yields carbon dioxide, water, and energy, whereas, in anaerobic respiration, hexose is incompletely decomposed into carbon dioxide, ethyl alcohol, and energy. Direct consequences of respiration are the loss of mass and gain in moisture content of the grain, rise in the level of carbon dioxide in the intergranular air space, and rise in the temperature of the grain. The respiration rates of freshly harvested grains are different from those of old grains and grains damaged by insects, fungi, and molds. The rate of respiration is high for the old grain and the pest-infested grain compared to the fresh grain. The molds, etc., respire at much higher rates compared to the grain itself.

The respiratory rate of stored grain depends largely on its moisture content and temperature. Figure 7.4 shows that at a certain temperature the respiratory rate of brown rice increases as moisture content increases. The moisture content at

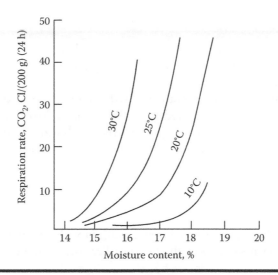

Figure 7.4 Effect of moisture content and temperature on respiratory rate of brown rice. (From Chikubu, S., *Training Manual for Training in Storage and Preservation of Food Grains,* **APO Project, Japan, 1970.)**

which the respiratory rate undergoes a sharp change may be called crucial moisture content and it moves toward lower level as rice attains the higher temperature.

Other factors involved in grain storage are the amounts of O_2 and CO_2 in the air. The respiration of grain under anaerobic condition usually weakens viability and induces quality deterioration. Storing cereal grains under vacuum or airtight conditions appears to be very effective, but anaerobic respiration of the grain at high moisture content makes it unsuitable. The checked and cracked grains have higher respiratory rates than the whole grains under the same condition.

Longevity

The viability period of a grain during storage can be short or long. The grain dies owing to the degeneration of protein which, in turn, is influenced by decay of components in the cell nucleus. Generally, the life of a stored grain is regulated by the grain type, the seed-borne microflora, and by the interaction between temperature and moisture.

Sprouting

Sprouting of the grain during storage occurs mainly owing to generation of heat as a result of infestation. A grain sprouts only when its moisture content exceeds certain limit of moisture content of 30%–35%.

Heating

The stored grain is sometimes heated up by itself without any external cause. This spontaneous heating can be attributed to the respiration of grain in combination with the respiration of infested pests.

Heating usually occurs when grain is stored in bulk. By respiration alone, however, the temperature may not exceed above 35°C. Under favorable conditions, the growth of microorganisms is very rapid and the combined effects of respirations become high, generating more heat. Of the total heating phenomena, 60%–70% of heating can be ascribed to the respiration of pests.

Biological Changes

The changes and damages of grain during storage brought about by insects, microorganisms, and rodents have been discussed earlier. These will be further discussed subsequently.

Moisture Migration

Moisture migration in stored food grains occurs owing to the changes in temperature. One of the most important factors that affect storage life of grain is its moisture content. Moisture migration takes place in a bin even if the cereal grain is at a safe storage moisture level of about 12% or so. But the moisture content of the top layer may go as high as 25%–30%. It results in spoilage of food grains from stored grains in a bin after about 1 year. Greater losses may occur in stored grains in bins for the following conditions: (1) when grains are stored at a high moisture content; (2) where there is an appreciable difference between ambient and grain temperatures; and (3) when a tall or deep bin is used.

Generally, grains are fed to the bins either in warm or in cold condition. The moisture accumulated at the top or bottom of the bin is attributed by the movement of convection air. Most of the times, grains are placed in the bin when the grains are warm. Under this condition, the air in the grain near the surface of the storage bin is comparatively cold and moves down along its circumference and then goes up near the center of the bin, where the air and grain are warm. The air moving through the center of the bin picks up moisture until it moves across the top of the bin. At this location, the grain surface is comparatively cold and the moisture is condensed on the grain, thus raising its moisture content (Figure 7.5a). In consequence, a large quantity of grain is spoiled at the top of the bin.

Sometimes, the grains are fed to the bin under cold condition as in the winter. The air currents rise along the circumference of the bin and move down through its center to the bottom, where moisture condenses on the cold bottom. Under this condition, spoilage of grain takes place at the bottom of the bin (Figure 7.5b) (Hall, 1957).

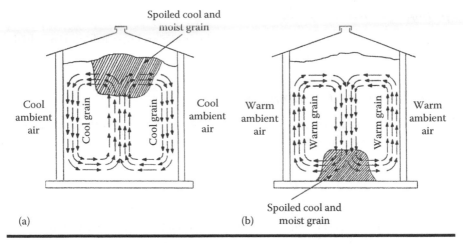

Figure 7.5   Moisture migration in (a) summer and (b) winter.

Grain Storage Pests and Their Control

Deteriorations Caused by Insects and Microorganisms

There are many causes for the qualitative and quantitative deteriorations of stored cereal grains. However, following are the most important: (i) deterioration in the quality of grains, caused by microorganisms; (ii) alternation in chemical constituents of grain owing to its metabolism; and (iii) insects and rodents causing damage by eating and contamination.

Insects produce uric acid, inoculate fungi and bacteria and larvae on the food materials, and develop a foul odor. Quinone and other harmful substances are also being produced on the infested products. Some fungi produce harmful mycotoxins.

Types of Grain Spoilage

Reduction in Mass

About 15 mg of rice kernel can be eaten away by a grain weevil during its growth from egg to adult. A female weevil breeding through three generations per year has the biotic potential of reproducing 1,500,000 offsprings, which are likely to consume 1,500,000 kernels of rice. It is also known that warehouse moth, *Ephestia elutella*, attacks the embryo of wheat kernels making it unfit for germination. These reveal the extent of damage caused by the grain pests and the need for their control (Mitsui, 1970).

Spoilage of Grains by Heating

Heating of grain sometimes is brought about by insect and pest infestation, which ultimately causes serious grain damage.

Reduction in Seed Germination

A seed grain attacked by a germ eater is not likely to germinate. In consequence, germination of seed grains will be seriously affected.

Contamination of Grains by Insects

Food grains, specially the milled products contaminated with dead bodies of insects and their excreta and secretions, often lead to a serious loss of grain quality.

Detection of Insect Infestation

Determination of infestation can be carried out by the following methods: (i) visual examination of surface holes; (ii) floatation method; (iii) staining method for detecting egg plugs; (iv) cracking floatation method; (v) gelatinization method; (vi) uric acid method; (vii) phenol method; (viii) ninhydrin method; (ix) aural techniques; (x) carbon dioxide as an index of infestation; and (xi) x-ray method.

Apart from these methods, milled grain can be analyzed for insects and fragments by the standard AOAC method and grain mites can be detected by any of the two methods, namely, liquid paraffin floatation method and saturated saline floatation method. These methods are not described here in order to keep the volume of the book within limit.

Grain Storage Pests

Among all the grain storage insects, the following 12 species are considered important: (i) the Khapra beetle; (ii) the borer beetle; (iii) the grain weevil; (iv) the rice moth; (v) the grain moth; (vi) the meal moth; (vii) the pulses beetle; (viii) the flour moth; (ix) the red rust flour beetle; (x) the long headed flour beetle; (xi) the saw-toothed grain beetle; and (xii) the flat grain beetle.

Of these 12 insect species, the first 8 are known as primary pests as these are able to damage all kinds of stored grains. The last four insects are unable to damage sound grains but these grow on broken grains or infested grains. That is why the last four species are known as secondary pests.

It may be noted that the major orders are coleoptera or beetles and lepidoptera or moths, which account for about 60% and 9%, respectively, of the total number of species of stored grain pests.

Some of the major stored grain pests are described as follows.

Important Insect Species

Lesser grain borer, *Rhyzopertha dominica* (family: Bostrichidae). These insects mainly eat grain kernels. These are also grown on milled grain products.

Khapra beetle, *Trogoderma granarium* (family: Dermestidae). Under tropical climate, the Khapra beetles often cause more serious damage to stored grain than the grain weevil or the lesser grain borer.

Grain weevils (family: Curcurionidae). The grain weevils are very destructive to stored rice grains. Of these, rice weevils and small rice weevils are the most important, which are distributed all over the world wherever grains are stored.

Saw-toothed grain beetle, *Oryzaphilus surinamensis* (family: Silvanidae). The saw-toothed grain beetle is commonly found in grain and grain products. As it can hardly attack sound grain kernels, it is usually found in grains damaged by primary insects.

Hour beetle (family: Tenebrionidae). The flour beetles are the most destructive insects for flour bran.

Warehouse moth, tropical warehouse moth (family: Phycitidae). These are the destructive insects for both stored grains and grain products.

Indian meal moth, *Plodia interpunctella* (family: Phycitidae). It generally grows and multiplies on grains and grain products, seeds, nuts, etc. It is especially destructive to wheat and brown rice. But it seldom occurs on milled rice.

Angoumois moth, *Sitotroga cerealella* (family: Gelechiidae). It is the only moth that develops within the grain kernel. It attacks both stored and field grains. It grows on barley, oat, rye, corn, and rice.

Rice moth, *Corcyra cephalonica* (family: Galleriidae). It attacks rice, wheat, green gram, cocoa, etc. In India, it grows well on milled jowar and millets also.

Control of Stored Food Grain Pests

Food grain pests can be controlled by preventive and curative methods (Mitsui, 1970).

It is always desirable that preventive measures are taken before the occurrence of infestation and subsequently curative measures have to be taken. The preventive measures are undertaken to avoid infestation by the pests, while the curative measures are used to wipe out any kind of infestation.

Preventive Measures

Preventive measures need consistent and thorough application at frequent intervals. However, preventive methods are divided into the following heads: (1) physical and mechanical measures; (2) chemical measures; and (3) hygienic measures.

Physical and mechanical control methods are (i) drying; (ii) cooling by aeration; (iii) airtight storage; (iv) low temperature storage; and (v) protective packaging.

Chemical control methods include (i) grain protectants; (ii) attractants; and (iii) repellants.

Hygienic measures: The entire storage is to be kept under clean and hygienic condition.

Curative Measures

Among the curative measures, the chemical methods are the most effective, which involve the use of insecticides, are toxic to man, and require special protective devices to safeguard the lives of the personnel applying these chemicals.

Curative measures are (i) physical methods; (ii) mechanical methods; (iii) chemical methods; and (iv) biological methods.

Physical Methods

Heating

All species of stored grain pests at any stage of development will be killed if these are exposed to a temperature of 60°C for more than 10 min or to 50°C for 2 h.

Radiation

Direct Method

Irradiation of insects by β-rays or γ-rays causes physical disorder such as loss of their reproductive power and lives even. The γ-rays have a strong penetrating power and have the potential to be the most effective method of control of grain storage pests. But any of these methods requires a very costly irradiation plant.

Indirect Method (Male Sterilization)

Adult males can be sterilized by irradiation. A continuous release of sterilized male adults for about 1 year can completely destroy the entire population of the pests in some specific areas.

In indirect and other chemical methods, reinfestation may occur from outside sources of insects brought in contact with the same stored grains. Hence, all other methods are only temporarily effective.

Mechanical Methods

Centrifugal force can be effectively utilized for this method. In flour mills, the impact of the flour against the rotating disks and housing of the entoleter is so great that all stages of insects and mites, including the eggs, are killed. The flour thus treated with the centrifugal force comes out of the machine.

Chemical Methods

The chemical control methods employ the following: (a) spraying; (b) fogging; (c) dusting; (d) vaporizers; and (e) fumigation.

Fumigation has been discussed in detail separately.

Chemosterilants: This is a simple method of spraying and dusting chemosterilants for sterilization of insects.

Biological Methods

Natural enemy: Wherever insect infestation occurs in stored grains, there exists almost invariably parasitoid wasps that are parasitic on them. This method is likely to be applicable for field pests, but its application to stored grain pests had not been successful.

Microbial control method: Various bacteria and fungi are parasitic in nature. Hence, these can perhaps successfully be applied to field pests.

Fumigation

Fumigation is an insect-controlling method of exposing stored grains to a lethal concentration of highly toxic gas long enough to kill the insects. Fumigants are the effective chemicals for killing stored grain pests. In the gaseous phase, fumigants can penetrate through stored grains anywhere in bags on stacks or in bulk and mill the hiding insects. These do not have any residual effect.

Insecticides

The insecticides for grain storage pests are also divided into preventive and curative insecticides. For control of additional reinfestation from outside sources after fumigation, a contact insecticide (grain protectant) must be applied immediately after the fumigation on the surface of the stored grains.

Principles of Fumigation

In any fumigation process, a fumigant acts in accordance with the following flow diagram (Mitsui, 1970):

Application of fumigant → Vaporization → Diffusion → (Leakage)

→ Sorption → Penetration → Lethal effect

After application of fumigant, generally it starts vaporizing.

The rate of vaporization mainly depends upon the kind of fumigant, method of application, temperature, and air flow rate.

Generally, the boiling points of the fumigants are approximately proportional to their molecular weights. Both liquid and gaseous fumigants are common. The low-boiling fumigant methyl bromide, which is in gaseous form at room temperature, is called gaseous fumigant.

Ethylene bromide, having comparatively a high boiling point, remains in a liquid state at room temperature and is known as a liquid fumigant. The methods of application are, certainly, different for the gaseous and liquid fumigants.

There are also solid types of fumigants such as aluminum phosphide tablet. They react with the atmospheric moisture to form hydrogen phosphide gas that has lethal effect on microorganisms.

Diffusion

The vaporized fumigant gases reach every nook and corner of any storage system by diffusion. The diffusion of gas takes place owing to convection of air. The rate of diffusion depends on (i) kind of fumigant; (ii) temperature; and (iii) method of application.

Sorption

Reduction in concentration of fumigant gases in any grain storage system takes place mainly owing to sorption of gases by grain and structural materials of the same storage system. The rate of sorption of fumigant gases by grains is dependent on (i) kind of fumigant; (ii) kind of stored grain; (iii) temperature; (iv) gas concentration; and (v) exposure time.

Penetration

The penetration of gas in between individual kernels of the stored grain mass is accomplished by diffusion. It is gradually done by molecular diffusion of fumigant gas due to concentration gradient of gas. Therefore, efficiency of the penetration depends upon (a) kind of fumigant; (b) kind of stored grain; (c) gas concentration; and (d) temperature.

Lethal Effect

The fumigant gas reaches the body of the insect through the aforementioned operations to render the insecticidal effect. The lethal effect mainly depends on (i) toxicity of chemical agent; (ii) dosage; (iii) exposure period; and (iv) temperature.

Applications of Fumigants

In India, fumigation with ethylene dibromide or aluminum phosphide is adopted for farm storage, and methyl bromide or ethylene dibromide or aluminum phosphide is employed in commercial storage of grains. In addition to fumigation, surface treatment with malathion or pyrethrum is employed in commercial grain storage. Due to the low cost of fumigation and convenience, the solid fumigant, namely, aluminum phosphide, has become popular.

Rodent Control

Exclusion of rodents from a storage system is possible if the storage unit as a whole is kept free from rodents and the rodent population is completely eradicated from the locality. Rat-proof construction is necessary to minimize the damage of food grains by rat. The metal bins can be tight enough to prevent entry of rodents.

Construction of a rat-proof food grain storage system is the main step in the direction of rodent control. Silos made of either steel or concrete are sufficiently rat proof. The flat godowns for bag storage are not always rodent proof.

Rat and Mice in Stored Grains

Rodents are one of the most destructive vertebrate animals on earth. The rats consume food grain and at the same time spoil it by nibbling it into brokens and contaminating it with their droppings of urine. Rats eat about 10% of their weight in food each day and contaminate a great deal, thereby rendering it unfit for human consumption. The following six species of rodents are common: (i) *Gerbillus* species (field rat); (ii) *Rattus rattus* (the black rat); (iii) *Rattus norvegicus* (the brown rat); (iv) *Mus musculus* (the house mouse); (v) *Bandicoota bandicoota* (the bandicoot); and (vi) *Bandicoota bengalensis* (the bandicoot).

The genus *Rattus* alone has about 570 named forms. The Norway rat (*R. norvegicus*), the black rat (*R. rattus*), and the house mouse (*M. musculus*) cause the most extensive damage. Two or more of these species are found in most of the countries of the world. In many countries, they are joined by other destructive native rodents such as the bandicoot rat of India and Sri Lanka. Jointly, they destroy many million tons of food grains each year.

Rodenticides (for Rats and Mice)

Hundred percent effective rat poison that meets all requirements under all conditions has not yet been developed. Toxicity dosage levels and relative effectiveness are the most important factors. Less often considered, but of equal importance, are degrees of acceptance and reacceptance and the development of tolerances.

Food Grain Storage Structures

Generally, food grains are stored in traditional or improved rural storage structures in urban warehouses/godowns and in advanced large silos. The material of construction and the size of the structures vary widely. The farmers usually store their produce in either aboveground or underground storage structures. In most cases, local agricultural wastes and minerals are used. These structures are made of bamboo, wood, earthenware, cement concrete, stones, or bricks. In India, capacities of these vary from 250 to 5000 kg. However, higher capacities of these are also available. Nowadays, some progressive farmers use plastic and metallic bins also.

Usually in India, the marketable surplus is being handled in jute bags. The godowns are constructed by the government and other organized sectors. The different kinds of godowns/warehouses vary in plinth, height, construction of walls, and also of the roofs. Their different names are as follows: (1) tubular truss, (2) flat roof, (3) gable type, (4) Nissen hut, (5) cubicles, (6) shell type, (7) reinforced cement concrete (RCC) circular bin, (8) ferrocement structures, (9) rectangular steel bin, (10) crib type, (11) air-warehouse, etc. ISI specifications of some of these structures are available. The metallic or RCC silos are the most advanced storage structures.

As regards improved storage structures such as improved khattis and kothis of India, plastic-lined metal bins of the United States, hemispherical grain silo, silo-cum-elevators, improved flat storage godown of FCI, India, and RCC structure of CBRI call for special attention. Recently, (1) low-density polyethylene (LDPE), (2) high-density polyethylene (HDPE), and (3) polyvinyl chloride, etc., have also been used for the manufacture of plastic bins. The ballooning technique of making stacks moisture proof and resistant to insect and rodent attacks is also unique. As a result, expensive flexible special lined bags are no more necessary for storage of the hygroscopic material (Food Industries, 1979).

Rural Storage Structures

Some of the common improved storage structures in India are described in subsequent paragraphs.

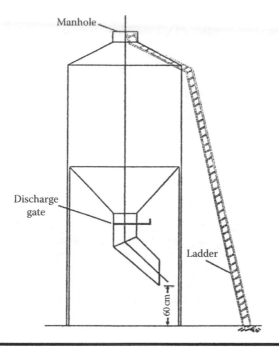

Manhole

Discharge gate

Ladder

60 cm

Figure 7.6   Steel bin.

Steel Bin

It is an outdoor structure of a prefabricated steel bin with a hopper bottom. It is made of a 16 gauge curved MS sheet. It has also metallic ladder and pulley arrangement for filling the grains (Figure 7.6).

Aluminum Bin

It is an outdoor structure, circular in shape.

The bin consists of a cylindrical body of several corrugated aluminum-curved sheets and a conical roof made of flat aluminum sheets. Bolts and nuts are used for assembly of the body and the roof. This bin is constructed on a platform or plinth of 60 cm height, which provides a permanent system against ground moisture. At the same time, the spout that is embedded in the platform allows for easy bagging of the grain. The bin is filled through a manhole at the roof. Both the manhole and the spout are provided with locking arrangements (Figure 7.7).

RCC Bin

The RCC storage structure is circular and waterproof. Its capacity varies from 1.5 to 6 tons. It can be built at the site. The bins having capacity less than 6 tons can be made with 30 cm high RCC rings. The joints of the rings are sealed with cement.

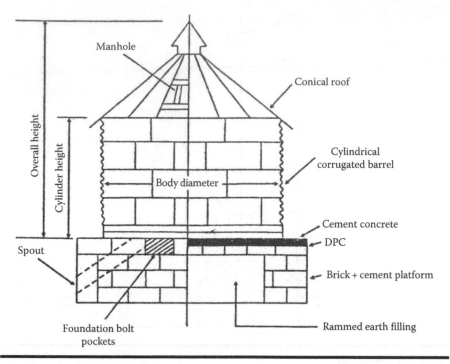

Figure 7.7 Aluminum bin.

A manhole of 60 cm diameter on the top and sliding door or a spout near the bottom are provided for filling and discharging grains. Slope at the bottom is preferable. The manhole and spout should have locking arrangements. Specifications of the bins are tabulated later (Figure 7.8).

Bag and Bulk Storage

In Indian godowns (Figure 7.9a and b), the food grains are stored in bags. Rice, "dhal" (milled pulses), and other milled products are stored in bags. Bag storage has occupied a dominant place in the country because of comparatively small capital investment for godowns and lack of sufficient bulk storage facilities.

Whether a bulk storage or bag storage should be used is based on the following: (i) type of grain to be stored; (ii) length of storage period required; and (iii) whether a particular grain or different kinds of grains are to be stored.

The relative advantages and disadvantages of these are summarized in Table 7.1.

The drawings of the common godowns are furnished in Figure 7.9a and b.

Stack Plan

The floor space is marked into equal rectangles to ensure uniformity in stacking and accounting. A minimum space of 1 m is to be left around each stack.

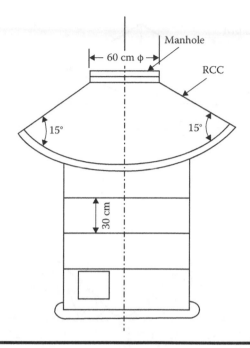

Figure 7.8   RCC bin.

Stack dimensions are on the basis of bag dimension (0.7 m × 0.6 m) and floor dimension. Bags are placed in blocks containing one layer lengthwise and another layer widthwise. The common dimensions of stack are 9.14 m × 6.09 m, 7.43 m × 12.2 m, etc. Various stack plans used for different capacities are shown in Figure 7.10a–d.

Dunnage

It comprises of wooden planks or polythene sheets. Wooden planks of 1.5 m × 0.9 m size are convenient to place below and are easy to carry. The height from the ground is 0.2 m. These should be strong to bear a load of about 0.07 ton/m². A polythene sheet of 0.03 mm thickness is used. It should be black in color. Stacking of bags are generally done in two rows. One is arranged lengthwise and the other is set breadthwise, which constitute a block. Cross stack is made as follows: lengthwise layer alternates with breadthwise layer in the height.

Silos

A silo is a tall storage structure, which is nothing but a deep bin. Reinforced concrete structure (RCC) has been in use widely in the construction of high capacity silos. Concrete is a durable and economical material, but it should be used in

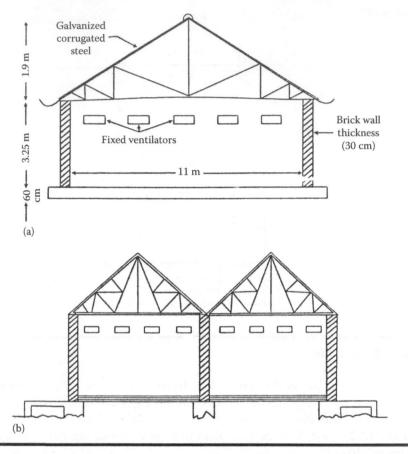

Figure 7.9 **A typical godown structure and conventional trusses. (a) Structure of a typical godown (***L* = 28.5 m, *H* = 3.25 m**). (b) Conventional trusses.**

prefabricated or precast form. Steel is perhaps the most common building material used for the construction of modern grain storage facilities. However, the use of steel silos has been limited in India and many other countries.

The schematic representations of the silos are shown in Figure 17.11a and b.

The major advantages of concrete silos are as follows: (i) These structures are moisture proof, vermin proof, and insect proof; (ii) these have high strength, durability, and workability; (iii) there is maximum space utilization per ton of stored material with a reasonable cost of storage; (iv) practically, no food grain loss takes place in filling, storing, and emptying; and (v) silos are completely fire proof and have a long life. Restructural details are available in Baikov (1978).

Steel silos have the following main advantages: (i) steel members have high strength and these can be used as prefabricated members; (ii) steel silos are gas- and

Table 7.1 Merits and Demerits of Bag and Bulk Storage

Item	Bag	Bulk
Land requirement	Double that of bulk storage (2 acres/5000 tons wheat)	Half of the bag storage
Cost of storage	Slightly cheaper	Comparatively costly
Maximum storage period	1–2 years	5–10 years
Feasibility of mechanical	Difficult	Mechanically operated
Period of construction	12–18 months	12–8 months
Feasibility of shifting the facility	Nil	Metallic structure can be shifted
Economics of handling	Cheaper	More costly
Fumigation	More costly	Cheaper
Cost of jute or other bags	Substantial	Nil
Storage loss (weight)	About 1%–1.5% in 1 year storage	Up to 0.2% for a longer storage period
Possible losses:		
(a) Rodents	It can be made rodent proof	Rodent proof
(b) Birds	Difficult to avoid	Bird proof
(c) Insect	Fumigated	Fumigated
(d) Effects of humidity and temperature	Can be minimized	Can be controlled by aeration
(e) Drainage	Can be controlled	Nil

watertight and have long service life; and (iii) steel can be readily disassembled or replaced.

The main drawback with steel silos is that these are susceptible to corrosion.

Expert System

The complex stored grain ecosystems involve several numbers of parameters such as temperature, humidity, water activity, biological activity, pest development, and pesticide degradations. Keeping these in view, computer programs

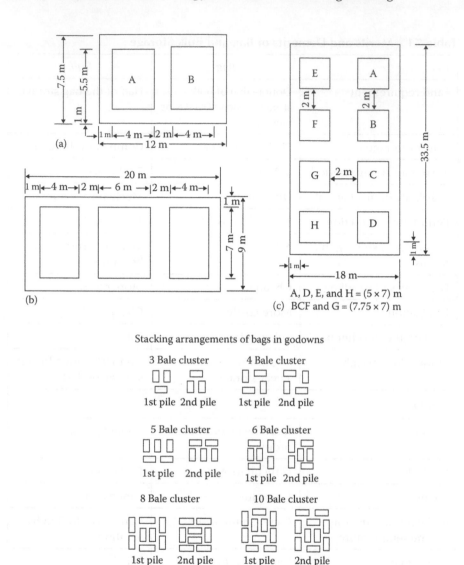

Stacking arrangements of bags in godowns

Figure 7.10 Stacking arrangement of bags in godowns/warehouses. (a) For 100 tons. (b) For 250 tons. (c) For 1000 tons. (d) Stacking arrangement of brown rice in Japan. (From Apo Proj/TRC/IV/68, 1970.)

known as expert systems have been introduced in Australia to forecast the outbreak of pests during storage and to take care of the role of a storage expert to offer solutions of the problems under the prevailing storage conditions. The computerized expert system can be interlocked with remote sensors indicating the grain temperature, moisture, etc., and accordingly signal for acoustic

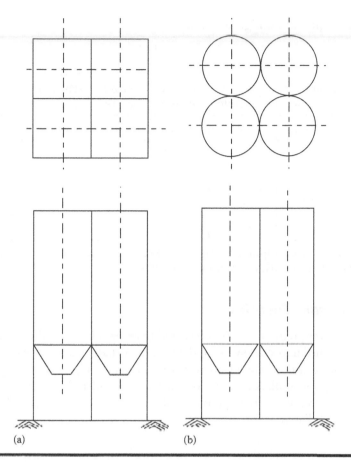

Figure 7.11   Plan and elevation of the silos. (a) Square cross section. (b) Circular cross section.

detection of insect pests and others during bulk storage in silos that finally enables automated control measures.

Optimal Configurations of Silos

The optimal configurations of silos of capacities ranging from 100 to 2000 tons for different *H/D* ratios (viz. 1–12) are furnished in Table 7.2. A small variation in the *H/D* ratio affects much on the weight of concrete and steel and overall cost of silo. The mean of the recommended values should be adopted for the practice. In general, the total cost of silo is the main criterion for the selection of optimal height to diameter ratio (Dubey, 1984).

Table 7.2 Optimal Configuration of Silos

Capacity (tons)	H/D Ratio					
	Based on Total Cost	Based on Cost of Concrete	Based on Cost of Steel	Based on Area	Based on Weight of Concrete	Based on Weight of Steel
100	9–11	3–5	10–12	7–9	8–10	9–11
500	9–11	5–7	10–12	6–8	7–9	5–7
1000	7–9	5–7	10–12	7–9	7–9	7–9
1500	7–9	5–7	10–12	6–8	5–7	7–9
2000	6–8	4–6	8–10	5–7	4–6	7–9

Source: Dubey, O.M., Optimum design of RCC silos for bulk storage of paddy, MTech thesis, IIT, Kharagpur, India, 1984.

Grain Pressure Theories

Theoretically, the grain in bulk is considered to be semifluid and the storage structures are classified into shallow bin and deep bin. A shallow bin is one in which the plane of rupture of the material stored emerges from the top of the filling before it strikes the opposite wall. A deep bin is a structure in which the plane of rupture of the material stored meets the opposite wall before meeting the top surface of the filling material (Figure 7.12a–c).

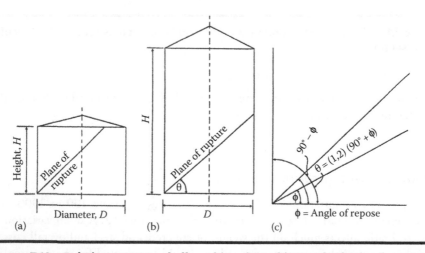

Figure 7.12 Relation among shallow bin, deep bin, and plane of rupture. (a) Shallow bin. (b) Deep bin. (c) Position of plane of rupture between vertical wall and angle of repose.

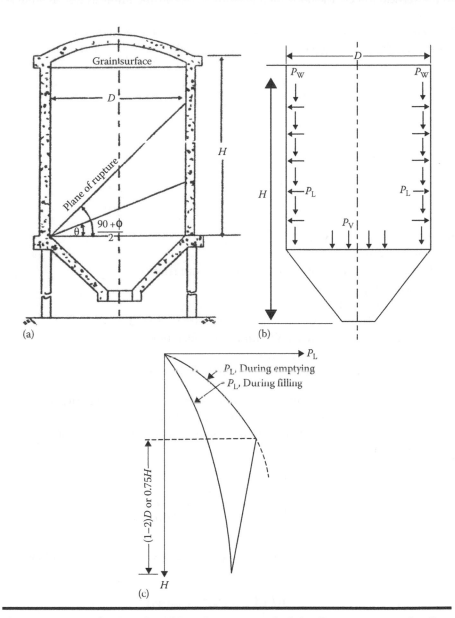

Figure 7.13 **(a) Silo (i.e., deep bin). (b) Stress analysis in silo. (c) Pressure distribution in silo during emptying and filling operations.**

The grains stored in silos (i.e., deep bins) exert pressure on its sides in addition to the vertical forces. The horizontal pressure varies during filling as well as emptying operation (Figure 7.13a–c). It also varies with the location of the discharge hole. Janssen studied the pressure in deep bins for the first time in 1878. On the basis of his theory, the following equations have been developed:

$$P_L = \left(\frac{WR}{m'}\right)\left[1 - e^{-(km'h)/R}\right]$$

$$P_w = WR\left[\frac{hR}{\mu'K}(1 - e^{-(K\mu'h)/R}\right]$$

$$P_v = \frac{P_L}{K}$$

where for a consistent system of units:

P_L is lateral pressure
P_w is vertical pressure transferred owing to wall friction
P_v is vertical pressure
W is specific weight of grain
K is ratio of lateral pressure to vertical pressure in the grain
h is depth of the grain to a point under consideration
R is area of the bin floor/perimeter
μ' is coefficient of friction between grain and bin wall

There are different theories to predict pressure in deep bins, but Janssen's equation is still most widely used in the design of deep bins.

Airy and Rankine developed grain pressure theories for shallow bins. Airy's equation is as follows:

$$P_L = Wh\left[\frac{1}{\sqrt{\{\mu(\mu + \mu')\}} + \sqrt{(1 + \mu^2)}}\right]^2$$

Rankine's equation as given in the following is still widely used in shallow bin design:

$$P_L = Wh\left[(1 - \sin\varphi)/(1 + \sin\varphi)\right]$$

where

μ is coefficient of friction of grain on grain
$\mu = \tan\varphi$, where φ is the internal angle of friction
$\mu' = \tan\varphi'$, where φ' is the angle of wall friction

Both these equations are based on the assumptions that the pressure is caused by a sliding wedge of grain and there is no surcharge. Rankine's equation further assumes that there is no active fractional force between the stored grain and the bin wall.

Economics of Storage

Storage in Godowns

The three types of storage of cereal grains practiced in India are (i) storage in village bins or godowns; (ii) scientifically built modern godowns; and (iii) most advanced silos. Nearly, 75% of cereal grains grown in India are still stored in the very old type of unscientific rural godowns and bins occurring very high losses of grain. Generally, in rural godowns, labor charges, maintenance cost, and cost of gunny bags are involved. Bagged storage occupies a much larger storage area for a given weight of stock and the deterioration of the grains will be more if the building is not properly constructed or adequately water and moisture proof.

The economics of paddy storage in both village godowns and silos reveals that silo storage is more economical than the traditional bag storage in godowns considering the usual 5% losses that occurs in ill-ventilated godowns. Though capital investment is much higher in silo storage, yet the paddy can be kept clean and safe after mechanical cleaning and dying. These facilities are not easily available in the conventional godown storage (Shivanna, 1971).

Storage in Silos

The modern methods of grain storage in scientifically built godowns and silos are definitely advantageous. The bulk holding of grain is simpler and requires much less floor area in silos. If bulk transportation of grain, either in lorries or in railway wagons, is resorted, the transport cost may become considerably cheaper. The paddy storage in silos posed some problems, one of which was the maintenance of quality during storage. Though 14% is the acceptable grain moisture level for safe storage, damage of grains has been noticed due to variation in the atmospheric relative humidity and temperature particularly during rainy seasons. For safe storage of paddy in silos, a regular aeration system should be there to counter the effects of high atmospheric temperature and relative humidity.

Cost of Storage in Traditional Godowns and Modern Silos

The traditional bag storage has an advantage that there is no apparent heavy capital investment.

The major factors augmenting the cost of paddy storage in silos were financial such as interest on loan taken for silo construction and purchase of paddy.

The cost of furnace oil consumption for drying was also an important factor. If husks were used as fuel in place of furnace oil, however, the drying cost would have been reduced considerably. Depreciation on building and machineries were comparable with bagged godown storage. But the cost of gunny bags, labor, etc., came in.

The silo storage had many advantages in saving manual labor, which was difficult to obtain during peak seasons of transplantation and harvesting coinciding with the procurement season. It also minimized the grain losses and the investment on the cost of gunnies and gave a definite advantage in the form of a long hygienic storage. But proper care had to be taken in the initial stages of precleaning, drying, and cooling of paddy before sending into silos (Shivanna, 1971).

PARBOILING
AND MILLING

Chapter 8

Parboiling of Grain

Parboiling is a hydrothermal treatment given to the grain for gelatinization of starch within the grain in order to improve its milling quality. During gelatinization, an irreversible swelling and fusion of starch granules occur that change the structure of the starch from a crystalline to an amorphous one. Due to the gelatinization of the starch, hardness of the kernel increases during parboiling. In consequence, it changes the physicochemical and organoleptic properties of grain kernel. Practically, parboiled rice is rice partially precooked in paddy form. Generally, the parboiling process consists of soaking grain in water to its saturation and steaming or heating the grain to gelatinize its starch. The grain is then dried before milling. Some varieties of paddy have inherently poor milling qualities due to their brittle kernels. To counteract this, premilling treatment, known as parboiling, originated in ancient India. Parboiled rice is still popular in India, Pakistan, Bangladesh, Sri Lanka, and Myanmar. It is consumed in some of the countries of Africa as well as some parts of Asia, like Malaysia and Thailand. Parboiling has been in practice in the United States and Italy on a commercial scale for the last several decades. About 20% of the world rice production is parboiled (Gariboldi, 1974). Thus, this practice is a widespread food industry in the world.

The main operations, namely, soaking, steaming, drying, and milling, involved in the production of milled parboiled rice are discussed with principles and practice in the subsequent paragraphs.

Advantages: Parboiling of paddy results in the following major advantages:

- Less broken grains in the milled rice due to hard texture
- An increase in the total yield of rice (milling yield)
- Greater nutrient status (particularly the B-vitamin contents)
- Translucent rice kernels
- Less loss of solid gruel during cooking

- Less susceptible to insect attack during storage
- Non-glutinous and nonsticky rice
- Higher percentage of oil in parboiled rice bran (about 25%–28% oil content)
- Relatively stabilized bran produced during parboiling
- Easier husking (leading to higher husking efficiency) during milling

Disadvantages: Some of the disadvantages of parboiling of paddy are as follows:

- Relatively darker-colored rice produced
- Off-flavored rice, produced by traditional parboiling process, due to prolonged soaking
- Longer cooking time compared to raw rice
- Chances of development of mycotoxins in the long traditional parboiling process
- Destruction of some natural antioxidants during heat treatment
- Requirement of more power for polishing
- Requirement of an additional initial capital

However, in spite of these disadvantages, the higher milling outturn of parboiled rice compared to raw rice and higher returns from high oil content parboiled bran ensures higher return to the miller.

Principles

Rice starch consists of two components: amylose and amylopectin. Amylose has a simple chain structure and is soluble in hot water. Some non-waxy varieties of rice starch may contain as high as 37% (by weight) of amylose. Amylopectin, with a branched chain structure, is not soluble in water and tends to form a viscous suspension, especially when heated. The cooking qualities of rice are largely dependent on the amount of its amylose content. An original form of starch granule exhibits the characteristics described as follows (Bhattacharya, 1985).

Birefringence: Orderly distribution of starch layers in the crystalline structure appears like a distorted spherocrystal when observed against plane polarized light under a microscope. This phenomenon is known as birefringence. Owing to the hot water–soaking treatment during parboiling, rice starch loses birefringence as a result of a change of structure from crystalline to amorphous.

The breakdown of the starch granule on heating in water occurs in three distinct phases: (1) swelling, (2) gelatinization, and (3) retrogradation.

1. *Swelling*: During the first phase, water is slowly and reversibly taken up, resulting in a limited swelling. The viscosity of starch suspension does not increase appreciably. The starch granules retain all their characteristics like appearance and birefringence on cooling and drying. This phenomenon controls the swelling of paddy during hydration in water at a temperature below 65°C.

2. *Gelatinization*: The second phase of swelling starts during hydration at a temperature above gelatinization temperature (GT) of around 70°C. When starch granules attain a certain level of moisture at this temperature, they suddenly swell in size with an absorption of a huge amount of water. It is caused by the rupture of hydrogen bonds between amylose and amylopectin components, exposing more surface area for water absorption by the starch granules. This is also characterized by rapid rise in viscosity of the starch suspension, as on cooling and drying, the granules alter in appearance, losing their crystalline structure and, consequently, their birefringence. This phenomenon of irreversible change of phase is known as gelatinization. The temperature at which it occurs is referred to as gelatinization temperature. The GT varies with the varieties of paddy.

3. *Retrogradation*: With further rise in temperature above 75°C or so, the starch granules become somewhat like formless sacs. The soluble part is dissolved in the solution. On slow cooling and drying, samples produce a hard gel-like starch resulting in retrogradation, which is due to a close alignment of the simple chain amylose molecules.

Soaking/Hydration

The process of water absorption along with the swelling of grain is known as soaking, steeping, or imbibition. It is mainly a diffusion operation. As a result of water absorption, the grain swells and the water moves inside the grain until the water vapor pressure inside the grain is less than that of soaked water. Its movement stops when equilibrium is reached. Soaking is the result of molecular absorption, capillary absorption, and hydration.

Some of the important hydration parameters, namely, soaking temperature, time, and pH affecting the efficiency of parboiling system and quality of parboiled rice are described as follows.

Soaking Temperature, Duration, and pH

While hot water accelerates the rate of soaking and reduces the time of soaking, the milled product is more likely to be discolored as a result. The loss of whiteness of the milled parboiled rice is also evident with the increase of either soaking temperature or duration. Both soaking temperature and time are responsible for the development of color in parboiled rice due to Maillard reaction. The color of parboiled rice may also be affected by the pH of the steeping water. The coloration is minimum if the pH of water is close to 5. There may be a loss of whiteness on either side of pH 5.

The volume of paddy increases due to soaking. The volume of soaked paddy is less than the sum of the initial volume of paddy and the volume of water absorbed

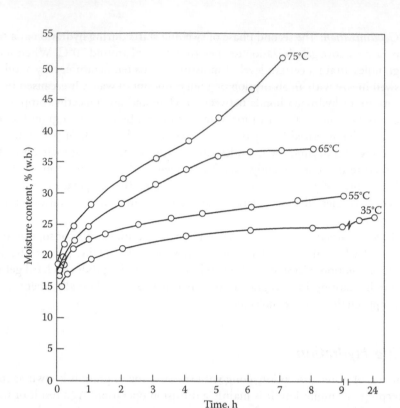

Figure 8.1 Soaking characteristics of paddy (Pankaj) at different temperatures. (From Pandeya, 1997.)

(Ali and Ojha, 1976). The soaking process is always accompanied by release of heat. A considerable amount of kinetic energy is lost as heat when the molecules are absorbed.

The soaking provides the starch with a quantity of water for gelatinization. The rate of soaking is always dependent on the temperature of soaking water (Figure 8.1). The initial rate of soaking is very high at all temperatures, but it decreases with the progress of soaking period and gradually tends to zero at any soaking temperature up to a level of 60°C. However, at GT and above, following gelatinization of the starch and bursting of the grain, the soaking rate further increases rapidly.

The initial high rate of soaking is mainly due to quick absorption of water by the hull and the filling of water in the space between hull and grain kernel. The absorption of water by grain is a diffusion process. At higher temperatures, the diffusion coefficient increases as does the soaking rate. Absorption of water takes place in two distinct phases. In case of paddy, the first phase of soaking continues till it reaches a moisture content of 24%–43% (d.b.), depending upon soaking temperature. At this point, the grain bursts and the rate of soaking becomes very fast. The moisture at which the second phase of soaking starts may be called the critical moisture content in soaking. Its value depends both on the soaking temperature and the variety of paddy. The soaking characteristics of different varieties of paddy are usually different because of the differences in the GTs. The GT varies normally from 68°C to 75°C.

Pneumatic Pressure and Reduced Pressure

In order to reduce the steeping time, vacuum (reduced pressure) and pneumatic pressure have been used, keeping the temperature of steeping water within limits that will not adversely affect the quality of parboiled rice. The steeping time can be substantially reduced, first by removing interstitial air and then by applying pneumatic pressure at a relatively low steeping temperature (Figure 8.2a and b).

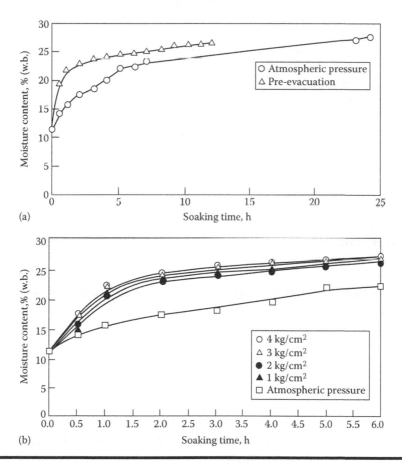

(a)

(b)

Figure 8.2   (a) Effect of pre-evacuation on cold water at room temperature soaking of paddy. (b) Effect of pneumatic pressure on cold water soaking of paddy.

Hydration Equations

The following rate equation has been developed with the help of a mathematical model of nonstationary state diffusion in solids of arbitrary shape for soaking wheat in water:

$$(\bar{x} - x_0) = k_m \sqrt{\theta} \tag{8.1}$$

where

$$k_m = \left(\frac{2}{\pi}\right)(x_s - x_0)\left(\frac{S}{V}\right)\sqrt{D} \tag{8.2}$$

x_0 is initial average moisture content (kg/kg, d.b.)
$\bar{x}$ is average moisture content for a soaking period (kg/kg, d.b.)
x_s is effective moisture content at the bounding surface at times greater than zero (kg/kg, d.b.)
S is exposed surface area of a solid grain (m²)
V is volume of solid grain (m³)
D is diffusion coefficient (m³)
θ is soaking period (s)

The Arrhenius equation for diffusion operation can be expressed as follows:

$$D = D_0 \exp(-E/RT)$$

where
E is activation energy, kcal/kg mol
R is gas constant
D_0 is diffusion constant

Following the Arrhenius type of diffusion equation, an empirical equation has been developed:

$$(\bar{x} - x_0) = \Delta x_i + k_m \sqrt{\theta} \tag{8.3}$$

where Δx_i is initial moisture gain of grain (kg/kg, d.b.).

The values of Δx_i and k_m can be estimated from the intercept and slope of $(\bar{x} - x_0)$ vs. $\sqrt{\theta}$ plot, respectively. The earlier equation is a linear relationship between $(\bar{x} - x_0)$ and $\sqrt{\theta}$ within a certain limit of moisture content (Figure 8.3).

A break in Arrhenius plot of k_m indicating two different values of activation energy for the two ranges of temperatures has been noted. The break for different

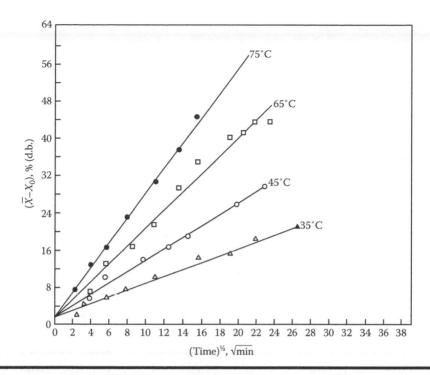

Figure 8.3 **Plot between gain in moisture ($\bar{x} - x_0$) vs. $\sqrt{\text{time}}$ for paddy (Pankaj).**

varieties of paddy has been found to occur at a temperature between 65°C and 75°C, due to gelatinization of starch granules.

Figure 8.4 clearly exhibits this phenomenon of break in the Arrhenius plots. For a more general expression, Equation 8.3 can be expressed as

$$(\bar{x} - x_0) = \left(\frac{2}{\pi}\right) k_m \sqrt{\theta} \tag{8.4}$$

where

$$k_m = (\bar{x} - x_0)\left(\frac{S}{V}\right)\sqrt{D} \tag{8.5}$$

Considering the moisture gain due to initial absorption, Δx_1 (kg/kg, d.b.), all soaking data can be expressed by

$$(\bar{x} - x_0) = \Delta x_i + k_m \sqrt{\theta} \tag{8.6}$$

The break in the Arrhenius plot of log k_m vs. the reciprocal of the absolute temperature around 65°C–70°C took place in the vicinity of GT as shown in Figure 8.4.

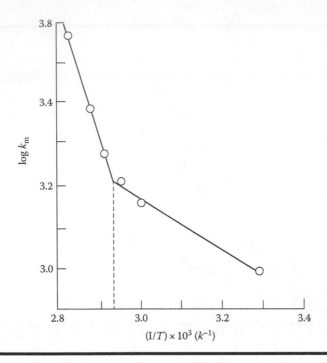

Figure 8.4 Log k_m as a function of the reciprocal of the absolute temperature for soaking of paddy.

Problem

In a batch soaking and steaming unit, raw paddy with an initial moisture content of 0.128 (d.b.) is soaked at 70°C under isothermal condition. If the initial moisture gain Δx_i for this paddy at 70°C is 0.045 and the value of k_m is given as 2.199 × 10^{-3}, calculate the time required to attain a moisture content of 41.9% (d.b.).

Solution

Following the equation $(\bar{x} - x_0) = \Delta x_i + k_m \sqrt{\theta}$,

$$\bar{x} = 0.419 \quad \text{and} \quad x_0 = 0.128$$

$$\Delta x_i = 0.045, \quad k_m = 2.191 \times 10^3$$

$$\sqrt{\theta} = \frac{\bar{x} - x_0 - \Delta x_i}{k_m} = \frac{0.413 - 0.128 - 0.045}{2.191 \times 10^3} = 117.00 \text{ s}^{1/2}$$

Therefore, $\theta = 3.8$ h

Steaming

The steam is preferred for gelatinization because of the following reasons:

1. It does not remove moisture from the grain, rather adds moisture by condensation.
2. It has high heat content.
3. It sterilizes the grain mass.

Steaming temperature and time are the two most important parameters controlling the degree of coloration and hardness of parboiled rice. Higher the values of these parameters, darker is the color and harder is the texture of the end product. However, a suitable combination can be found out to produce translucent parboiled rice of good quality. During steaming, the granular texture of the endosperm becomes pasty due to gelatinization; any crack in the caryopsis is sealed; the endosperm becomes compact and translucent; most of the biological processes are completely annihilated; and the enzymes are also inactivated. If the starch is not fully gelatinized, white cores will remain in the endosperm. So steaming should be just sufficient to cause complete gelatinization. If there is a higher degree of gelatinization, it will be reflected in the hardness of the grain and the deepness of the color. The coloring may be due to reaction between reducing sugars and amino acids. The time and temperature of steaming are to be controlled in order to get the desirable characteristics in the final product. It shows that large increase in volume occurs between steam temperatures of 100°C and 120°C.

In India, saturated steam at 4–5 kg/cm² pressure for a period of 20–30 min is employed for steaming the soaked paddy in a batch process. The requirement of steam per ton of paddy parboiled in a modern parboiling plant using steam at 4–5 kg/cm² is about 120 kg for soaking, 60 kg for steaming, and 20 kg for losses.

Drying

After soaking and steaming, paddy contains about 45%–50% (d.b.) moisture. It is required to be dried to a moisture content of about 17.3% (d.b.) for safe storage and proper milling.

In India and many other Asian and African countries, a part of drying is still carried out in the sun on large paved yards in rice mills. The paddy is dried to a moisture content of about 20% in a day and then heaped and tempered and then again dried for a few hours to bring down the moisture to 14%–15%.

The drying is also carried out in mechanical dryers with hot air. The most important aspect in drying is that the process should be carried in two stages. If the drying is continued in one stage below a moisture of 18%, there is considerable amount of breakage. But if it is conditioned at that level and again dried to 14%–15% moisture, the breakage is considerably reduced no matter how fast is the drying (Bhattacharya, 1985).

If parboiled paddy is uniformly dried by shade, sun, or hot air following the preceding two-pass drying procedure with conditioning/tempering in between two passes, there is practically no breakage. While drying, parboiled paddy two points are to be borne in mind. First, breakage does not occur throughout the drying process but only when the moisture content reaches and then crosses 18%, no matter how fast is the drying operation. After that, breakage increases sharply (Bhattacharya and Indudharaswamy, 1967). Second, the cracks are not developed during drying but cracks develop over a period of 2 h after drying. Therefore, drying of parboiled paddy should be carried out in two passes to avoid the breakage of rice during milling.

In mechanical drying, hot air is forced through the grain, which evaporates and carries away the moisture from the grain. LSU or any other columnar continuous flow dryers can be used for drying of parboiled paddy. In India, a large number of rice mills are now using the rice husk as a fuel for generating the steam.

During the first stage of drying, the moisture content is reduced from about 40% to 25%. This is followed by 8–10 h of tempering to equalize the moisture within the kernels. The tempered paddy is then dried to 14%–15% moisture content during the second pass. To achieve faster moisture removal, drying air at a temperature as high as 120°C is being used in commercial parboiling and drying plants. To remove the bulk of the surface moisture during the first drying pass, the drying air may be kept at a lower temperature of 95°C–100°C. However, during the second drying pass when moisture from inner part of the grain is to be removed, the temperature of the drying air may be kept at a low level of 75°C–80°C.

Milling

Breakage of rice during milling is due to inherent cracks and fissures present in the kernel. Parboiling seals the cracks and hardens the grain. The breakage also depends on the milling conditions and the type of milling machineries used. Short rice varieties are usually more resistant to breakage as compared to long slender varieties. The head yield of rice is considerably improved by parboiling, particularly for the varieties having poor milling qualities.

Parboiled rice takes longer time compared to raw rice for the same degree of polish. Moreover, parboiled rice requires three to four times as much abrasive load as raw rice to attain the same degree of polish (Raghavendra Rao et al., 1967). However, to achieve an equal degree of surface bran removal, parboiled rice must be subjected to less polishing than raw rice. If the degree of polish is done to a level required for minimum consumer acceptability (80% of bran removal), parboiling increases the total yield from 1% to 1.5% and the yield of marketable rice (containing brokens not larger than one-fourth the size of the original rice) by about 3.5%.

Product Qualities

During parboiling, many nutrients are diffused inside the endosperm and hence loss of vitamins during polishing is minimized. As much as 70% vitamin B is retained in the parboiled rice produced by improved method, whereas the traditional parboiled rice retains only 30%:

- Development of color in rice during parboiling is mainly due to the nonenzymatic Maillard-type browning that can be inhibited by bisulfite. The husk pigment may also contribute color by diffusing into the endosperm.
- The cooking quality of rice is expressed in terms of cooking, swelling capacity expansion ratio, color, solids in gruel, and pastiness. Raw rice takes about 15–20 min to become fully cooked in boiling water, whereas parboiled rice requires about 30–40 min for a comparable degree of softness. Raw rice cooked beyond 20 min becomes pasty (Mahadevappa and Desikachar, 1968).
- Swelling capacity is expressed as the ratio of the final to the initial volume or weight of rice. The water absorption capacity, as expressed by the swelling ratio, is significantly lower for parboiled rice than for raw rice cooked for the same period. But if raw rice and parboiled rice are cooked to an equivalent degree of softness, then parboiled rice can absorb more water without losing its shape (Mahadevappa and Desikachar, 1968).
- Expansion ratio is the ratio of the dimensions of cooked and uncooked rice, both along the length and breadth. At the equivalent degree of softness, parboiled rice expands more along the breadth than raw rice, whereas the expansion along the length is not significantly different.
- The loss of solids into the gruel is greater in raw rice than in parboiled rice.

Methods of Parboiling

Parboiling is a premilling treatment of paddy, which is actually a method of precooking rice kernel within the husk cover. It requires both water and heat and involves mass as well as heat transfer simultaneously. As discussed earlier, the process mainly involves soaking, steaming, and drying before milling. The parboiling methods can be broadly discussed under two heads, namely, traditional and improved methods.

Traditional Methods

In India as well as in other countries, a number of traditional premilling treatments known as "sela," single boiling/steaming, double boiling/steaming methods, etc., are still in practice with some variations. In "sela" treatment, paddy is steeped in water under ambient conditions for 24–48 h and then gently roasted in hot sand for gelatinization. Finally, the roasted paddy is further dried in the sun before milling. However, the following two commercial methods are popular in India and some other countries:

- Single boiling/steaming
- Double boiling/steaming

Single Boiling/Steaming Process

The method is called "single boiling/steaming" because soaked paddy is steamed only once. The paddy is soaked in cold water at ambient condition for around 2–3 days in the cement tanks of about 5 tons capacity each. The soaked paddy is steamed for a few minutes in small mild steel cylindrical kettles and then dried in the sun. During prolonged soaking of the paddy, the rice prepared out of it produces a foul odor, as a result of fermentation in the course of soaking.

Double Boiling/Steaming Process

The "double boiling/steaming method" is thus called as paddy and is steamed twice in its sequence of operations. The raw paddy is first steamed for few minutes in the steaming kettle. There are two generally vertical mild steel steaming kettles. The size of each kettle is generally 700–900 mm in diameter and 1.2–2.0 m in height with a conical bottom fitted with a sliding gate. The kettles are provided with steam pipe. The direct steam from the boiler at a pressure of 6–7 kg/cm² is injected through the steam pipe. The steamed paddy is discharged from the bottom and quickly dumped into the masonry water in the soaking tank. The water level in the tank is maintained in such a way that the steamed paddy is completely submerged in warm water during the soaking period. Because of large exposed surface area, the temperature of soak water comes down to around 40°C–45°C within a few hours. The paddy is allowed to soak for about 36–72 h after which the water is drained off. The soaked paddy is steamed in the steaming kettles for a sufficient period for the second time for actual gelatinization of the kernel. The parboiled paddy is discharged from the kettles and allowed to dry in the sun in the drying yard or in a mechanical dryer. In this process, light yellowish rice of good cooking quality is produced. But the foul odor of rice persists owing to fermentation during long soaking.

Improved Methods

Realizing the drawbacks and limitations of the traditional methods, many improved commercial methods have been developed for efficient parboiling. Of these, the following are well known:

- Rice conversion process (United Kingdom)
- Schule process (Germany)

- Crystal process (Italy)
- Malek process (United States)
- Jadavpur University method (India)
- Central Food Technological Research Institute (CFTRI) method (India)
- Pressure parboiling method (India)

Rice Conversion Process (United Kingdom)

This batch process of parboiling was patented in the United Kingdom and adopted in the United States. Steeping is carried out in an autoclave after the paddy is deaerated by applying vacuum. This facilitates rapid water absorption. Next, hydrostatic pressure is applied by introducing compressed air to the steeping water. The combined effect of vacuum and pressure reduces the steeping time to a considerable extent. Steaming is done in a rotating steam-jacketed autoclave at about 1 kg/cm² for sufficient time, after which a vacuum is applied to make the grain free from excess water. Drying is first done in the same autoclave under vacuum keeping the paddy hot by contact with steam-heated surfaces and then completed in a rotary dryer using heated air.

Schule Process (Germany)

This batch process has been developed by Schule Co., a rice-processing machinery manufacturer in Germany. Only water is heated by steam. The paddy is fed to a pressure vessel where it is soaked for 2 or 3 h at a moderate temperature of water, which is kept in circulation. Then a hydrostatic pressure of 4–6 kg/cm² is applied in the tanks with compressed air. The cooking is performed by releasing pressure and re-admitting very hot water to achieve the gelatinization of starch. The water is drained off and the paddy is first dried in a vibratory dryer under semifluidized condition and then in a series of columnar dryers under mild conditions with proper tempering.

Crystallo Process (Italy)

The Italian "crystallo" process has been in use from the early 1950s. In this process, paddy is first soaked in cold water to remove the impurities and lighter grains. Steeping is done in a stationary autoclave where vacuum is applied first after which hot water is injected and then a hydrostatic pressure of about 6 kg/cm² is maintained at a controlled temperature. Steaming and drying are done in a rotary autoclave, which is also fitted with a steam jacket. The drying is done under vacuum. The product is then allowed to pass through a cooling tower and then it is stored in a storage bin.

Malek Process (United States)

This is a semicontinuous process. The steeping is done at a relatively high temperature for 3–6 h in batch tanks. It is then steamed into a vertical cylindrical

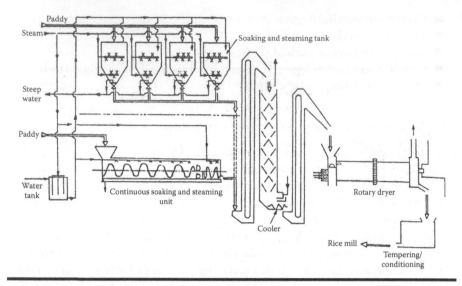

Figure 8.5   Jadavpur University parboiling method.

autoclave, which has a conical base. The drying is done continuously in a steam-heated rotary cylindrical dryer and then in a vertical through flow columnar drier with medium hot air. The parboiled rice so produced is amber-colored and fully gelatinized. The grains are sufficiently hard.

Jadavpur University Method (India)

This is a semicontinuous process. Steeping is done at about 70°C within 3½ h in vertical cylindrical batch tanks. Steaming is carried out in the same tank having perforated steam pipes and is continued only for a few minutes. After steaming, the paddy is rapidly cooled in vertical cup-and-cone-type air-cooling arrangements. The cooling helps to maintain a lighter color in the finished rice. After cooling, the paddy is dried in a steam tube rotary dryer at a temperature of about 90°C. The paddy is then tempered and stored before milling. Simultaneously, a continuous soaking and steaming unit has also been developed (Figure 8.5).

CFTRI Method (India)

The CFTRI batch method was primarily aimed at improving the yield and quality of rice with a lower capital investment. Both the soaking and the steaming are done in the same upright mild steel cylindrical tank, each holding 3–4 tons of paddy. Steam enters through the perforated pipe at the center and through other perforated pipes. The base of the tank is cone shaped and is closed at the bottom by a watertight hatch. At the side of the hatch, there is a valve for draining the steeping water (Figure 8.6).

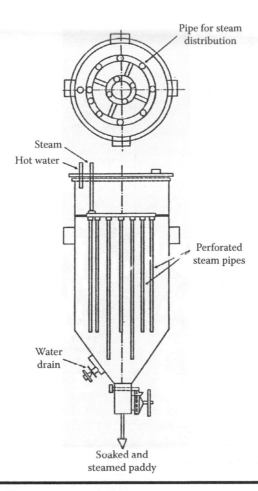

Pipe for steam distribution

Steam

Hot water

Perforated steam pipes

Water drain

Soaked and steamed paddy

Figure 8.6   Details of a parboiling tank.

During parboiling, the tank is filled with water heated by steam injection to 85°C–90°C. The paddy is then fed inside the tank. The temperature of water drops to 70°C–75°C. After 3½ h of steeping with occasional recirculation of hot water, the water is drained off. Steam is then passed through the perforated pipes until the husks just begin to crack open. After steaming, the hot paddy is unloaded through the bottom hatch to spread over the drying yard for sun or mechanical drying. If the drying is done mechanically, soaked and steamed paddy is discharged onto a belt conveyor and then fed to the dryer by an elevator (Figure 8.7).

Pressure Parboiling Method (India)

This batch method had been developed in India in 1969 (Shivanna, 1971a). The paddy is soaked for nearly 40 min at a temperature of 80°C–90°C in a vertical

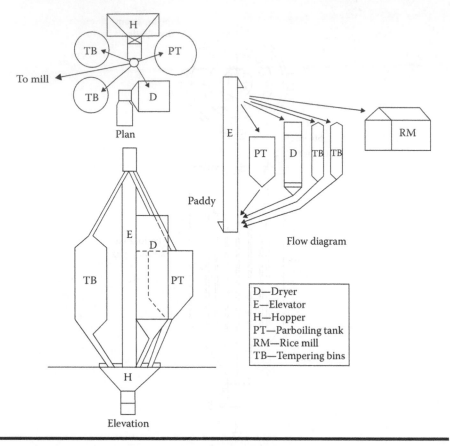

Figure 8.7 Layout of a 12 tons/day parboiling plant.

closed tank and then steamed under pressure for about 18 min. It requires short time for drying also. The whole parboiling is completed within a short time. The rice has a pleasing, slightly yellowish, uniform color.

The main advantages of the aforementioned method are as follows:

■ A reduction in soaking and drying time
■ An increase in shelling efficiency (nearly 80% of paddy-husk splits during steaming)
■ An increase in milling outturn because the grains are resistant to breakage

The deep yellowish colored rice produced requiring longer time to cook is the main drawback of the process. However, some pressure parboiling plants have been installed in India.

A design problem: Design of a CFTRI-panboiling tank matching with a 2 tons paddy/h rice mill (Figure 8.8).

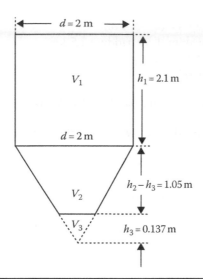

Figure 8.8 Dimensions of the parboiling tank.

CFTRI method of parboiling of paddy has the following requirements:

Operation	Time Required (Approximately) (min)
1. Filling paddy into parboiling tanks (PT)	30
2. Filling hot water at 90°C in PT	30
3. Soaking paddy at 70°C in PT	210
4. Draining of steep water from PT	30
5. Steaming of soaked paddy in PT	30
6. Unloading	30
Total cycle time (approximately)	360 min = 6 h

$$\text{Batch size} = (2 \text{ tons/h}) \times 6 \text{ h} = 12 \text{ tons}$$

Parboiling operations and necessary data

Soaking of paddy in hot water at 70°C for 3½ h
Steaming at about 4 kg/cm² pressure for about 30 min
Bulk density of raw paddy = 560 kg/m³
Allowance for swelling of soaked paddy = 33%
Nos. of parboiling tank of 3 tons capacity each = 4

Diameter of steam pipes = (25–37) mm
Water required for soaking 1 kg paddy = 1.5 kg
Steam required = 250 kg/ton of paddy
Capacity of PT = 3000 kg

$$\text{Volume of 3 tons paddy} = 3000/560 = 5.36 \text{ m}^3$$

Total volume of soaked paddy after swelling (33%) = 5.36 × 1.33 = 7.125 m³

$$10\% \text{ free board} = 7.125 \times 0.1 = 0.7125 \text{ m}^3$$

$$\text{Total volume} = 7.837 \text{ m}^3$$

$$\text{Volume of cylindrical portion, } V_1 = (\pi/4) \times (d^2 h_1)$$

Assuming $d = 2$ m

$$V_1 = (\pi/4) \times (2)^2 \times (h_1) = \pi(h_1) = 3.142 \ h_1 \text{ m}^3$$

Volume of the conical hopper bottom, $V_2 = 1/3(\pi/4) \times (d^2) \cdot (h_2)$.
Assuming $d_2 = d = 2$ m and angle of the hopper bottom = 50°

$$\text{Since, } h_2 = (d/2) \times (\tan 50°) = 1.192 \text{ m}$$

$$\text{Hence, } V_2 = (\pi/12) \times (4) \times (1.192) = 1.248 \text{ m}^3$$

$$\text{Volume of } V_3 = 1/3(\pi/4) \times (d_3)^2 \times (h_3)$$

Assuming $d_3 = 0.23$ m

$$h_3 = (d_3/2) \times (\tan 50°) = 0.137 \text{ m}$$

$$\text{Hence, } V_3 = 1/3(\pi/4) \times (d_3)^2 \times (h_3) = 1.89 \times 10^{-3} \text{ m}^3$$

Actual volume of the tank, V

$$V = V_1 + V_2 - V_3$$

$$= 3.142 \ h_1 \text{ m}^3 + 1.248 \text{ m}^3 - 1.89 \times 10^{-3} \text{m}^3 = 7.837 \text{ m}^3$$

Therefore, $h_1 = 2.1$ m (approximately)
Therefore,

■ Diameter of the parboiling tank, $d = 2$ m
■ Height of the cylindrical portion, $h_1 = 2.1$ m (approximately)
■ Height of the conical bottom = $h_2 - h_3 = 1.055$ m

Chapter 9

Parboiling of Wheat

Debranned (polished) and cracked parboiled wheat is known as *bulgur*. Bulgur is a special product of wheat. Various cooked foods, quick cooking products, breakfast cereals, etc., can be prepared using bulgur. It is mainly used as a substitute of rice. Parboiled wheat resembles rice when it is debranned and split into two or more pieces. It takes about 25–30 min to cook fully.

Bulgur is consumed in many countries, namely, Australia, Argentina, Canada, and the United States. Bulgar wheat has been successfully produced on a commercial scale in India also.

The process of bulgur wheat production is more or less the same as that of parboiled rice. It consists of soaking, followed by steaming, drying, and milling (i.e., debranning and cracking). However, in this chapter, only parboiling and drying of wheat have been discussed in detail and others in brief.

Principles of Parboiling of Wheat

Soaking

For complete gelatinization of starchy endosperm by open steaming, the moisture content of raw wheat has to be raised to the level of 80% (d.b.) by soaking prior to steaming.

Figure 9.1a and b shows the hydration characteristics of a local variety of wheat at a soaking temperature of 75°C. It may be seen from this figure that at 75°C, it takes about 2.5 h to increase the moisture content of wheat from 15% to 83% (d.b.). Soaking temperatures should not be above 75°C so as to minimize the bursting of kernels and leaching losses as well.

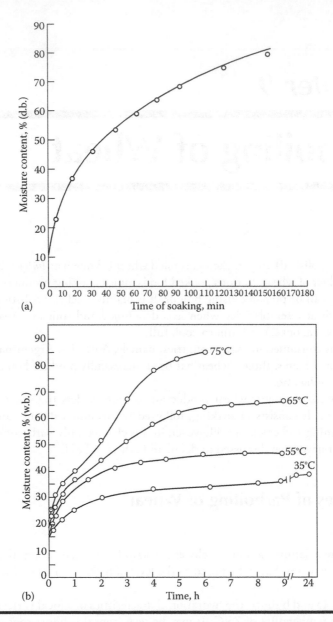

Figure 9.1 (a) Hydration characteristics of wheat (temperature of water—75°C). (b) Soaking characteristics of wheat at different temperatures. (From Pandeya, A., Drying and milling characteristics of various types of parboiled paddy and wheat, MTech thesis, IIT, Kharagpur, India, 1997.)

Steaming (Cooking)

Open steaming for about 15–20 min is necessary for complete gelatinization of the soaked wheat of 83% moisture content (d.b.). It may be pointed out that only 3%–5% moisture is added during steaming of soaked wheat containing 83% (d.b.) moisture.

Drying

The effects of drying air temperatures on the thin layer (one grain thick), drying characteristics of parboiled wheat, and the quality of the dried product are shown in Figures 9.2 and 9.3, respectively.

It should be kept in mind that at a constant air temperature, the rate of drying decreases significantly as the thickness of grain increases. Figure 9.3 shows that the percentage of multicracks increases with a corresponding increase in drying air temperature. Multicracks are not desirable when the dried debranned parboiled wheat is to be split into two pieces to appear to resemble rice. Moreover, multicracks in the grain always lead to the production of higher percentage of fines in the product after debranning and cracking. From the standpoint of drying time and quality (percentage of multicracks), the drying air temperature of 75°C appears to be optimum.

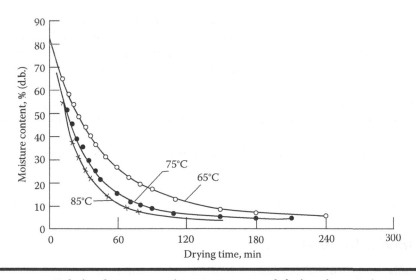

Figure 9.2   Relation between moisture content and drying time at air temperatures of 65°C, 75°C, and 85°C at an air velocity of 12 m/min.

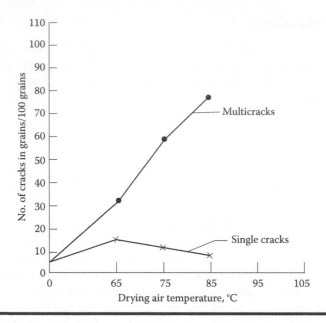

Figure 9.3 Effect of drying air temperature on development of cracks in dried parboiled wheat.

Methods of Parboiling and Production of Bulgur

The methods of production of bulgur wheat can be grouped into three major heads, namely, traditional batch method, modern method, and chemical lye peeling method.

The *traditional batch method* consists of soaking and open steaming bulgur wheat in a kettle, drying in dryers, partial debranning in a polisher, and cracking under runner disk type or any other type cracking machine.

The *modern method* involves soaking and pressure cooking (steaming) the wheat in modern parboiling units, drying in dryers, debranning, and cracking in modern milling machines.

The *chemical lye peeling method* includes continuous soaking and pressure cooking, lye peeling, drying, and cracking the wheat.

A few special processes developed in India (Method I) and the United States (Methods II–IV; Roger, 1970) have been described in the subsequent paragraphs.

Method I: Batch Method

In India, bulgur wheat has been produced on a commercial scale successfully under the supervision of Jadavpur University, Calcutta. The process is described as follows.

The clean raw wheat is soaked at 65°C–70°C for a certain period of time to raise the moisture content to a level of about 85% (d.b.), and then the soaked wheat is steamed with live steam for about 20 min in a batch-type soaker–steamer (i.e., a paddy parboiling tank developed in India). The parboiled wheat is then dried to the proper moisture content with heated air in an LSU dryer.

The clean and dry-parboiled wheat is partially debranned in a Satake-horizontal abrasive-type polisher. The debranned grain is then allowed to pass through the small clearance between the two rotating steel rolls of a Satake-type husker for cracking. Now the bulgur is ready for consumption.

Method II: Preheat Treatment Process

In this process, the clean and washed wheat is first preheated in an oven at a temperature between 93°C and 105°C and the preheated grain is then introduced into a first water quench. During this operation, the moisture content of wheat is raised from 10% to about 50% (d.b.). The partially soaked wheat is led to the second reheat oven, whereas the grain is again heated to a temperature between 65°C and 75°C. The reheated wheat grain is then fed to the second water quench to increase its moisture content to a level of about 90% (d.b.). The total time required for the aforementioned operations is ~1 h.

The soaked wheat is then cooked by steaming at a pressure of 2 kg/cm². The cooked grain (gelatinized grain) is dried to 10% moisture content (d.b.) by the dryers.

The dried parboiled wheat may then be conveyed to the mill for debranning and cracking. The flow diagram for the heat treatment process is shown in Figure 9.4 (Roger, 1970).

Method III: Continuous Process

In a continuous process, the cleaned and washed wheat is treated with a spray of water while moving through a series of three screw conveyors. Steam is also injected into each of the three conveyors so as to raise the temperature of wheat to 65°C and increase the moisture content to the desired level. The heated and moistened wheat from the first screw conveyor is discharged into the first steeping tank. The wheat is partially steeped while moving from the top to the bottom of the tank. The partially steeped wheat is then conveyed to the top of the second steeping tank by the second screw conveyor wherein it is again subject to a spray of water and open steam. In the final step, the wheat is moistened and heated in the third screw conveyor and steeped to the proper moisture level in the third steeping tank. The earlier sequence of operations requires about 12 h.

The steeped wheat is blanched in the blanchers and then introduced into a continuous steamer wherein it is cooked for about 2 min by steam under a

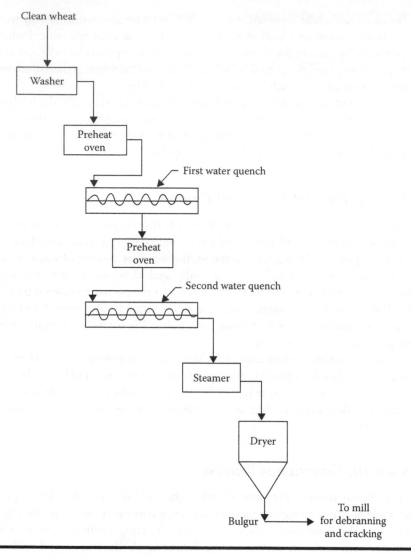

Figure 9.4 Flow diagram of preheat process.

pressure of 2 kg/cm². After cooling in a cooler, the cooked (gelatinized) wheat is dried to 10% (d.b.) with hot air in a series of two columnar dryers.

The dried product may be debranned and cracked as usual. The outturn of the cracked bulgur and fines are about 85% and 15%, respectively.

The flow diagram for the continuous process is shown in Figure 9.5 (Roger, 1970).

The process is particularly suitable for soft white wheat berry.

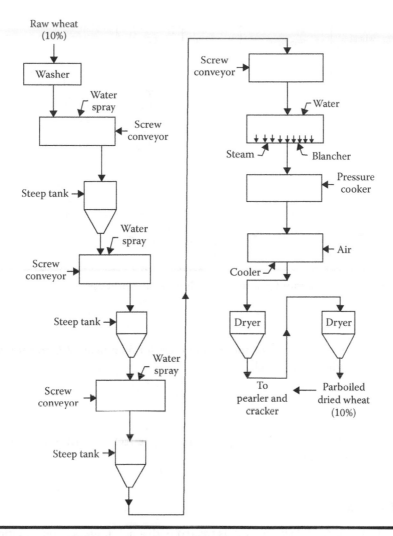

Figure 9.5 Flow sheet for continuous process.

Method IV: Chemical Lye Peeling Process

The lye peeling process has been discussed in the following (Roger, 1970). The process is illustrated in Figure 9.6.

Soaking

Raw wheat is allowed to pass through a screw conveyor filled with hot water. Heating jackets are provided to maintain the temperature of grain and water mixture at 82°C at the exit. The soaking period in the conveyor is adjusted to 1 h.

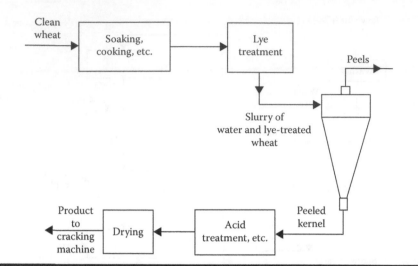

Figure 9.6 Flow diagram for continuous lye peeling process.

Tempering

The soaked wheat is fed to a bin and held in there for 30 min at 82°C to equilibrate moisture throughout the kernel.

Cooking

The soaked and tempered wheat is passed through a reel in which it is steamed at 100°C for a few minutes.

Lye Treatment

The cooked wheat, while in the reel, is treated with a hot dilute aqueous solution of caustic soda (NaOH) at 85°C and the lye-coated wheat is held in the reel for about 3 min. Then the lye solution is removed from the lye-treated wheat by a spray of cold water in a rotating screen.

Peeling

The lye-treated wheat along with water is pumped through a hydrocyclone at a pressure of 3 kg/cm². The peeled kernels issuing from the discharge tip are washed thoroughly in water.

Acid Treatment

The washed and peeled wheat kernel is then treated with 1% aqueous acetic acid solution at 50°C for 5 min in a worm on neutralize, the last trace of caustic soda adhering to the lye-peeled wheat kernels.

Drying

The lye-peeled wheat kernel is then dried from 10% to 14% moisture content (d.b.) by dryers.

Advantages

- The light-colored product resembles rice in appearance.
- The product is suitable for quick cooking cereals and breakfast cereals and is as an ingredient for baby food.

Disadvantage

- The production cost is comparatively high.

Chapter 10

General Grain Milling Operations

Food grains are naturally endowed with outer protective husk/bran layers composed of rough, fibrous, pigmented, and waxy substances that are undesirable for edible purposes. They also consist of oily germs that are undesirable for storage purposes. Removal of these parts constitutes the most fundamental prerequisite in grain milling or flour milling technology of cereals. In grain milling, the outer husk/bran layers are removed from the grain while retaining its shape, whereas in flour milling, flour without or with negligible bran content is prepared without the grain shape.

In general, milling refers to the size reduction and separation operations used for processing of food grains into an edible form by removing and separating the inedible and undesirable portions from them. Milling may involve cleaning/separation, husking, sorting, whitening, polishing, grinding, sifting, etc.

To increase the milling quality of the food grains or to improve the quality and quantity of their end products and facilitate milling operations for the desired products, food grains are sometimes subjected to hydrothermal treatment prior to milling. This process is called conditioning.

Cleaning/Separation

Classification of Separation Methods

Any mixture of solid materials can be separated into different fractions according to their difference in length, width, thickness, density, roughness, drag, electrical, conductivity, color, and other physical properties.

Each of the various types of separators employed in flour and grain milling is designed on the basis of the difference in the following physical characteristics of grain: (1) width and thickness of the grain for sieves, screen cleaners, sifters, thickness graders, grading reels, inclined sifters, etc.; (2) length of the grain for indented-type or disk-type pocket separators; (3) aerodynamic properties for husk aspirators, cyclone separators; (4) form and state of the surface for separators for coarse grain, spiral separators, and belt-type separators; (5) specific gravity and coefficient of friction for separating tables and stone separators; (6) ferromagnetic properties for magnetic and electromagnetic separators; (7) electrical properties for electrostatic separators; and (8) color for electronic separators.

Separation according to Aerodynamic Properties

The pneumatic separation is based on the difference in aerodynamic properties of the different components. The aerodynamic properties of a particle depend on the shape, dimensions, weight, state, and position of the particle with respect to the air current.

The aerodynamic properties have been discussed in detail in another section.

Separation according to Specific Gravity

If the components of a mixture are different in densities and subjected to a reciprocating movement of an inclined table or screen, then the components of the mixture are readily separated into different fractions according to their densities.

This type of separator works on the principle of self-sorting or stratification. The heavier particles of the mixture sink to the bottom by the to and fro movement of the inclined table and are then separated by any suitable method.

In a mixture of components of the same density, the finer particles sink, while in a mixture of components of equal dimensions but of different densities, the heavier particles sink to the bottom. This is the principle behind the operation of stone separators.

Composite stone separators of pneumatic grading table type are used to enhance the settling of stone. The stone separator has been discussed in detail in Chapter 12. The effectiveness of operation of a stone separator depends on many factors. Of them, kinematic parameters are most important. Continuous and uniform feeding of mixture are also important.

Separation according to Magnetic Properties

Metallic impurities in the grain accelerate the wear and tear of the milling machinery. Moreover, even minute quantity of metallic impurities present in the milling products can make them unfit for human consumption.

The magnetic impurities like steel, pig iron, nickel, and cobalt particles present in a grain mixture can be separated on the basis of their differences in magnetic properties.

Since the effective removal of metallic impurities depends on the force of attraction of the magnet, an electromagnet is preferred to an ordinary permanent magnet as the force of attraction of the former can be increased just by increasing the strength of electric current. Magnetic separation consists of three steps: (1) distribution of feed over the magnet; (2) collection and retention of the ferromagnetic impurities by the magnet; and (3) cleaning of the magnet from the impurities.

Separation according to Electrostatic Properties

When different particles are charged with static electricity and are passed through another electric field, the action of the outer electric field on the electric field of the charged particles produces some mechanical work, which is used for separation (Figure 10.1).

The electrical separation consists of two stages: (1) preliminary charging of the particles with electricity and (2) separation of the charged particles by electrostatic forces in accordance with the magnitude and nature of the charges on the particles.

Magnitude of the preliminary charge is determined by the following factors: electrical conductivity, dielectric constant, and other properties of the particles such as particle size and shape, specific gravity, and design of the separators.

Separation according to Colors (Electronic Separators)

Difference in color can be used for separation. With the help of an electronic separator, some fruits, vegetables, and cereal grains can be sorted out from the discolored or defective ones in accordance with their differences in color.

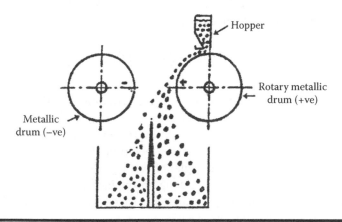

Figure 10.1   Electrostatic separator.

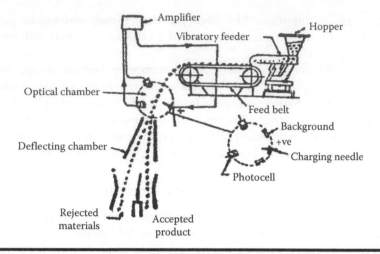

Figure 10.2   Flow diagram of an electronic color sorter.

In an electronic separator (Figure 10.2), the seeds are uniformly fed to the optical chamber. Two photo cells are set at a certain angle in order to direct both beams to one point of the parabolic trajectory of the seeds.

A needle connected to a high voltage source is placed on the other side. When a beam through photoelectric cells falls on a dark object, a current is generated on the needle. The end of the needle receives a charge and imparts it to the dark seeds. The grains are then allowed to pass between two electrodes with a high potential difference between them. As a result, two fractions of a mixture are separated according to difference in colors.

Electronic separators have been used in various industries for a long time. In recent times, electronic sorters are being used for the separation of discolored grains in some advanced countries. But the sorting capacity of these units is limited, resulting in very high cost of separation.

Separation according to Surface Properties (Frictional Separators)

The frictional properties can be utilized for the separation of a mixture of grains of almost the same size. Sizes of oats and hulled oats, millets, and bind weed are almost the same. These mixtures car be separated by frictional separators (Figure 10.3) consisting of an inclined plane surface. The operation of the separators is based on the differences between the friction angles of two types of grains.

When grains are allowed to move along an inclined plane, frictional forces of different magnitudes act upon these particles. Therefore, different particles move on an inclined surface at different velocities.

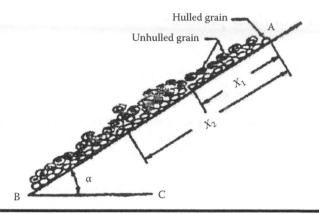

Figure 10.3 Inclined plane factional separators.

In this case, heavier particles will sink at the lower layer and move at a lower velocity, while lighter particles will float at the top and move downward at a higher velocity.

The velocity of the particle at any point on the inclined plane can be found as follows:

$$v_1 = \sqrt{2g\,K_1 x_1}$$

where

$K_1 = \sin \alpha - \mu_1 \cos \alpha$
x_1 is the distance traveled by a body that has initial velocity zero
α is the angle of inclination
μ is the coefficient of friction

When coefficient of friction of the dehusked grain is greater than the coefficient of friction of the unhusked grain, the movement of the lower layer of husked grain will be retarded. The unhusked grain will move above the lower layer with higher velocity, v_0,

$$v_0 = v_1 \left(1 + \frac{K_2}{K_1}\right)$$

v_0 is the absolute velocity of the upper layer
$v_0 > v_1$ by the factor K_2/K_1

These properties of free flowing particles can be used in separation of husked and unhusked grains.

Effectiveness of the Separation

$$\varepsilon_A = \frac{(X_F - X_D)X_C(X_F - X_C)(1 - X_D)}{(X_D - X_C)^2 X_F(1 - X_F)}$$

where

ε_A is the separating effectiveness for a mixture of two components A and B

F is the rate of feeding to the separator, kg/h

C is the rate of separating the product, kg/h

D is the rate of reject, kg/h

X_F, X_C, and X_D are fractions of component A in the feed, separated product, and reject, respectively

Husking/Scouring/Hulling of Grain

In general, husking refers to the removal of outer seed coat from the grain kernel. The terms "hulling" and "scouring" are also used in cereal milling. In grain milling, husk is removed from the grain, retaining its original shape, whereas in flour milling bran is removed from the grain to produce flour without any emphasis on its shape. Husking and scouring are the most important operations in grain milling or flour milling technology of cereals.

Different types of machines are employed for husking and scouring operations because of the differences in anatomical structure, type of bonds, and strength properties among husk, bran, and kernel of different grains.

Methods of Husking

The operation of husking and scouring machines can be divided according to the following three basic principles (Figure 10.4):

1. *Compression and shear*: Compression and shear can compress, split, and strip off the husk from grain. Concave type of husking machine, rubber roll husker, etc., are designed on the basis of this principle.
2. *Abrasion and friction*: Hollanders are based on the friction of grain on an abrasive surface (emery).
3. *Impact and friction*: Husk can be stripped off by the action of impact and factional force. Centrifugal-type paddy sheller comes under this group.

Concave-Type Husker

This machine consists of a horizontal rotating cylinder, called roll, and a stationary cylindrical surface known as concave (Figure 10.5).

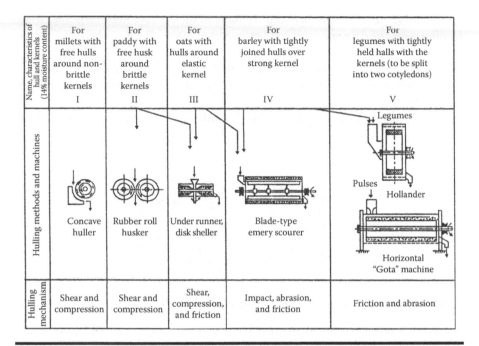

Name, characteristics of hull and kernels (14% moisture content)	For millets with free hulls around non-brittle kernels	For paddy with free husk around brittle kernels	For oats with hulls around elastic kernel	For barley with tightly joined hulls over strong kernel	For legumes with tightly held halls with the kernels (to be split into two cotyledons)
	I	II	III	IV	V
Hulling methods and machines	Concave huller	Rubber roll husker	Under runner, disk sheller	Blade-type emery scourer	Legumes Pulses Hollander Horizontal "Gota" machine
Hulling mechanism	Shear and compression	Shear and compression	Shear, compression, and friction	Impact, abrasion, and friction	Friction and abrasion

Figure 10.4 Classification of husking/scouring/hulling methods.

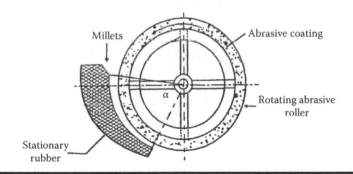

Figure 10.5 Principle of hulling by concave-type huller.

The husk/bran layer can be removed from buck wheat and millets keeping the original shape of the kernel by applying mild shear and compression. These types of machines are fairly efficient and require low power.

On feeding into the mill, the grain is first caught up by the rolls and passed through the husking zone between the roll and the concave where it is subjected to shear and compression simultaneously. One part of the husk is sheared by the rotating roll while the other part is pressed against the stationary concave and subjected to breaking forces.

The minimum clearance between the roll and the concave must be greater than the dimensions of the grain kernel; otherwise, the kernel will be crushed. The radius of curvature of the concave is usually the same as the radius of the roll.

Composition of roll and concave varies with the type of grain to be husked. For example, roll made of abrasive material and concave made of commercial rubber are used for millets.

Usually the following specifications of rolls and concaves (Kuprits, 1967) are used:

Diameter of concave	50–60 cm
Length of concave	19–30 cm
Angle of contact	40°–70°
Peripheral speed	10–15 m/s

Husking by the Action of Rubber Rolls

In case of paddy, deformation caused by shear and compression of the two differentially rotating rubber surfaces is sufficient to split and separate the husk from the grains. The paddy is passed through the clearance between two rubber rolls, rotating in opposite directions at different speeds. The clearance between them is smaller than the mean thickness of the paddy. One part of the husk is subjected to shearing forces whereas the other part in contact with the slower roll is under compression and is thus subjected to breaking forces. Husking is done by the action of these forces.

Suppose a grain having thickness b (Figure 10.6) is brought into contact with the rolls at points A and A_1. The grain is then acted upon by the rolls until it leaves the working zone. If α_1 is the angle between the line connecting the

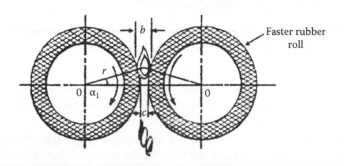

Figure 10.6   Husking principle of rubber roll husker.

centers of the rolls and the radius drawn through the point at which the grain contacts the roll,

$$\text{then} \quad \cos \alpha_1 = \frac{r + (C/2)}{r + (b/2)} = \frac{d + C}{d + b}$$

where
  r is radius of the rolls, cm
  d is diameter of the rolls, cm
  C is clearance between the rolls, cm
  b is size of the grain, cm

$$\text{Hence,} \quad \alpha_1 = \cos^{-1} \frac{(d + C)}{(d + b)}$$

The full length of the husking zone from the point at which the grain is caught by the rolls to the point where the grain is no longer in contact with the roll will correspond to the length of the arc l_h and angle $2\alpha_1$.

Length of the husking zone l_h can be expressed as follows:

$$l_h = \frac{2\pi d}{360} \cos^{-1} \frac{(d + C)}{(d + b)}$$

The length of the husking zone l_h depends on the diameter of the rolls, the thickness of grain to be husked, and the linear distance between the centers of the rolls. When the size of the grain, b, and the clearance, C, are fixed, then the length of the husking zone increases with the increase in diameter of the roll:

$$\text{Period of husking, } t = \frac{l_h}{V_f} \text{ (faster roll)}$$

$$\text{Period of husking, } t = \frac{l_h - l_{ad}}{V_s} \text{ (slower roll)}$$

where l_{ad} is the difference between the paths of faster and slower roll:

$$l_{ad} = \frac{l_h (V_f - V_s)}{V_f}$$

This equation shows that the difference between the paths of the two rolls depends not only on the geometrical parameters but also on the peripheral speeds of the rolls, their relative velocities, and the differential speeds.

If V_f = 15–17.5 m/s and V_s = 12.5–15 m/s, then $V_f - V_s$ should not be less than 2.5 m/s (Kuprits, 1967).

Husking by Under Runner Disk Husker (Disk Sheller)

Paddy as well as oats may be husked by the under runner disk sheller. Previously, it was used for millets.

Principle

The under runner disk husker consists of two horizontal and coaxial cast iron disks, partly coated with an abrasive material. The top disk is stationary and provided with an opening at the center for feeding, while the bottom disk is rotating. The clearance between the two disks can be adjusted by changing the position of the lower rotating disk vertically.

The grains that fall on the lower disk through the central opening are pulled along by its rotation and move outward by centrifugal force. During rotation around the axis of the disk and simultaneous movement toward the edge, grains tend to take a vertical position. At this point, their tips strike against the hard surface of the top disk, which exerts pressure and a slight friction resulting in splitting of husk apart.

The effective width of the abrasive coating should not be more than one-sixth of the stone diameter and the optimum peripheral speeds should be about 14 and 20 m/s for paddy and oats, respectively.

The trajectory of the grain with respect to the working zone of the stationary and rotating disks is a curve directed from the inner radius r of the rotating disk to its outer radius R. The grain moves with increasing velocity:

$$\text{Velocity, } V = \omega \times r'$$

where
ω is the angular velocity of the rotating disk
r' is the variable radius vector for the moving grain
$(r < r' < R)$
Centrifugal force $C = m\omega^2 - r'$

Under the influence of these forces, the grain describes a curve stretching from the center to the periphery of the disks. The force acting on the grain increases as it travels from the center to the periphery, since the peripheral speed and the relative velocity are simultaneously increased. This increase in velocities leads to complex deformations of the husk (deformation by compression, shearing, and slip). As a result, the husk is split and stripped off from the kernel.

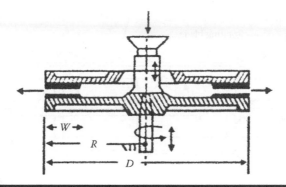

Figure 10.7   Under runner disk husker.

Construction

Figure 10.7 shows the constructional features. The main parts of this machine are the feed control device, the upper stationary disk, the lower rotating disk fitted to the shaft, and the pulley mounted on shaft. The shaft can be moved up and down.

Advantages

1. Construction and operation of the machine are simple.
2. Life of the machine is long.
3. Running cost is low.
4. The performance of the machine is satisfactory for raw and parboiled paddy of uniform size.

Disadvantages

1. Compared to rubber roll husker, the head and total yields are less.
2. Damage to the rice kernel is done due to the formation of scratches on it by the hard abrasive coating.

Husking by Scourers and Blade-Type Huskers

Impact huskers can be used for barley, wheat, and oats having moisture content of about 13%.

The husker consists of a rotating horizontal shaft fitted with blades encaged in a cylinder. When grains are fed to the cylinder (Figure 10.8), they come in contact with the revolving blades and are thrown to the internal abrasive surface of the

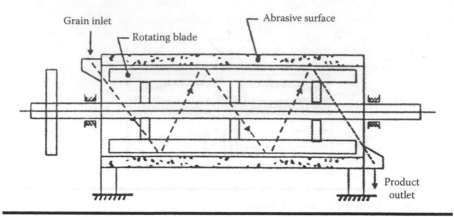

Figure 10.8   Principle of husking by blade-type emery scourer.

cylinder that rebounds them. The grains impinge on the blade again. Two impacts per revolution between the grain and the abrasive cylinder and friction amongst the grains result in removal of husk from the grain.

The holding time of the grain in the scourer depends on the peripheral speed of the abrasive cylinder and the blade angle. The effective inclination of the blade is 3–6 s only.

Advantages

1. The machine is easy to operate.
2. It does not require much attention.

Disadvantage

1. Yield of cracked and crushed grain is high.

Hulling by an Abrasive Drum in a Cylindrical Steel Shell

Grains with hull firmly attached to the kernel such as barley, wheat, etc., can be husked by a more drastic treatment.

Hollander (Figure 10.9) can be used for the aforementioned purpose. It comprises rapidly rotating abrasive roll encaged in a slowly rotating steel shell. In a batch huller, the grains enter into the space between the abrasive roll and the perforated cylindrical shell and take a considerable period to travel from one end to the other. The shell rotates slowly in the opposite direction. In Hollanders,

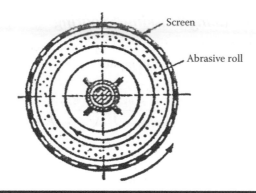

Figure 10.9 Working principle of batch-type Hollander.

the hull is removed from the grain because of the friction among the grain, abrasive roll, and perforated steel shell and friction between grain and grain.

On account of the following disadvantages of batch-type Hollanders, continuous Hollanders are being increasingly used:

1. The machine is heavy and bulky.
2. It consumes high power.
3. Due to batch system, continuous production is not possible and operational schedule becomes complicated.

Factors Affecting the Effectiveness of Hulling/Husking/Scouring

The effectiveness of hulling depends on the properties of the grain and the type of huller.

The amount of desired and undesired products obtained from cereal milling depends on the following grain parameters:

1. Type of grain and its special properties
2. Bond strength between kernel and husk, strength of the kernel, and strength of the husk.
3. Sound or cracked grain, grain size, and uniformity of size
4. Moisture content of grain and difference in moisture content between husk and kernel
5. Extent of hydrothermal treatment given to the grain
6. Proportion of husked kernel in the grain
7. Ease of separation

Effectiveness of Hulling/Husking/Scouring

The coefficient of hulling E_{hulling} is defined by the percentage of husked grain obtained from the total amount of grain input:

$$E_{\text{hulling}} = \left(\frac{n_1 - n_2}{n_1} \right) = \left(1 - \frac{n_2}{n_1} \right)$$

where

n_1 is the amount of grain before hulling
n_2 is the amount of unhulled grain after hulling

The coefficient of wholeness of the kernels $E_{\omega k}$ is defined by the proportion of whole kernels to the total amount of kernels (kernel + crushed + grain + mealy waste) extracted by the given system:

$$E_{\omega k} = \frac{k_2 - k_1}{(k_2 - k_1) + (d_2 - d_1) + (m_2 - m_1)}$$

where

$k_2 - k_1$ is the yield of whole kernels
$d_2 - d_1$ is the yield of crushed kernels
$m_2 - m_1$ is the yield of mealy waste in the product
k_2 is the content of the whole kernels after hulling
k_1 is the content of the whole kernels before hulling
d_2 is the content of the crushed kernels after hulling
d_1 is the content of the crushed kernels before hulling
m_2 is the content of the mealy waste in the product after hulling
m_1 is the content of the mealy waste in the product before hulling

The efficiency of operation of hullers can be expressed by an overall coefficient, which takes into account both the qualitative and quantitative aspects of the operations carried out:

$$\eta_{\text{hulling}} = E_{\text{hulling}} \times E_{\omega k}$$

Grinding

Cereal grinding system can be divided into two groups: plain grinding and selective grinding.

In plain grinding, hard bodies are ground to a free-flowing material consisting of particles of sufficiently uniform size. This material is either the final product or a product ready for further processing.

In selective grinding, the grinding operation is carried out in a number of stages successively using differences in structural and mechanical properties of the components of the body.

It should be noted that the power consumption for grinding is about 50%–80% of the total power required for all operations.

Therefore, the following important points are to be considered for power design of the grinding system: rational utilization of the raw material, quality of the products, size and color of the products, efficiency of the grinders, specific power consumption, and costs of production.

The characteristics of grinding operation are affected by the following grain parameters: (i) type of cereal grain; (ii) variety; (iii) moisture content; (iv) extent of hydrothermal treatment given to the grain; and (v) mechanical properties.

Each type of grain has an optimum moisture content for efficiency of grinding. The maximum formation of new surface for each type of wheat is related to the optimum moisture content. Moreover, power consumption is also dependent on the moisture content. The starch granules are separated from the proteins due to deformation of hard wheat.

Hardness of rice and barley endosperms are $11.5–14.5\ kg/mm^2$ and $6.0–9.0\ kg/mm^2$, respectively (Kuprits, 1967).

Effectiveness of Grinding

The main criteria for evaluation of effectiveness of grinding of any solid body including grain are the degree of grinding, specific power consumption, and specific load of the initial product on the working tool of the grinder. Degree of grinding, i, is expressed as follows:

$$i = \frac{S_a}{S_b}$$

where
S_a is the overall surface area of the product after grinding, cm^2
S_b is the overall surface area of the product before grinding, cm^2

The term "overall extraction" E_X expressed in percent refers to the difference between the percentage of undersized particles C_1 in the final ground product and the percentage of undersized particles I in the initial feed entering the grinder:

$$E_X = C_1 - I$$

$$C_1 = \left(\frac{u_1}{u_1 + O_1} \right) \times 100$$

$$I = \left(\frac{u_2}{u_2 + O_2} \right) \times 100$$

where
  u_1 is the weight of undersized particles obtained by sifting the ground product
  O_1 is the weight of oversized particles obtained by sifting the ground product
  u_2 is the weight of the undersized particles in the initial product to be ground
  O_2 is the weight of oversized particles in the initial product to be ground
  E_X depends upon structural–mechanical properties of the material, dimensions of the roll (i.e., diameter, etc.), geometry of the surface of the rolls, kinematic parameters of the rolls, and specific load on the rolls

Machinery Used in Cereal Grinding

Depending on the objective of grinding and mechanical properties of cereal grain, effects of the following are utilized for grinding: compression and simultaneous shear of the material, impact of the material, etc. The classification of grinding machines is shown in Figure 10.10.

Grinding of Grain in Roller Mills

The roller mill consists of two cylindrical steel rolls revolving in opposite directions at different speeds. In the roller mill, the grain or its parts are ruptured in a space that is narrowed toward the bottom.

A certain degree of grain rupturing starts at above the line connecting the centers of the rolls. The slow roll holds the grain during the action of the fast roll. In the grinding zone, the grain or its parts are simultaneously subjected to compression and shear resulting in deformation of grain. Rupturing of endosperm without much grinding the bran of wheat under grinding conditions is the characteristic of the first breaking system.

Factors Affecting the Effectiveness of Roller Mills

Clearance between rolls: Even a small variation in rollers' clearance leads to considerable variations in the products.

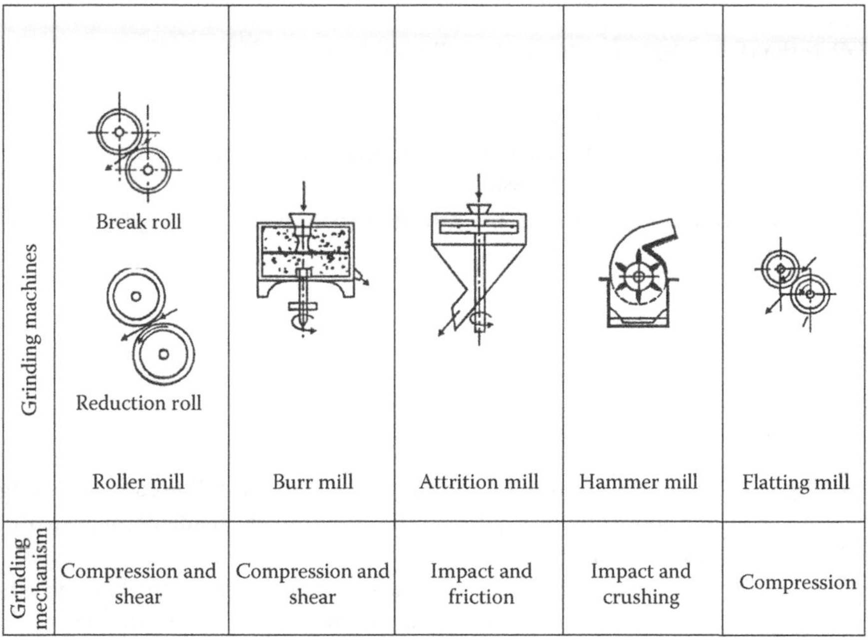

Grinding machines	Break roll / Reduction roll				
	Roller mill	Burr mill	Attrition mill	Hammer mill	Flatting mill
Grinding mechanism	Compression and shear	Compression and shear	Impact and friction	Impact and crushing	Compression

Figure 10.10   Classification of grinding machinery.

Geometrical parameters of rolls: Conditions of rupture of the particles to be ground depend upon the roll diameter, clearance, and initial size of the particle. Shape, number, slope, mutual position, and shape of the cross section of corrugations of the rolls have significant effects on the quality and yield of flour, total output, and specific power consumption of the roller mills. Position of the corrugation edges also has the same effects on the product.

Kinematic parameters of the rolls: The important kinematic parameters of the rolls are speed of the fast roll, V_f, and ratio of the speeds between fast and slow rolls (i.e., $k = V_f/V_s$). The efficiency of grinding depends upon the kinematic parameters.

Capacity of the Roller Mills and Power Consumption for Their Operation

The capacity of a roller mills is the amount of product in kilograms, ground per unit time.

The theoretical capacity of a pair of rolls can be determined from the following formula (Kuprits, 1967):

$$Q_r = 3.6 \times \rho_1 \times l \times V_z \times b\psi$$

where

Q_r is the theoretical capacity of the roll pair, kg/h

ρ_1 is the bulk density of the product before grinding, kg/cm^3

l is the length of the rolls, cm

V_z is the mean calculated velocity of the product in the grinding zone, cm/h

b is the clearance between the rolls, cm

ψ is the coefficient of filling of the grinding zone, where $\psi = Q_a/Q_r$ (Q_a is the actual capacity of the roll pair)

The capacity of a roller mill is also dependent upon type of grain and the moisture content of the grain.

Grinding Grain in Hammer Mills

Grinding in a hammer mill involves impact on the material followed by crushing. The main working tools of this type of mill are hammers made of high-quality steel, screens, and metal lines.

The output of hammer mill depends on the peripheral speed of the hammer, the clearance between the screen and hammers, the area of the screen, the size of the screen openings, and the structural–mechanical properties of the material.

Symbols

E_{hulling}	Coefficient of hulling/husking, decimal fraction
E_{wk}	Coefficient of wholeness with respect to hulling/husking, decimal fraction
E_{X}	Overall extraction with respect to grinding, percent
i	Degree of grinding ratio
Q_r and Q_a	Theoretical and actual capacity, respectively, of a pair of grinding rolls, kg/h
V_1	Velocity of a particle at a point on an inclined plane, m/s
V_f and V_s	Velocity of a faster and a slower rubber roll, respectively, m/s
ε_{A}	Separating effectiveness for a mixture of two components A and B, ratio
η_{hulling}	Efficiency of hulling/husking, decimal fraction

Chapter 11

Hydrothermal Treatment/Conditioning of Cereal Grains

Sometimes, cereals, pulses, and other food grains are subjected to conditioning or hydrothermal treatment. Hydrothermal treatment of grains refers to the addition of moisture and heat to the grains to improve the quality and quantity of their end products or to facilitate different milling operations for the desired products. Hydrothermal treatment is commonly called conditioning and is considered a pre-milling treatment. It can be carried out either at room temperature, at a little elevated temperature, or even at high temperature.

Conditioning or hydrothermal treatment is used for various purposes. It can be used for improving the shelling efficiency, nutritional quality, and milling quality of paddy and for facilitating degermination and dehulling of corn and wheat. Pulses are conditioned by alternate wetting and drying. It helps in the dehusking and splitting of the kernels during milling. Even toxins of *khesri* pulses, which cause paralysis, can be removed by soaking in hot water. Green vegetables are blanched with hot water for retention of the green color and removing any disagreeable odor. The disagreeable odor of soybean can also be removed by blanching. Therefore, simultaneous addition of moisture and heat to the grains and other crops is an important step in cereal milling and crop processing.

Hydrothermal treatment brings about several changes in the properties of cereal grains. They are grouped into three major heads, namely, changes in physicothermal, physicochemical, and biochemical properties. These changes and the effects of basic parameters on them have been discussed here.

Physicothermal Properties

Strength/Hardness

Generally, hardness of cereal grains decreases with the increase of their moisture content. Among all crops, the hardness of corn is highest, followed by highly vitreous wheat variety.

Density and Hardness

Bulk density of grain decreases with increase in moisture content. With the increase of moisture, the softness of grains increases. Therefore, increase of moisture is related to the strength properties of the grain and to the power consumption for grinding as well. Hence, the role of hydrothermal treatment of grains is very important.

Hysteresis

Hysteresis of grain plays an important role in grain milling. When grains are subjected to alternate wetting and drying processes, their primary hardness decreases. Husk is also loosened from the kernel, resulting in higher efficiency of shelling.

Thermal Properties

Thermal properties, namely, specific heat, thermal conductivity, and thermal diffusivity of grains, increase with the increase of either temperature or moisture content or both. The rate of hydration and chemical reactions also increases as temperature increases.

Biochemical Properties

Biochemical properties of grains are largely dependent upon their protein and amylose contents. In case of wheat, both qualitative and quantitative changes in crude gluten take place during hydrothermal treatment. Changes in the proteolytic and amylolytic complexes determine the physical properties of the dough and the bread-making properties to a considerable extent.

Thus, changes in biochemical properties affect primarily the bread-making properties, i.e., color and structure of the crumb and especially the volume of the bread produced. During parboiling of paddy, steaming inactivates the lipase and stabilizes the rice bran.

Physicochemical Properties

The germ and bran layers can be removed quite easily without much reduction in their sizes if the grains are subjected to hydrothermal treatment prior to milling.

Changes in the physicochemical properties of the grain during hydrothermal treatment of wheat mainly affects the flour milling properties, i.e., whiteness of flour and specific power consumption for grinding, etc.

Effects of Various Factors on the Changes of Different Properties

The three important factors, namely, grain moisture content, heating temperature, and heating time, have significant effect on the aforementioned changes.

There is an optimum moisture content for each type of grain at which it should be milled. This optimum moisture content is determined by the initial and final properties of grain and the atmospheric conditions.

The maximum temperature rise and heating time of grain are determined mainly by its biochemical properties, which are ultimately dependent upon its protein complex.

Effects of Moisture Content

The moisture content of any grain for milling is determined on the basis of its physical properties (mainly hardness). The physical properties of the grain are dependent upon the type, variety, region, harvesting condition, biochemical properties, and the type of grinding machines used. The optimum moisture contents for milling of parboiled paddy and soft and hard varieties of wheat are about 15%, 14%, and 17%, respectively.

Husk, germ, and endosperm are hydrated and swelled by hydrothermal treatment, but plastic deformation takes place only in the case of husk and germ, making them more rubberlike and less brittle. As a result, during degermination and bolting husk and germ are easily separated from the endosperm and specific power requirement for grinding is also minimum at the optimum moisture content of grain.

Effects of Temperature

The best results are obtained when heat is added along with the addition of moisture. For example, wheat should be soaked to the desired moisture level at about 40°C–60°C for a desired period of time only.

Thermal Properties

Moisture moves in the same direction as the flow of heat. This phenomenon is known as thermal moisture conductivity. It is characterized by a value numerically equal to the moisture transfer at a temperature difference of 1°C through arbitrary sections under constant operating conditions. Both specific heat and thermal conductivity of grain increase with the increase of temperature.

Chemical Kinetics

The rate of moisture absorption, moisture equilibration, and chemical reaction increases as temperature increases. Therefore, duration of hydrothermal treatment depends on temperature.

This explains why the effect of steam is considerable. It affects temperature and moisture gradients simultaneously.

Coefficient of Expansion

During hydrothermal treatment, expansion of cellulose of husk is different from that of proteins and starch of the endosperm. As a result, an internal slip occurs, which has a favorable effect on the separation of husk, formation of middlings, and power requirement during grinding.

Biochemical Properties

Proteins: Temperature has a great influence on the quality and quantity of gluten. When soaked wheat is conditioned at a temperature between 40°C and 50°C, the yield of crude gluten is higher than that of wheat conditioned at ordinary temperature.

The most remarkable changes that occur in the qualitative characteristics of the gluten are flexibility and extensibility. At a higher temperature, denaturation of protein takes place.

Enzymes: Within limits, enzymatic activities increase with temperature. But most of the enzymes are inactivated at temperatures above 80°C. Catalase is one of the heat-sensitive enzymes.

Optimum Conditions

Therefore, it is necessary to find out an optimum combination of heating temperature and time and moisture content for successful hydrothermal treatment of each type of grain for effective milling and desired bread-making quality, if any.

Chapter 12

Rice Milling

Rice milling machinery used in different countries ranges from crude hand-pounding equipment and small-scale hullers to highly sophisticated and capital-intensive units. However, the rice milling machinery can be broadly classified into two groups: traditional and modern rice milling machinery.

Traditional Rice Milling Machinery

Traditional rice mills include hand-pounding equipment, single huller and battery of hullers, sheller-cum-huller, and shelter mills.

Single Huller

Construction

The common type of Engleberg huller consists of an iron-ribbed cylinder mounted on a rotating shaft on ball bearings and fitted in a concentric cylindrical housing (shown in Figure 12.1). The inner ribbed cylinder has helical ribs up to one-fourth of its length and four to six straight ribs for the rest of the length. The cylindrical casing can be divided into two halves. The bottom half of the cylinder is replaceable and made of slotted sheet so that the bran removed during milling may pass through the slots by the pressure generated in the cylinder. The ribbed cylinder is rotated at a speed of 600–900 rpm.

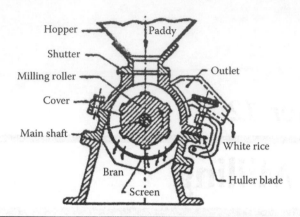

Figure 12.1 Engelberg-type huller.

Principle

The paddy is husked and whitened in the huller by friction and pressure generated in the milling chamber. Both husking and whitening are done by the same machine. The milling can be carried out either in a single pass or in multipasses.

Paddy-handling capacity of the common huller varies from 0.25 to 1.00 ton/h. The total rice outturns are about 56% and 64% for raw and parboiled paddy, respectively.

Advantages

- The initial investment and operating cost are small.
- The huller can be manufactured locally and operated with unskilled labor.
- The huller can be utilized for whitening of parboiled rice as it produces uniformly whitened rice.

Disadvantages

- Both total and head yields are low (in comparison to any other milling machine) as the degree of whitening cannot be adjusted to a low value.
- The huller bran, contaminated with large amount of husk, cannot be utilized for oil extraction and for feed purposes.

The huller bran is sometimes used for boiler fuel. From the conservation of food materials and other points of view, the huller is an uneconomic and wasteful machine. Therefore, the huller should be discarded in all countries as soon as possible.

Sheller Mill

As far as basic milling operations are concerned, the shelter mill is almost identical to a modern mill, except that the husking operation is done by the under runner disk husker (sheller) in place of modern rubber roll husker (Figure 12.2).

This type of mill consists of (a) a cleaner; (b) one or more disk shellers; (c) aspirator (to remove husk); (d) one or more paddy separators; and (e) one or more cone-type rice whitener. Commonly, the capacities and total power consumption vary from 1 to 2 tons/h and 35 to 50 BHP, respectively.

Among all traditional rice mills, total and head yields are highest in a sheller mill. However, in huller and sheller-cum-huller and sheller mills, a certain amount of husked rice, smaller and lighter paddy escape along with the husk due to the defective design of the aspiration system, and small brokens are also lost with the bran. Moreover, immature paddy cannot be separated in these mills and is mostly lost in the pile of husk.

Advantages

- Sheller bran can be used for oil extraction.
- Operating cost is low.
- Life of the machinery is long.
- The emery cone can be recoated at the mill sight.
- It works well with parboiled paddy.
- Shelter mill can be manufactured locally.

Disadvantages

- Head and total yields are less than the modern rice mills.
- Some scratches may be formed on the rice kernel, which is undesirable for long storage.

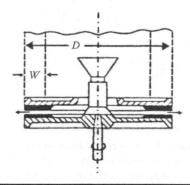

Figure 12.2   Particulars of the under runner disk sheller.

Modern Rice Milling Machinery

The modern rice milling machinery can be divided into two major groups: (1) rice milling machinery developed in Japan and (2) rice milling machinery developed in Europe.

Both European and Japanese modern rice milling machines have been described in this chapter.

The major operations performed by modern rice mills (Figure 12.3) are as follows: (1) storage; (2) cleaning; (3) husking; (4) separation; (5) whitening; and (6) grading.

Cleaning

The paddy procured from the farmer may be dried, if necessary, and stored in storage bins, and then it is cleaned with the help of paddy cleaners. The removal of impurities from the grains is essential to protect the subsequent milling machinery from unusual wear and tear as well as to improve the quality of the final product.

If the procured paddy contains excessive amounts of foreign materials or if paddy is parboiled prior to milling, then the procured paddy is sometimes precleaned in an open double sieve–type precleaner or enscalper installed outside the mill room.

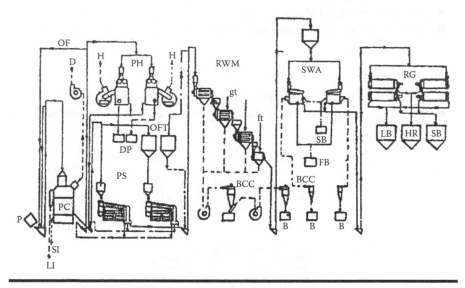

Figure 12.3   OF—Overflow, RWM—rice whitening machines, RG—rice graders, D—dust, H—husk, PH—paddy huskers, SWA—sieve with aspirator, OFT—overflow tank, DP—defective paddy, SB—small broken, LB—large broken, HR—head rice, FB—fine broken, B—bran, BCC—bran collecting cyclones, PS—paddy separators, P—paddy, PC—paddy cleaner, LI—large impurities, gt—grinding type, ft—friction type, SI—small impurities.

General Principles of Cleaning

Differences in physical characteristics such as size, specific gravity, weight, and sometimes length of the impurities compared to the paddy grain are being utilized in cleaning operations. Light impurities can be removed either by aspiration or by sieving. Impurities larger and smaller than paddy are removed by sieving, whereas impurities of the same size but heavier than paddy are removed by gravity separation.

Iron parts of particles can be removed with the help of sieve or magnetic separators.

Open Double-Sieve Precleaner

In many rice mills in India prior to parboiling, precleaning is performed through open single or double layer oscillating sieves. The precleaners are driven directly by an eccentric drive from the main transmission shaft. Sometimes, single sieve–type precleaners are equipped with suction fans also, for aspiration of light impurities.

Advantages

- Price is low.
- It can be manufactured locally.

Disadvantages

- It is an open sieve, so dust formation in the mill premises is considerable.
- Without having a self-cleaning device the bottom sieves with small perforations are often clogged, resulting in lower separation efficiency.
- The separation of impurities of about the same size as the paddy grains is not possible.

Single Scalper Drum Cleaner (Japan)

This machine can be used for precleaning of either harvested paddy or rough paddy from the impurities such as straws, ears, chaffs, big stones, dust and light impurities, sands, etc. (Satake, 1973).

Construction and Operation

Basically, this machine consists of a horizontal rotating scalper drum, an aspirator, and vibrating sieves (Figure 12.4).

The dispersing plate (1) disperses the paddy evenly on the whole width of the vibrating screen. The paddy input is controlled by a feed roll (2) and a valve (3).

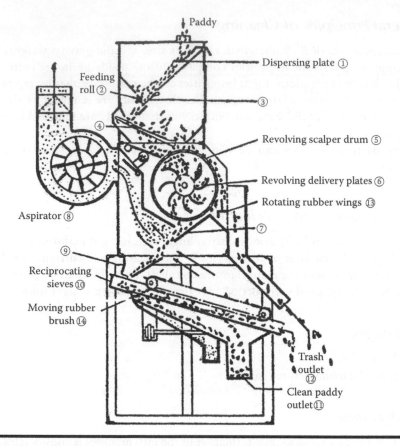

Figure 12.4　Single scalper drum-type precleaner (Japan).

An inclined vibrating screen (4) helps the paddy grain in getting them into loose form and removing large impurities. The paddy is then fed to the rotating scalper drum (5). The scalper drum, covered with hexagonal slotted screens, removes the large impurities such as straw, ear chaffs, big stones, etc. The paddy with small impurities first enters through the hexagonal slots of the scalper drum and is then discharged from the drum uniformly with the help of the device (6). The speed of the falling grains is equalized by an inclined plate (7). When they fall from the inclined plate as a film, an air stream is sucked through the grain film by a suction fan (8) to aspirate the lighter impurities like, dust, etc., from the grains. The paddy with remaining impurities is fed through hopper (9) to the reciprocating sieve (10). The reciprocating sieve is a double-layer type. The top sieve of large mesh removes large impurities through the outlet (12), while the bottom sieve of small mesh separates heavier and smaller impurities like sand. The clean paddy overflows and is discharged from the outlet (11). The scalper drum is continuously cleaned by cleaners (made of rubber) and the wide vibrating screen is cleaned by a special moving scraping device. These two devices prevent them from clogging.

Features

1. Straws and light impurities are removed effectively.
2. It is easy to operate and maintain.
3. Sieve cleaning device prevents clogging.

Stoner

Stones of about same size as paddy grains can be separated by the gravity separation method.

Stoner with Aspirator (Japan)

The main purpose of this machine is to remove dust and stone from paddy effectively. It separates immature paddy also. The system can be used independently or in combination with other cleaning systems (Satake, 1973).

Principle

The machine consists of an inclined reciprocating tray having convex slots all over the surface. A large amount of air is blown from underneath through the slotted separating tray. When a mixture of grain, stone, etc., is fed at the top of the tray, the stones having higher specific gravity slowly go down and occupy the bottom layer of the mixture and thus come in contact with the reciprocating tray.

The heavier stones are carried to the top of the tray by its movements. Since the direction of movement is more inclined than the inclination of the tray, the heavier particles in contact with the reciprocating tray are lifted. Paddy, being lighter, floats on the stone and moves downward by way of gravity. The strong air current through the slots of the tray directed toward the movement of the paddy further facilitates in lifting and shifting them downward. Any paddy grain in contact with the stones is separated and returned or blown back to the tray with the help of another smaller blower installed underneath the tray. The stones collected under a flap can be unloaded either manually or by an automatic device.

Construction

The machine is shown in Figure 12.5. The grain with impurities is distributed uniformly over the whole width of the machine by a dispersing plate (1).

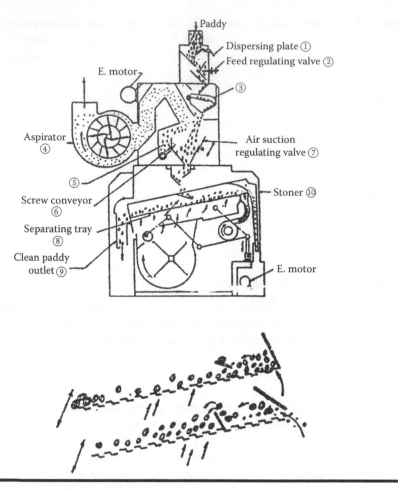

Figure 12.5   Stoner with aspirator.

The flow of the mixture is controlled by a feed control valve (2). With the help of a device (3), the speed of the flow is reduced and equalized, light and small impurities in the grain are aspirated by an aspirator (4). The immature grain is separated/settled in the expansion chamber (5) and discharged by a screw conveyor (6). The airflow rate is controlled by the air-controlling device (7). Then the grain is delivered to the separating tray (8), in which stones and a few grains move upward while almost all grains move downward. The paddy grains are discharged through the grain outlet (9), while stones are discharged through the stone outlet as described earlier.

Operation

The stone separator is a very delicate machine. Its satisfactory performance can be achieved by a careful operation. The following suggestions may be given for satisfactory performance:

1. The machine is to be labeled properly so that the separating tray is in the horizontal position perfectly.
2. The grain flow should be uniform and in proper quantity.
3. The inclination of the separating plate should be adjusted according to the moisture content of paddy.
4. Blowers underneath the separating tray should be installed in the correct position.
5. The chrome-galvanized separating tray is to be kept in a rust-proof condition.

Advantages

- Stones, immature grains, and light impurities are removed effectively.
- It can be used either independently or with other types of cleaners.
- It can be used for brown rice also.
- The inclination of the separating tray is adjustable.

Disadvantages

- It can separate a small percentage of stone content from a mixture. Hence, the capacity of the stone-separating part is limited.
- Careful maintenance and operation are necessary.

Paddy Cleaner with Stoner (Japan)

The purpose of this machine is to remove large impurities, light impurities, sand, stones, and metallic particles from paddy.

Construction and Operation

The machine consists of a rotating drum covered with a sieve having hexagonal slots (rotoscalper), a suction fan, and a slant vibrating tray with convex perforation. Under the separating tray, a blower is installed. Large impurities are removed by the rotoscalper and light impurities are aspirated by the suction fan. Stones and other heavy impurities of the same size, etc., move upward toward the upper part of the tray and are separated. Cleaned paddy moves downward by gravity and is discharged.

Advantages

- The rotoscalper and the aspirator are capable of removing about 30% impurity content.
- Large impurities, light impurities, stones, sands, and metallic particles are separated by the same machine.

Disadvantages

- The separating capacity of the stone-separating tray is limited up to 2% stone content only.
- Careful operation and maintenance are necessary.
- The separating tray has to be imported from the manufacturer only.

Paddy Cleaner (West Germany)

Figure 12.6 illustrates the constructional features of a grain cleaner.

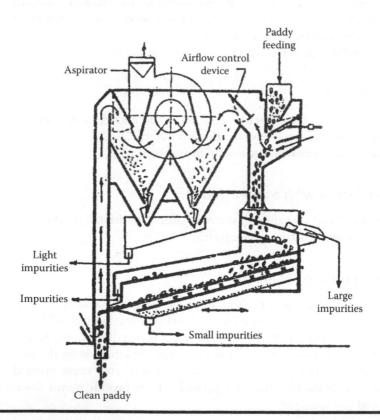

Figure 12.6 Closed-type double action aspirator precleaner (West Germany).

Unclean grain enters the machine through a balanced feed gate that ensures even distribution of grain. A suction fan situated in the upper portion aspirates off lighter impurities and dust. Then the grain has to pass through an aspiration channel where the air current lifts off chaff and other lighter particles that are drawn into the settling chambers and delivered through gate flaps to inclined chute. Lighter immature grains are also collected in the chamber, which otherwise go along with air current. The mixture is then passed through the vibrating screen where large impurities such as straws, big stones, clods of dirt, etc., are removed. The following grain sieve overtails seeds and other admixtures larger than grains. On the sand sieve, the grain tails over while sand, etc., passes through it. Before being discharged, the last traces of light impurities are removed from the grains when they are subjected to a strong aspiration through a deep aspirating leg.

Husking

The purpose of a modern husking machine is to remove husk from paddy without damage to the bran layer and rice kernel.

Husking machines are known by different names such as huskers, dehuskers, shellers, and sometime hullers also.

Impact-Type Paddy Husker (Japan)

Principle

The working principle of the impact or centrifugal-type of husker is based on the utilization of impact and frictional force for husking of paddy. In the impact-type husker, paddy is thrown against a rubber wall (Figure 12.7) by a rotating disk. The impact on the rubber wall due to the centrifugal force of the rotating disk causes splitting of hulls with a minimum damage to the kernel.

Relations between husking percentage and rpm of the disk, husking percentage, and angle of inclination of the husking plate with *Japonica* varieties of paddy are shown in Figures 12.8 and 12.9. Figure 12.9 shows that the husking percentage increases sharply from 2000 rpm and reaches its maximum at 4300 rpm. But the head yield decreases as rpm increases. Therefore, optimum rpm of the disk for different varieties of paddy is very important. It was also observed that the most important factor affecting husking ratio was not the diameter of the rotating disk but the circumferential speed.

Construction

A common type of centrifugal husker consists of a rotating disk of diameter 28.5 cm within a stationary rim of synthetic rubber 30 cm in diameter centered at the same

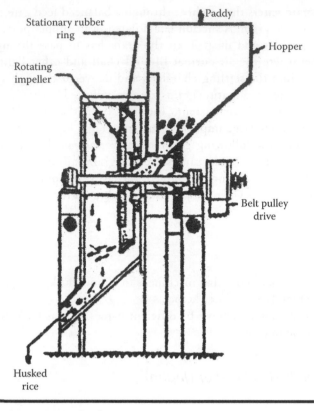

Figure 12.7   Impact-type husker.

axle (Figure 12.7). The thickness of the rubber rim is about 2.5 cm. The diameter of the rotating disk varies from 23 to 31 cm. The shore hardness of rubber rim is about 85°. The husking capacity of a machine with a disk of 30 cm in diameter is about 300 kg/h.

Advantages

- Price is reasonably low.
- Operating cost is low due to small power requirement (1 HP/500 kg paddy/h) and long life of the rubber rim.
- Total and head yields are higher in comparison to huller.
- Husking efficiency and head yields are quite satisfactory for parboiled paddy.
- The machine is compact and portable.
- It can be manufactured locally.

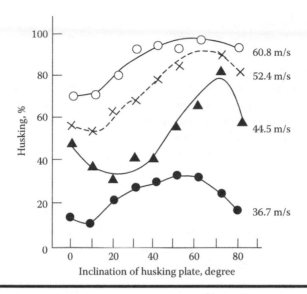

Figure 12.8 Relation between husking percent and the angle of inclination of husking plate. (From Ezaki, H., Paddy husker, Group training course, Fiscal Institute of Agricultural Machinery, Japan, 1973.)

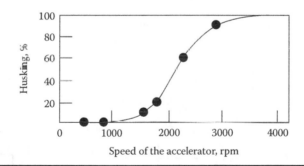

Figure 12.9 Relation between husking percent and the speed of accelerator plate. (From Ezaki, H., Paddy husker, Group training course, Fiscal Institute of Agricultural Machinery, Japan, 1973.)

Disadvantages

- Critical adjustment is required for different varieties and moisture contents of paddy.
- Breakage percentage increases with the increase of husking ratio.
- It is not possible to increase the husking capacity of the machine unless the number of husking disk on the same rotating shaft is increased.

Rubber Roll Husker (Japan)

Principle

If paddy is passed between two resilient surfaces rotating at different speeds in different directions, its husk will be split and stripped off. Figure 12.10 shows the principle of husking. If paddy is allowed to pass through the gap (smaller than the thickness of paddy) between two rubber rolls rotating at different speeds, then it makes contact with two rolls for different periods of time. The contact of the faster revolving roll is longer than the slower revolving roll. As a result, the paddy is sheared and compressed and its husk is stripped off and removed.

Construction

Figure 12.10 shows the constructional features of the Satake husker. It consists of two major parts: the upper part is the husking section and the lower part is the husk aspirating section (Satake, 1973).

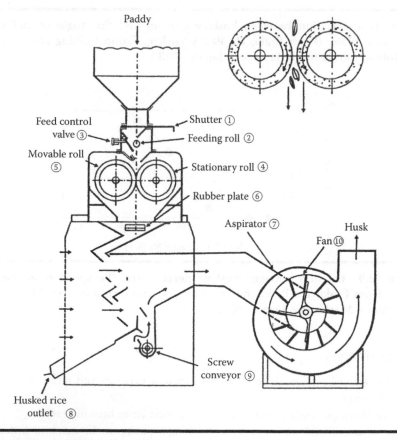

Figure 12.10 Rubber roll husker (Japan).

Clean paddy is delivered through a shutter (1) and a feeding roll (2) fixed below the shutter. Feed rate is controlled by a feed-regulating valve (3). The paddy is then passed through the gap between a fixed (stationary) roll (4) and an adjustable (movable) roll (5) and husked. After reducing the grain speed with the help of a resistance (rubber) plate (6), the mixture is allowed to pass through the aspiration section (7) to aspirate husk. Mature grain being heavy is collected in the husked rice outlet (8). Immature grain being light is collected and discharged by a screw conveyor (9). The husk is aspirated and thrown out of the mill building by a fan (10).

Both rolls have the same diameter. Diameter of the rubber rolls varies from 150 to 250 mm depending on the capacity of the husker. Their differential speeds are shown as follows:

Roll Diameter (mm)	High Speed (rpm)	Low Speed (rpm)
200	1200	900
250	1000	740

The wear of the rubber is considerable and with the reduction of the roll diameter, the capacity is also reduced. The main reason for the capacity reduction is the decrease in the relative speed of the two rolls.

In general, only 85%–90% of the paddy fed is husked due to variation in moisture content, size, degree of maturity of grain, eccentricity of the revolving rubber rolls, and uneven wearing of rubber roll surfaces.

The clearance between the two rolls can be adjusted manually. A sophisticated pneumatic device for automatic adjustment of the clearance has been introduced in place of manual adjustment.

Nowadays, rubber roll huskers are equipped with a blower to blow air on the rubber rolls' surface and bring down their temperature rise due to friction between husk and roller during husking.

Operation

When the husker is in operation for a certain period, the faster revolving roll wears more than the slower revolving roll. As a result, the diameter of the former becomes smaller than the latter and the difference in peripheral velocity between the two rolls becomes less. Though the speed ratio is kept constant, it causes lowering of husking efficiency. To encounter this difficulty, application of high pressure is not the proper means as it would break the grains. However, the rubber rolls are to be interchanged at a regular interval of time.

Advantages

- Highest percentage of sound and whole husked rice is produced as the risk of breaking the kernel is small and the chance of forming scratches on the kernel is also nil.
- The mixture of different sizes and varieties of paddy can be used without any significant increase of brokens in husked rice.
- Husking ratio can be increased to 0.9 without reduction in head yield.
- It does not remove germs.

Disadvantages

- Operating costs are high due to wear of rubber rolls.
- Storage life of rubber roll is limited as storage deteriorates its quality and in consequence, shortens its working life.
- If the paddy separator fails to separate paddy completely and the husked rice is returned to the husker along with the paddy, the constituent of the rubber may impart color and odor to the rice.
- It requires skilled labor to operate the machine efficiently.
- Sometimes on account of uneven grain distribution and uneven thickness of rubber, the rolls surface wears out unevenly, which adversely affects the efficiency and capacity. If the roll surfaces are corrected by turning them, the life of the rubber rolls will be reduced considerably.
- In general, the husking capacity of the rubber rolls in tropical countries is low due to (a) high temperature and humidity of the atmospheric air; (b) structure; and (c) larger surface area of the long grain husk in contact with the rubber rolls.

Power Transmission System

Figure 12.11 shows the power transmission system.

This type of power transmission system is most commonly used for the rubber roll husker. In the rubber roll husker with 25 cm rubber roll, a hexagonal belt is usually used (Figure 12.11).

Rubber Roll Husker (Europe)

It consists of (a) a drive pulley; (b) a box containing wheels and gears, etc.; (c) a special mechanism for adjusting the position of one roller; (d) a casing containing the two rollers; (e) a base on which the aforementioned parts are fixed; and (f) a steel underframe (Figures 12.12 and 12.13).

Paddy enters through the transparent feed pipe, which is fitted with an adjustable gate to stop the flow. The one baffle is fixed while another baffle is movable.

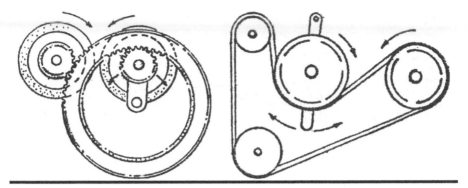

Figure 12.11   Power transmission system for rubber roll husker (Japan).

The feed roller driven by the pulley ensures even feeding. To adjust feeding, the hand wheel is turned and this moves the movable roll, the position of which can be seen on the graduated scale. The door is used to see the inside of the chamber (Gariboldi, 1974).

There is also an arrangement for air cooling the rubber rolls and removing the dust as well. In this type of husker, the husk aspiration system is installed separately. Usually, the husk separator is placed between the husker and paddy separator.

Separation

The husked rice is separated from the mixture of paddy and husked rice by the paddy separator.

Paddy Separator (Japan)

Principle

The paddy separator (shown in Figures 12.14 and 12.15) consists of a number of identical inclined trays with dimples (1) over the surface. These dimpled trays are in reciprocating motion. When a mixture of husked rice and paddy is delivered at the upper corner (2) of the tray, the husked rice being heavier occupies the bottom layer and comes in contact with the dimpled tray, while the paddy, being lighter, floats on the husked rice. The size of the dimples is slightly bigger than the brown rice but smaller than the paddy. The downward movement of brown rice is thus partially restricted by the dimples while the paddy is free to move downward. With the reciprocating movement of the tray, the husked rice, in contact with the tray, picks up the movement and moves to the side (3),

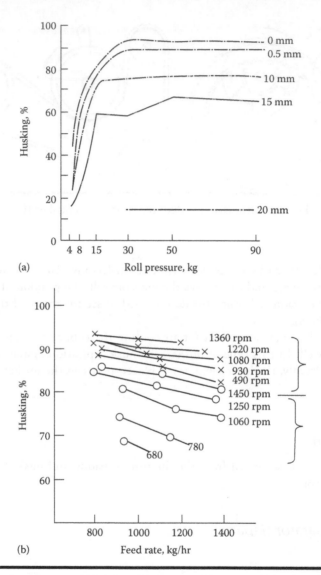

Figure 12.12 **(a)** Relation between husking percent and roll pressure. **(b)** Relation between husking percent and feeding rate of paddy. (From Ezaki, H., Paddy husker, Group training course, Fiscal Institute of Agricultural Machinery, Japan, 1973.)

while the paddy flows downward fast by gravity (4). At the same time, paddy and husked rice gradually move to the right side of the tray separating the husked rice at the upper part (5) and the paddy at the lower part (7) and leaving the mixture at the central part (6). The tray overflow is, therefore, received by the three compartments divided by flaps (8). The two flaps divide the whole receiving

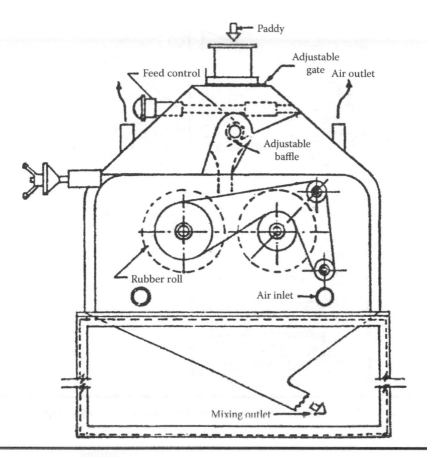

Figure 12.13　Rubber roll husker (Europe).

chamber into three compartments. The position of the flaps is adjustable so that pure brown rice can be received in the upper compartment, paddy in the lower compartment, and the mixture of paddy and brown rice in the middle compartment of the chamber (Satake, 1973).

Construction

The machine is made of steel. It consists of three to seven identical indented/dimpled trays, mounted one above the other with a spacing of about 5 cm. The whole tray assembly moves up and forward, making a typical reciprocating motion during operation. The dimples on the whole surface of the trays provide necessary frictional coefficients for separation of the mixture. The amount of mixture, fed from the distribution box placed on the left top side of the machine, is adjustable. Each tray receives an equal amount of the mixture. The trays are

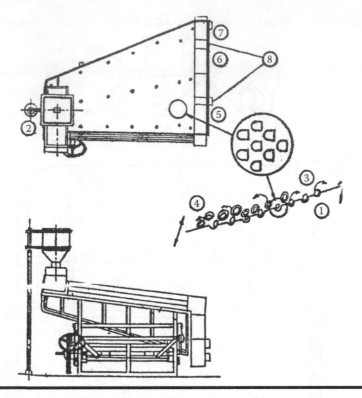

Figure 12.14   Construction of indented tray–type paddy separator (Japan).

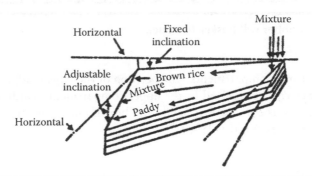

Figure 12.15   Inclination of the indented tray–type paddy separator (Japan).

inclined in both latitudinal and longitudinal directions. The latitudinal inclination is fixed at about 5°, while longitudinal inclination can be varied within
20° + 6°. The inclination adjustment handle located on the frame is used to
adjust the longitudinal inclination according to variety, moisture content, and
other conditions of paddy.

Operation

When the amount of mixture to be delivered becomes small, the grain should be distributed only to the upper plate. At the beginning of separation, paddy is likely to be mixed up with the husked rice outlet (5) until all the separating plates are covered with grain. In this case, all grains should be circulated in the machine until enough grain flows through the machine. Similarly, too much grain flow causes mixing of husked rice in the paddy outlet (7).

When the machine is not used, the separating trays should be coated with rust-proof grease.

Advantages

- Perfect separation can be achieved if different varieties are not mixed up.
- Separating condition of the mixture on the dimpled tray is always visible and can be adjusted accordingly.
- Operation is very simple.
- The unit is very compact and requires minimum space.
- Power requirement is also small.

Disadvantages

- It is difficult to manufacture the dimpled trays locally.
- Difficulties in perfect separation are encountered when the size and shape of the dimples do not correspond to the length of paddy and brown rice, when moist or dirty grains are separated, and when a mixture of different varieties of grain are separated.

Paddy Separator (Europe)

Compartment Separator

Principle

When apparently similar grains move over an inclined plane, the downward movement of the different grains by gravity is related to the specific gravity, shape, contact area, and coefficient of sliding friction. The husked rice, being small, heavy, and smooth, sinks down whereas light, flat, and rough paddy floats up by stratification.

The mixture to be separated is distributed over an inclined surface of the machine and given each single component equal and intermittent, obliquely upward thrust. The thrust is so regulated that it cannot push up the husked rice. Thus, it acts as a brake, causing them to slide down slowly. On the other hand, flatter and rougher paddy grains of lower specific gravity can, however, move up the inclined surface and fall off on the opposite sides.

Construction and Operation

The machine is usually made of wood. It consists of the following main parts: (a) the feed box; (b) the table enclosing the compartments, the collecting troughs for the separated products and their discharge outlets; and (c) the framework carrying the table, reciprocating movement drive mechanisms and speed variator.

The main part of the separator is the oscillating compartment assembly where the separation of paddy and husked rice takes place.

Figure 12.16 shows how the compartment separator works. The table (1) is divided cross-wise into several compartments (2), the shapes of which form a zigzag. The number of compartments varies from five to eight, depending on the capacity. The bottom of these compartments is slightly inclined. The whole assembly consists of one, two, three, or four decks.

The mixture of husked and unhusked grains is fed to the compartments in equal quantities, through the hopper. The table (1) moves forward and backward horizontally along the direction of line A. The movement for thrusting the grains toward the upper edge of the slant table is obtained by means of eccentric motion. Normally, the stroke length is about 20 cm and not adjustable. The grains move cross-wise inside the compartment and alternatively hit the zigzag walls. These walls are oblique and the grains rebound obliquely.

The paddy receives the oblique upward thrust. The husked paddy moves down slowly whereas the unhusked paddy moves up the inclined plane. The two components are thus separated and discharged from the two opposite sides (Gariboldi, 1974).

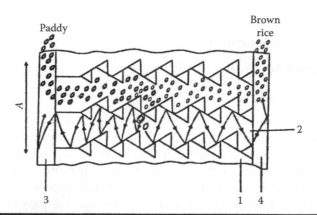

Figure 12.16 Internal construction of the compartments and operational principle of a compartment separator.

Advantages

- Running cost is minimum as there is no part to be changed or subject to serious wear.
- Life of the machine is very long due to its sturdy construction.
- If it is operated properly, the separation is perfect.

Disadvantages

- The machine is expensive to buy and transport.
- It needs strong foundation at the ground level.
- It requires skilled labor.
- The wooden parts of the machine are subject to termite attacks.

Whitening

The term "whitening" refers to the operation of removal of germ, pericarp, tegmen, and aleurone layers from husked rice kernels. It is also called "polishing" or "pearling" or "scouring."

There are three major kinds of whitening machines used in the modern rice processing industry. They are (1) vertical abrasive whitening cone; (2) the horizontal abrasive whitening roll; and (3) the horizontal metallic friction-type roll.

Vertical Whitening or Pearling Cone (Europe)

Principle

Basically, the machine consists of an inverted cast iron frustoconical rotor covered with abrasive material mounted on a vertical spindle, revolving inside a crib. The crib is lined with steel wire cloth or perforated metal sheets and it is equipped with rubber brakes that are placed vertically and spaced equally and protruded into the gap between the cone and crib. The pressure inside the machine can be adjusted by pushing in or pulling out the rubber brakes. The number of brakes to be used depends on diameter and is determined by the formula: $n = (D/100) - 2$, where D is cone diameter, mm. The peripheral speed of the cone should be 13 m/s. The husked rice enters the gap and is dragged along by the rough surface of the rotating cone. The rubber brakes tend to stop it and cause it to pile up against their side. While pressed up against the brakes, the grains undergo a strong swirling and revolving movement because of their oval shape and smooth surface (Gariboldi, 1974).

Each grain is scoured by the abrasive surface of the cone. It also rubs against the surrounding grains and the rough lining of the crib as well.

The grains meet almost the same conditions all the way round the cone until they sink lower and lower by gravity and are finally discharged at the bottom of the cone.

The bran is finally ground due to scouring and rubbing and escapes through the lining of the crib and regularly removed.

As the vertical abrasive roll is an inverted truncated cone, the peripheral speed at the upper part is higher than the lower part. But the density and pressure at the lower part are higher than at the upper part. As a result, the grinding action is predominating at the upper part while frictional force is stronger at the lower part. In the strict sense, this machine is a combination of abrasive and friction types of machines.

Generally, a series of two to four whitening cones are used to whiten rice successively in two to four passes.

Construction

The machine consists of a rotating vertical conical cast iron cylinder covered with an abrasive material. The entire rotating cone is encased within a fixed perforated metal sheet known as crib. The gap between the abrasive surface and the crib is about 10 mm. It is provided with rubber brakes, placed vertically and spaced equally, which protrude into the gap between the abrasive cone and the crib. The clearance between the rubber brake and crib is about 2–3 mm. The pressure inside the whitening chamber can be adjusted by pushing in or pulling out the rubber brakes.

The construction of the machine is shown in Figure 12.17. The whole abrasive cone can be lifted up or down to adjust the clearance between the cone and the rubber brake as and when rubber brakes are worn out a little.

Advantages

- The parts of the machine subject to wear are the abrasive surface, the crib lining, and the rubber brakes; these can be made and replaced locally.
- It can deal with any variety of rice. Its performance is excellent with parboiled rice.
- The degree of whitening is adjustable.
- The life of the machine is very long.
- The operational cost is low.

Disadvantages

- The machine is big and heavy.
- The machine requires skilled operators and attention.
- Power requirement is high.
- Power transmission system is troublesome.
- As the abrasive roll is not vitrified (i.e., not treated at high temperature), sharpness of the abrasive particles is not much.
- Pressure distribution along the vertical line is not uniform.

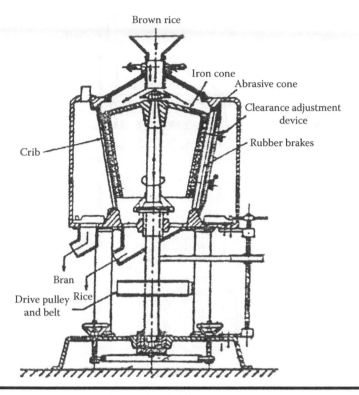

Figure 12.17 Vertical abrasive whitening cone.

Horizontal Rice Whitening Machine (Japan)

To remove bran layers from the husked rice, a combination of two different types of whitening machines, namely, grinding type (speedo type or horizontal abrasive type) and friction type (or pressure type), are employed by Satake Engineering Co. Ltd. (Satake, 1973). As the surface of husked rice is smooth and slippery, the grinding type of machine is used at the initial stage of whitening in order to shave and grind the bran layer into smaller particles and impart roughness to the grain surface. The thrust (mainly) and frictional force of a revolving abrasive roll are used for whitening. The friction-type machine is employed subsequently to peel off the remaining bran layer (in flakes) easily by the friction between rice and rice by the high pressure created by a rotating ribbed steel roll. Functions of these machines are shown in Figure 12.18A and B.

If the smooth husked rice kernels were directly whitened by the friction-type machine only, it would require much higher pressure and power for whitening. The yield of brokens would also be more.

Grinding-type machine is characterized by the increase of milling efficiency with the increase of speed of abrasive roll and decrease of milling efficiency with the increase of pressure.

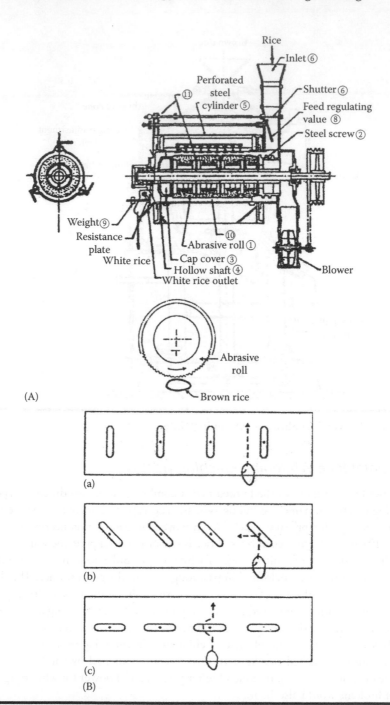

Figure 12.18 **(A) Function of grinding-type machine. Horizontal abrasive-type whitening machine. (B) Adjustable steel bar (resistance pieces).**

The differences in principle and action between the two types of machines are further illustrated in the following table:

	Grinding Type		Friction Type	
Peripheral Speed	Over 600 m/min		Below 300 m/min	
Pressure	Below 50 g/cm²		Over 200 g/cm²	
	Grinding Type		Friction Type	
System/Item	Initial Stage	Middle Stage	Initial Stage	Middle Stage
Efficiency	High	Low	Low	High
Breakage	Small	Small	Much	Small
Whitening degree	High	High	Low	Low
Glossiness (shine)	Low	Low	High	High
Moisture absorption	Fast	Fast	Slow	Slow
Deformation	Partially deformed	Partially deformed	Full	Full
Embryo removal	Easy	Easy	Easy	Difficult

In general, the first and second stages of whitening operations are done by the grinding-type machines and the last stage of whitening is done by the friction-type machine. Therefore, the friction-type machine is used to give uniform polish and surface finish to the rice. These machines are arranged in series so that the rice is whitened successively. The following combination can be used for *Indica* varieties of paddy: two to three grinding-type machines and one friction-type machine.

Grinding is caused mainly by impact and partly by friction with the abrasive surface. Grinding-type machine is characterized by the following:

1. Impact should be high and friction should be low.
2. Milling efficiency decreases with the increase of pressure or friction.
3. Milling efficiency increases with the increase of speed of the abrasive roll.

Construction

The horizontal abrasive roll–type whitening machine is shown in Figure 12.21. The machine consists of an abrasive roll (1) fitted with steel screw (2) and cap cover (3) clamped on a horizontal hollow shaft (4) rotating at a high speed (1000 rpm) in a slotted steel cylinder (5). An inlet hopper (6), shutter (7), and feed control gate (8) are provided above the feeding screw. An adjustable steel plate with weights (9) is fitted to the rice outlet spout. This plate with weight acts as a valve. The counterpressure of this valve, which is adjusted by the weight, controls grain flow rate mainly and milling pressure partially.

The outer cylinder, i.e., housing, can be separated into two halves. The bottom half is slotted with rectangular slots. The slots are inclined at an angle of 80° to the rotating direction of the roll so that grains move slowly toward the outlet. Over the full length of the cylindrical housing, two to three rows of adjustable steel resistance pieces are protruded into the annular space between the roll and the cylindrical housing. The resistance pieces can be turned from 0° (axial) to 90° (radial) (Figure 12.18B).

After feeding through the steel screw, the grain passes through the space (10) between the rotating roll and the slotted steel cylinder. The bran layer of the grain is shaved, cut, and ground into small particles by the fine and sharp edges of the abrasive roll. The bran passes out through the slots of the steel cylinder. An air current from the hollow shaft flowing through the gaps of the abrasive roll accelerates the separation of bran and cools the grain. The whitened rice is discharged through the outlet (Satake, 1973).

The abrasive roll is made of carborundum (SiC) particles bound by some ceramic substances.

Operation

1. Peripheral speed should be more than 600 m/min.
2. Resistance is slightly varied by weight. Pressure should be less than 50 g/cm².
3. *Flowing quantity*: The progress of whitening action in the abrasive roll–type machine is in inverse proportion to the flowing quantity.
4. *Function of resistance pieces* (Shibano, 1973): As the abrasive roll is horizontal, rice grain tends to stay in the lower part of the whitening cylinder. In order to overcome this problem, the resistance pieces should be set properly so that the grains are rebound by them.

   When the resistance pieces are set at 0°, they give minimum resistance and minimum thrust to the moving grain so that the grain moves along the direction of the slots. When the resistance pieces are set at 45°, they give maximum thrust and medium resistance and when they are set at 90°, they give maximum resistance but no thrust to the moving grain.

   If the resistance pieces are set at correct position, the grain density and the pressure will be uniform in the whitening chamber and bran will be dropped equally from all slots of the cylinder. This ensures perfect and effective whitening operation.

   By adjusting their inclination, the residence time of the grain inside the whitening chamber can also be adjusted. Thus, it helps in controlling the degree of whitening also.
5. By proper selection of pressure, mesh of the abrasive roll, peripheral speed/rpm of the abrasive roll, this type of machine can produce whitened rice of any desired shape. This is illustrated as follows.

Shape of the Whitened Rice			
Parameters	*Round*	*Bar Like*	*Flat*
Peripheral speed	Large	Medium	Small
Mesh	Large	Medium	Small
Pressure	Small	Medium	Large

Renewal of Different Parts

The various components are replaced under the following circumstances:

1. When the diameter of abrasive roll is reduced by 15 mm
2. When the lower half of the outer cylinder is broken
3. When the resistance pieces are worn out by one-third of their original size

Advantages

- Total and head yields are high.
- Bran removal is easier and uniform due to jet air system introduced in the machine.
- The edges of the heat-treated carborundum roll are very sharp and hard.
- The degree of whitening can be adjusted.

Disadvantages

- Since the machine is placed horizontally, the rice would tend to settle at the bottom part of the machine, if the angle of inclination of the resistance pieces is not adjusted correctly, according to variety and condition of rice resulting in nonuniform whitening of rice.
- The machine needs careful attention.
- Clearance between the screen and abrasive roll cannot be adjusted.
- Once the diameter of the abrasive roll is reduced by about 1.5 cm, it has to be replaced. It is difficult to manufacture carborundum roll (32 grits) of high quality locally.

Friction-Type Whitening Machine (Japan)

Principles

The purpose of the friction-type whitener is to remove the remaining bran easily and uniformly from partially whitened rice produced by the horizontal abrasive whiteners. When rice is allowed to enter into the space between the rotating

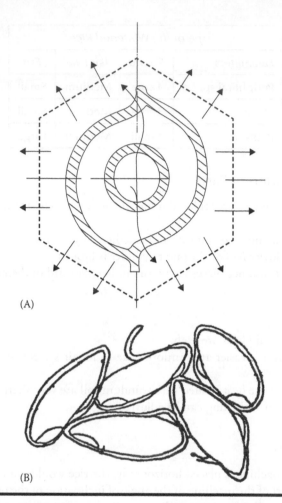

(A)

(B)

Figure 12.19 **(A) Slotted hexagonal cover and hollow shaft. (B) Function of friction-type whitening machine.**

steel roller and the hexagonal screen of the whitener, mutual rubbing of the rice kernels takes place under pressure.

Figure 12.19A illustrates the space variation between the rotating roll and the hexagonal screen.

The peripheral speed of the steel roller is below 300m/min. The space between the hexagonal screen and steel roller is different at different points. Mutual rubbing of grain occurs (Figure 12.19B) at a position where the annular space is narrow, and bran is removed from the rice kernel. After passing through the narrow gap between the extruding part of the roller and screen, the grain enters into the wider space where it receives a strong air stream that facilitates the separation of bran from the grain. It also helps in discharging bran through title slots of the hexagonal screen. The whitened rice produced by this machine is free from bran, slightly shiny, and cool.

The inner surface of the slotted hexagonal screen has a number of small projections that enhance the rubbing action. Pressure is applied by the resistance plate with weight. Whitening is mainly performed by mutual rubbing of grains and partly by rubbing of grain with the screen. Higher pressure is necessary to accelerate the whitening.

Construction

The machine (Figure 12.20) consists of a rotating steel cylinder having two friction ridges (milling roll) (1) and feeding screw (2) and a lock nut (3) mounted on a horizontal, partly hollow, perforated shaft (4) encaged in a hexagonal chamber (5). There are two lengthwise openings behind the ridges for the passage of air. The hexagonal chamber is made of slotted screen and has small projections on the inner surface. The inlet (6) is located above the feeding screw. An adjustable steel plate (7) with weights (8) is fitted at the rice outlet spout (9) to adjust the variable pressure. This pressure can be varied by adjustment of the counter weight. Strong air stream is blown by a centrifugal blower through the long slots of the cylinder and holes of the shaft to help in separating the bran and dissipating the heat generated by the friction between rice and rice (Satake, 1973).

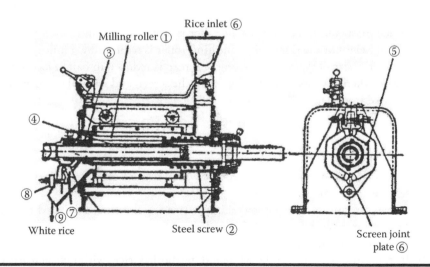

Figure 12.20 Friction-type whitening machine.

Operation (Shibano, 1973)

1. *Rpm of roller*: The rpm of the steel roller should be about 700 (standard). In order to generate effective frictional pressure on the grain, rpm should be so adjusted that the peripheral speed of roller is <300 m/min. The peripheral speed varies with the diameter of milling roller.

   Whitening efficiency increases as pressure increases, but the milling pressure decreases as the rpm of the roller increases.

2. *Flowing quantity*: In order to increase the milling efficiency, it is necessary to keep the pressure on the grain in the whitening chamber sufficiently high and density of grain even. Adjustment of milling pressure should be made mainly by the flowing quantity and partly by the weight at the outlet. The milling pressure above 200 g/cm² is effective.

3. *Jet air*: Proper quantity of air should be supplied; otherwise, germ separation may be difficult.

4. *Resistance by weight*: Generally, resistance to the rice can be divided into fixed resistance and variable resistance. A friction-type machine is characterized by the proportion of the two kinds of resistance.

Renewal of Different Parts

Due to high pressure, the rate of wear of different components of the machine is higher compared to the abrasive roll–type machine. The following components are to be replaced at proper time to maintain the whitening efficiency of the machine:

1. When diameter of the screw of the roller is reduced by 2 mm
2. When height of the ridges of the milling roller is reduced by 3 mm
3. When the size of the projection on the screen is reduced to half of its original size or when new holes are developed on the screen and when the screen joint is broken

Advantages

- Production of uniformly whitened rice with a shining appearance is the main feature of this machine.
- Strong air current helps in lowering the rice temperature, generated by friction and removing from the whitened rice.
- Pressure is adjustable.
- It can produce uniformly whitened parboiled rice without increase of brokens.

Disadvantages

- Broken percentage may be increased if the machine is not operated properly. However, its performance is excellent with bold/short *Japonica* varieties of rice.
- In case of parboiled paddy, the screen may be choked with the oily bran if higher percentage of polish is given particularly in the humid atmosphere.
- The slotted screen and the steel cylinder are subjected to wear and are replaceable.

Water Mist Whitening/Polishing

A water mist polisher has been developed in the rice whitening system in Japan to improve the luster and appearance of milled rice. This machine employs a fine spray of water mist to remove the last trace of fine bran and rice polish from the surface of the milled rice. Particularly, bran and polish remaining deep inside the grooves of the rice kernels can be removed as well. Moreover, a combined application of water mist and subsequent rice to rice rubbing action of the friction-type whitener imparts a high degree of polish and a luster to the milled rice. In consequence, the dull milled rice grains can be transformed into glossy rice kernels with an attractive appearance.

Modern Combined Vertical Milling System

Recently, the combined vertically oriented milling systems have minimized the floor space and overcome some of the drawbacks/problems of horizontal milling machines. In the present system, the combination of vertical abrasive-type and friction-type milling machines is used to improve pressure development within the milling chamber. It increases milling efficiency and the quality of both *Japonica* and *Indica* rice. Its other feature is that the water mist system can be fitted to the last friction-type machine to cool the rice, and remove the traces of adhered bran, and impart translucency to the polished rice.

Grading

After whitening the rice, it has to be graded into different grades by separation of broken from the head or whole rice. The rice grain above 4 g size is considered head rice. The brokens below 3/4 grain size can be further graded into coarse and fine brokens.

Traditionally, it is done by a sieving machine using different sizes of screens, which is not an efficient system.

An improved metallic indented cylinder grader is made of indented sheet. It is slightly inclined and is allowed to rotate at 30–40 rpm with a mixture of broken and head rice to be graded. The brokens small enough to lodge into the indents are

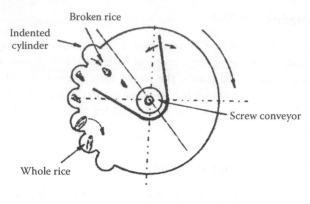

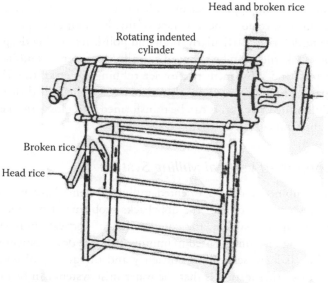

Figure 12.21   Rice grader.

raised to certain points after which they fall owing to gravity in a trough, longitudinally mounted into the rotating cylinder. This trough has a worm to convey the brokens to the broken outlet. The head rice tails over the bottom of the cylinder and reaches the broken outlet by sliding (Figure 12.21). Each cylinder can separate the mixture of brokens into two grades only. Hence for more broken grades, more number of cylinder graders are to be employed.

For efficient performance of this unit, cleaning of indents at certain intervals is necessary; otherwise, efficiency would fall.

Chapter 13

Milling of Corn, Wheat, Barley, Rye, Oats, Sorghum, and Pulses

Corn Milling

Introduction

Corn is one of the world's most versatile seed crops. Botanically named *Zea mays*, corn is used as food and feed. Corn can be processed into various food and feed ingredients, industrial products, and alcoholic beverages. But the modern corn milling technology developed for the aforementioned products is mostly confined to some of the developed countries only. However, modern corn milling technology is to be suitably adopted for producing the types of products required for other countries. At present, there are two modern methods of milling of corn: dry milling, and wet milling. Besides germ for corn oil extraction and hull and deoiled germ, etc., for feed, grits (mainly used for breakfast cereals) are the main products of corn dry milling, whereas pure starch, germ, and feed are the major products of wet milling.

Composition and Structure

The mature corn kernel is composed of four major parts: (1) endosperm (82%); (2) germ (12%); (3) pericarp (5%); and (4) tip cap (1%).

Pericarp: The pericarp is mainly composed of four successive layers, namely, outermost thick layer of tough cells, spongy layer of cells, seed coat or testa layer, and aleurone layer. The spongy layer is continuous with spongy cells of tip cap.

Germ: The germ is mainly composed of scutellum and embryonic axis. The major parts of lipids and proteins are reserved in the scutellum.

Endosperm: The endosperm of corn is composed of floury and horny parts. The proportion of horny to floury parts varies widely from dent corn to floury corn variety. The ratio of horny of floury endosperm is about 2:1 in dent, whereas the floury corn contains little horny part. As the horny parts are hard and floury parts soft, the latter is milled into flour easily during rolling. The horny regions have 1.5%–2% higher protein content than the floury regions.

Corn Dry Milling

Corn dry milling system can be divided into two groups: the traditional non-degerming system and modern degerming system. In the non-degerming system, the whole corn is ground into a meal of high fiber as well as high protein contents by a stone grinder without removing germ. After grinding, certain amount of germ and hull can be removed from the meal by sifting.

In the degerming system, the corn is moistened with a little amount of water and tempered for moisture equilibration. After degerming, the stock is dried, milled, and classified into different products. The purpose of all dry degerming corn milling methods is to remove hull, germ, and tip cap from the corn kernel as far as practicable and primarily produce corn grits with some meals and flours. The germ is then used for oil extraction and deoiled germ, hull, etc., are used as feed that is known as hominy feed. The yields of endosperm products and hominy feed are about 70% and 30%, respectively.

Tempering–Degerming Method of Dry Milling

The major objectives of this method are (a) to remove essentially all germ and hull so that endosperm contains as low fat and fiber as possible, (b) to recover a maximum amount of the endosperm as large clean grits without any dark speck, and (c) to recover a maximum amount of germ as large and pure particles.

Description of the Tempering–Degerming System

The basic operations/processes involved in the tempering–degerming (TD) method are as follows:

1. Cleaning of the corn
2. Conditioning of the corn by addition of control amount of moisture either at ordinary temperature or at an elevated temperature to toughen the germ and husk and facilitate their removal from the endosperm

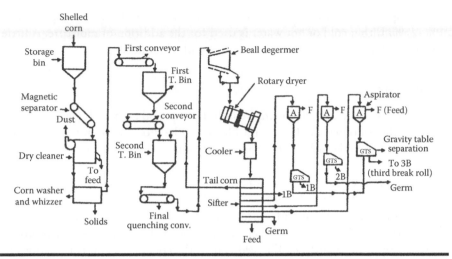

Figure 13.1   Flow diagram for TD-type corn mill.

3. Releasing hull, germ, and tip cap from the endosperm in a degermer
4. Drying and cooling the degermer products obtained from the degermer
5. Fractionating degermer stock by multistep milling through a series of machines, namely, roller mills, sifters, aspirators, gravity table separators, and purifiers, to separate and recover the various products
6. Further drying of the products as and when necessary
7. Blending and packaging of products

The flow diagram of the system is shown in Figure 13.1.

Cleaning of Corn

Thorough cleaning of corn is essential for subsequent milling operations.

Pieces of iron, etc., are removed by magnetic separators. Dry cleaners consisting of sieves and aspirators and sometimes a wet cleaner consisting of a washing destoning unit and a mechanical-type dewatering unit, known as whizzer, are used for cleaning of corn.

Hydrothermal Treatment/Conditioning

Predetermined amount of moisture is added to the corn in the form of cold or hot water or steam in one, two, or three stages with appropriate tempering times after each stage. The tempering times (rest periods) vary according to the hydration methods. Also, tempering temperatures vary from room temperature to about 50°C accordingly.

The optimum moisture content for degerming in the Beall degermer is 21%–25%. Either cold or hot water is used for the addition of moisture. A little heat in the form of open steam is added as and when necessary.

Degerming

The purpose of degerming is to remove hull, tip cap, and germ as far as practicable and leave the endosperm in large grits. However, the products from degermer consist of a mixture of kernel components, freed from each other to varying degrees, with the endosperm particles varying in sizes from grits to flour.

The Beall degermer consists of a rotating cast iron conical roller mounted on a horizontal shaft in a conical cage.

Part of the cage is fitted with perforated screens and the remainder with plates having conical projections on its inner surface. The rotating cone has similar projections over most of its surface. The feed end of the cone has spiral corrugations to move the corn forward whereas the large end has corrugations in an opposite direction to retard the flow. The product leaves the unit in two streams.

The major portions of the released germ, husk, and fines as well as some of the grits are discharged through the perforated screens. Tail stock containing large amount of grits escapes through an opening fitted with the large end of the cone. A hinged gate with an adjustable weight adjusts pressure inside the chamber and controls the flow of the stock.

Drying and Cooling of Degermer Stock

The degermer products are to be dried from 15% to 18% moisture content for proper grinding and sifting.

Generally, rotary steam tube dryers are used for drying the product. Rotary Louver-type dryer can also be employed. The stock is heated to about 50°C.

Counter-flow or cross-flow rotary vertical gravity or fluidized-bed types of cooler can be used for cooling the dried products.

Rolling and Grading

Recovery of various primary products is the next step. Further release of germ and husk from the endosperm products occurs during their gradual size reduction roller mills.

The germ, husk, and endosperm fragments are then separated by means of sifters, aspirators, specific gravity table separators, or purifiers.

Sifting is an important operation and is variously referred to as scalping, grading, classifying, or bolting, depending upon the means and purpose. Sifting is actually a size separation operation on sieves. Scalping is the coarse separation made on the product leaving a roller mill or degermer. Grading or classifying is

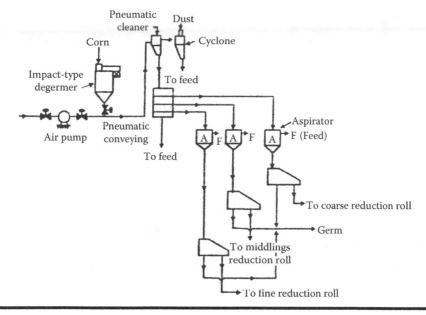

Figure 13.2   Flow diagram for Miag process.

the separation of a single stock (usually endosperm particles) into two or more groups according to particle size. Bolting is the removal of hull fragments from a corn meal or flour.

Another modern corn dry milling method, known as Miag process, is shown in Figure 13.2. Only the flow diagram of the Miag process is given in this figure but detailed process has not been described here.

Corn Wet Milling

It has been discussed earlier that pure starch, pure germ, and feed are the basic products of corn wet milling. But a few hundreds of by-products can be produced from these three main products.

The raw corn for wet milling should contain 15%–16% moisture and it should be physically sound. Insect- and pest-infested, cracked, and heat-damaged corns (treated at temperature around 75°C during drying) are unsuitable for wet milling. The heat-damaged corn affects the quality of oil extracted from its germ.

Sufficient amount of moisture is added to the corn during steeping in the wet milling process in order to prepare the corn for subsequent degerming, grinding, and separation operations.

The wet milling process consists of the following steps: (a) cleaning; (b) steeping; (c) germ separation and recovery; (d) grinding and hull recovery; and (e) separation of starch and gluten.

Cleaning

All impurities such as dust, chaff, cobs, stones, insect-infested grain and broken grain, and other foreign materials are removed from the corn by screening and aspirating. The clean grains are conveyed to the storage bins.

Steeping

The major objectives of steeping are (1) to soften the kernel for grinding; (2) to facilitate separation of germ; (3) to facilitate separation of gluten from the starch granules; and (4) to remove solubles, mainly from the germ.

Water impregnated with SO_2 (i.e., acidulated water with H_2SO_3) is used for steeping. It helps in arresting certain fermentation during the long steeping process.

The steeping is carried out at about 50°C for a period varying from 28 to 48 h in different plants. The steeped corn attains a moisture content of about 45%.

The flow diagram of corn wet milling process is shown in Figure 13.3.

Germ Recovery

The wet and softened corn kernels containing about 45% moisture are conveyed to the degerminating unit. This machine consisting of a metallic stationary plate and a rotating plate with projecting teeth is employed only for tearing the soft kernels apart and freeing the germs without grinding them.

The pulpy mixture containing germs, husk, starch, and gluten is passed through hydroclones, where the germ, being lighter, is separated from other heavier ingredients, by way of centrifugal force. Only modern starch plants employ hydroclones for germ separation. Otherwise, the floatation method of germ separation is still in use in old types of mills.

Milling and Fiber Recovery

After separation of germ and screening of the coarse particles, the mixture contains starch, gluten, and hulls.

Mainly horny endosperm and hull are generally ground by either traditional Burstone mill or modern entoletor impact mills to release the rest of the starch. Material to be ground enters the machine through a spinning rotor and is thrown out with great force against the impactors at the periphery of the rotor and also against a stationary impactor resulting in considerable reduction in particle size. Here only the starch is readily released, with a very little size reduction of hulls.

The milled slurry, containing the ground starch, gluten, and hulls, is passed through a series of hexagonal reels where the coarser hulls and fibers are removed.

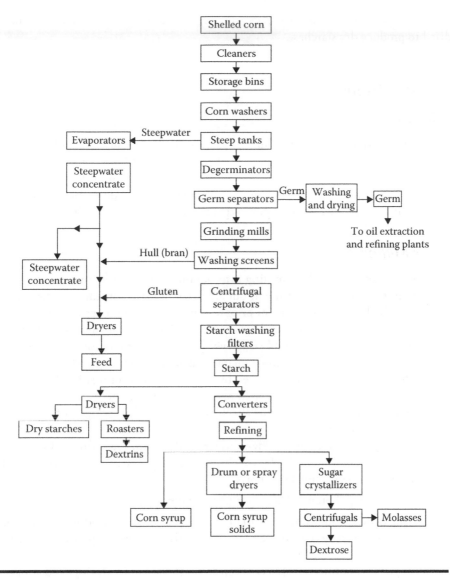

Figure 13.3 Flow diagram of corn wet milling and refining processes.

Starch–Gluten Separation

In the modern process, the slurry containing starch and gluten is concentrated and then the lighter gluten particles are separated from the relatively heavier starch particles by the centrifugal force in high-speed centrifuges. The centrifuging of starch is carried out in two stages. In many modern plants, the second stage of centrifugation is performed by a number of hydroclone type of equipments.

The starch obtained from the second stage of separation is filtered and then dried to produce dry starches.

Wheat Milling

Introduction

Wheat is the principal food grain in many countries of the world. One of the most important cereals, it is used as staple food in the form of flour. In India, a large proportion of wheat is used as the familiar *atta* and *maida* (wheat flour). The hard wheats are also ground into *suji* (semolina). Whole wheat is ground into atta by the traditional stone grinder without prior separation of bran and germ from it.

Flour Milling

The objective of modern flour milling is to obtain the maximum amount of white flour from the wheat endosperm without any bran or germ content. Conditioning of wheat by hydrothermal treatment prior to milling helps in the separation of bran and germ from the endosperm. If wheat is conditioned by hydrothermal treatment, bran and germ become rubberlike while the endosperm becomes soft. It also eliminates the difference in grinding characteristics between soft and hard wheat. When the conditioned wheat is sheared by the corrugations of first break roll during the milling operation, it splits open releasing small endosperm pieces and thus exposing the remaining endosperm, which could be carefully scraped off the bran in successive break rolls.

The yields of white flour and by-products (called mm reed) from white flour milling are about 70% and 30% by weight, respectively. The mill feed is composed of 12% bran, 3% germ, and 15% shorts.

Wheat consists of bran (12%), germ (3%), and endosperm (85%).

Milling of wheat into flour essentially involves a series of grinding and sifting operations with in between purification of granular endosperm particles (Sarkar, 1993, 2003).

Modern flour milling consists of seven steps: (1) receiving, drying, and storage of wheat; (2) cleaning; (3) conditioning; (4) milling into flour and by-products; (5) packaging and storage of finished products; (6) blending; and (7) milling by-products. Utilization of the most important operations, namely, cleaning, conditioning, and milling have been discussed here.

Cleaning

Wheat is thoroughly cleaned to remove all fine impurities and the dirt sticking to the surface of the grain. To remove loose fine impurities, a set of cleaners is employed. Small pieces of stick, stones, sand, etc., are removed by sieving and light impurities like chaff, etc., are removed by aspirations. Then the wheat is allowed to pass over powerful magnetic separators to remove pieces of ferromagnetic materials. The seeds of other food grains, defective grains, and weed are removed by disk separators.

The next step in the cleaning operation is the removal or dirt sticking to the surface by scouring. Usually, wheat is moved by paddles against stationary emery-coated surface. Then the dirt and loose outer coating are aspirated off. The scratches and cracks formed in wheat during scouring help in increasing the rate of moisture absorption at the time of washing and conditioning.

The final cleaning step is washing by water, which allows the dirt and bits of metal to sink. The moisture content of wheat increases by about 1% during washing.

Conditioning/Hydrothermal Treatment

The conditioning of wheat can be done either at room temperature, elevated temperature, or at high temperature. But the temperature of wheat grain should not be raised above 47°C; otherwise, the gluten quality will be affected, which deteriorates the baking quality of the flour.

Generally, the moisture contents of soft and hard wheats are increased 15%–17% and 16%–19%, respectively, by soaking and then the moisture of the grain is equilibrated by tempering for 18–72 h in the tempering bin.

In a modern system, conditioning of wheat is performed in four stages. The conditioner mainly consists of three sections, namely, preheating section, moistening section, and cooling section. In the first section, wheat is preheated to the proper temperature; in the second section, wheat is moistened to the desired moisture level; and in the third section, soaked wheat is cooled to the room temperature. Finally, the treated wheat is kept in a separate tempering bin for 18–72 h.

Hydrothermal treatment of grain by direct steaming has been popular for the last few years. It has many advantages over heating by air because both moistening and heating are carried out simultaneously in a single operation. Moreover, the grain is heated within 20–30 s to about 47°C. But the grain temperature above 47°C may adversely affect the quality of the flour. The rapid rate of heating weakens the intermolecular bonds in various parts of the grain to a considerable extent resulting in easier separation of bran, more effective grinding of endosperm, and stronger action on proteins and enzymes.

Grinding (Milling)

Milling of wheat is carried out by roller mills. The roller milling system is mainly divided into the break roll and reduction roll systems. In addition, most of the flour mills keep a standby system known as scratch system. The scratch system is nothing but an extension of the break roll system. The break rolls and the reduction rolls are differentiated with the variation in their surface conditions. The surface of the reduction roll is smooth whereas the surface of break roll is corrugated. In the break rolls, the bran is cracked, and the kernel is broken open. The endosperm adhering to bran is milled away successively in a few steps. Generally, a series of four sets of break rolls are used. Each set of rolls take stock from the preceding one. After each break, the mixture of free bran, free endosperm, free germ, and endosperm still adhering to the bran is sifted and separated. The endosperm adhering to bran is passed through the next break roll while the middle size endosperms called middlings are sent to the reduction rolls for proper size reduction to flours. Therefore, the break rolls are mainly used for the production of middlings and the reduction rolls are used for grinding of free middlings into proper flour size. After each reduction of endosperm (middlings), the flour is sifted away from the bigger size middlings and the remaining middlings are passed to the next reduction rolls. The aforementioned operations are continued until the desired products are obtained. As many as 12–14 reduction rolls are used in most flour mills. But all reduction rolls are not used for all break products (Matz, 1959). The flow diagram of the flour milling system is shown in Figure A.1.

Storage of Finished Products

The flour and the mill feed (bran, germ, and shorts) are bagged in waterproof bags, stitched, and stored in cold dry condition in flat godowns.

Components of a Wheat Mill

Break Roll

Break roll consists of twin pairs of corrugated steel rolls. One roll of a pair revolves faster than the other, differential speed being in the proportion of 2.5 to 1.

Break Sifting System

This can be divided into two parts—plan sifter and purifier.

> *Plan sifter.* Plan sifter is a scalping system removing large bran pieces adhering with endosperm at the top. The next series, which are finer, remove the bran and germ. The next layer of still finer sieve removes the endosperm middling and the bottom rough flow.
> *Purifier.* The middling containing finer bran particles are removed by the purifier before they move to reduction roll.

Reduction Roll

The reduction roll comprises two smooth rolls. The rolls in the reduction system are further divided into coarse and fine rolls, depending on the clearance between the rollers.

It is possible to grind flour into very fine particles by gradual grinding. But under high grinding pressure, the starch is ruptured and this should be avoided.

Reduction Sifting System

The same plan sifting system is used here.

After each reduction, the product is separated by plan sifters where the finished flour is sifted by 120 mesh sieve (silk) and removed and oversized material is sent back to the reduction rolls for further processing.

Scratch System

If the mill is functioning properly, i.e., a good release of endosperm is obtained on the break rolls, the scratch system can be bypassed; if not, the scratch system is employed to maintain proper release of endosperm from bran. The scratch system is an extension of the break system and thus used as a standby system only.

Barley Processing and Milling

Barley is used primarily as a source of carbohydrate. A whole barley grain contains about 12% hull, 58% starch, 12% protein, 2% fat, 6% fiber, and 3% ash. Barley is used for feed, brewing, and food. It is consumed as barley flakes, popped barley, sprouts, barley starch, etc.

Regarding animal feed, barley grains are ground and mixed with other concentrates or grains for feeding. By-products from malting, brewing, and distilling industries can be used as feed for livestock. Pot and pearl barley and barley flour are used for baby foods. The main industrial use of barley is the production of malt. Food uses of malt are many.

Barley malt is a good source for the production of fermented beverages. To produce barley malt, barley seed is steeped in water for sprouting. The sprouted barley with an enzyme amylase is dried mildly so that its enzyme remains active. The sprouted dried barley, known as malt, is used in the brewing process for rapid fermentation. A malt flavor contributes to the distinctive flavor of brewed beer. The malt adds a characteristic flavor to breakfast cereals also.

As regards barley pearling, a very small amount of barley is being pearled in comparison to its other uses as feed or malt. The barley hull is so tightly adhered to its kernel that barley grain is difficult to hull. That is why it is generally pearled to abrade away the outer surfaces of the grain with the aid of a rotating abrasive surface for removal of the hull and other undesirable layers. Pearled barley can further be reduced to flour with the help of a series of break and reduction rolls like wheat flour milling. Barley flour is an ingredient in many baby foods, while pearled barley is generally used for the preparation of soups.

Rye Milling

In some parts of the world, rye grain is used more as food than feed. When rye flour is added to the wheat flour, it imparts some flavor and taste to the bread. Rye grain can be fermented to produce alcoholic beverages. As regards industrial uses, rye starch is the main ingredient of some adhesives and binders. Rye is also used as a part of a feed mixture with other grains.

Rye grain contains about 71.0% nitrogen-free extract, 12.5% protein, 1.8% fat, and 2.5% fiber. It also contains some poisonous and some nutritional inhibitors such as ergot, resorcinols, phytates, and enzyme inhibitors.

There is similarity between rye milling and wheat milling. In the United States and European countries, rye is thoroughly cleaned before milling using magnetic and milling separators, disk separators, entoleters, scourers, aspirators, and stoners (Figure 13.4). The cleaned rye is steeped in water to increase its moisture to a level of 14%–15% depending on the quality requirement. Both break rolls and reduction rolls are employed in rye milling. Generally, an American rye milling plant may consist of five break rolls, one bran buster, one sizing roll, seven reduction rolls, and one toiling roll. Purifiers may not be used in rye milling. West European mills

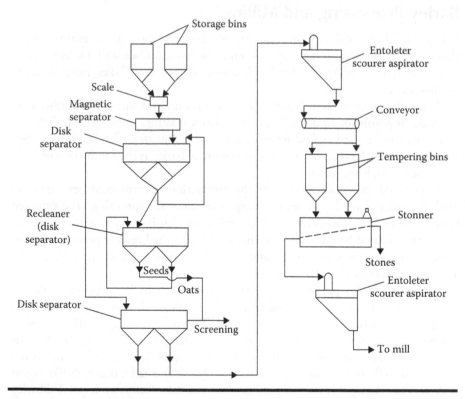

Figure 13.4   Flow diagram of a rye-cleaning system.

also use similar machines. The rye grains are gradually ground with the break and reduction rolls. Subsequently, bran and germ are separated from the products to produce rye flour.

Rye flour is used in combination with wheat flour for the production of rye bread. Rye flour as such is not useful for this purpose as its protein cannot form sufficiently strong films to support an expanded bread structure. Moreover, the rate of gas production in the dough may be high but the gas-retaining capacity is low. Various types of rye breads and rolls are produced.

Rye is also flaked in the United States to prepare breakfast cereal.

Oat Milling

At maturity the caryopsis, usually called the groat, is tightly enclosed within lemma and palea. The whole oat usually consists of 65%–75% caryopsis and 25%–35% hull. An oat kernel contains about 56%–68% starchy endosperm. Oats are classified by color. The milling oats are commonly white in color. After separation of hulls, the oats are called groats, which apparently resemble wheat kernels. The oat groats are characterized by their high fat and protein contents. These also contain an active lipase enzyme that is to be denatured or inactivated; otherwise, rancidity in the milled products may develop as they have a short shelf life.

The operations carried out for milling of oats are (i) cleaning; (ii) drying and cooling; (iii) hulling; and (iv) cutting and flaking.

After preliminary cleaning, oats are specially cleaned to remove other grains and seed. It follows a milling separation combining coarse and fine screening with aspiration. For further cleaning and grading, disk and indented cylinder separators, gravity separators, and a paddy separator are usually employed to separate seeds/materials of same size, shape, and specific gravity. Pin oats and light oats are removed and utilized as feed ingredients.

The clean oats are heat treated and partially dried by indirectly heating the oats with steam at about 90°C–95°C for about 1 h in large open pans to bring down their moisture to a level of 7%–10% and the dried products are then cooled. The heat treatment imparts a desirable roasted flavor to the oats, inactivates lipase, as well as makes it easy to remove the hull.

The treated oats are then separated by size with the help of disk separators. These are dehulled using an impact type of huller. The oats are fed to the center of an impeller (rotating at 1400–2000 rpm), which throws the oats against a rubber line attached to its cylindrical case. The lighter hulls, separated from the kernel by impact force, are removed by aspiration. The groats are polished by abrasion with a horizontal scourer. The groats are separated from the unhulled ones by using disk and paddy separators.

Usually, groats are cut into two to four uniform sizes before milling. The groats are cut into pieces and steamed with live steam at about 100°C for 10–15 min.

Steamed groats are allowed to pass through the rolls rotating at 250–450 rpm in order to roll into flakes for the production of old-fashioned rolled groats or thin flakes. Open steaming before flaking helps in reducing breakage during flaking and also denatures or inactivates enzymes in the products. The flakes are then cooled by blowing air, which also helps to remove the remaining hull.

To make oat flour, the same cleaning, drying, cooling, and cutting procedures are followed. The steamed cut groats are ground with impact-type hammer mill and the ground products are then sifted to have the desired oat flour.

Chemistry and technology of oats are detailed in Webster (1986).

Sorghum Milling

Grain sorghum is the staple food for a large number of people in the world. It is the main source of proteins and calories to the poor people of the arid and semiarid tropical areas. But the major use of sorghum grain in developed countries is for feed purposes. However, it is also a drought-resistant crop that can serve dual purposes of food and feed. Along with other millets, sorghum is said to be a coarse grain due to its outer tough and fibrous bran layer. It is mainly consumed in the form of bread (*bhakri, roti,* etc.), porridge, and other cooked products.

The sorghum consists of 80.6%–84.6% endosperm, 7.8%–12.1% germ, and 7.3%–9.3% bran (Hubbard et al., 1950). It mainly contains 55%–75% starch, 4%–21% protein, 2%–7% fat, 1%–3% crude fiber, 0.5%–4% sugars, and 1%–3% ash, but it also contributes some antinutritional factors like polyphenols and tannins in varying amounts (Subramanian and Jambunathan, 1980). Apart from these, an amino acid imbalance due to excess leucine has been implicated as the etiopathogenic factor of pellagra in sorghum eaters.

On account of the aforementioned prohibitory factors, sorghum processing has not been fully developed on a commercial scale. However, the objective of modern dry milling is to produce endosperm, bran, and germ fractions from the sorghum kernel and separate them from one another. The preliminary operations involved in dry milling are (1) removal/separation of all foreign materials like other seeds, sticks, chaffs, stones, dirt, etc., and (2) conditioning (hydrothermal treatment) of grain prior to milling. The conditioning of grain helps in separating germ from endosperm, removing pericarp in large pieces, and efficiently separating the desirable products.

In a roller milling system, a series of break and reduction rolls are employed. The corrugated rolls break open the grains. Then the exposed endosperm is gradually ground by a series of smooth reduction rolls. The flours are separated and the coarse fractions are fed to the next set of rolls for further grinding. The bran separated by sifting and aspiration are used as valuable by-products.

Other modern methods, namely, pearling, attrition, abrasive milling, etc., have also been attempted for sorghum processing/milling.

However, traditional stone mill and pestle–mortar systems are still in vogue in Asian and African countries.

Milling of Pulses

Introduction

Pulses are rich in proteins and are mainly consumed in the form of dehusked split pulses. Pulses are the main source of protein in vegetarian diet. There are about 4000 pulses mills (*Dhal mills*) in India. The average processing capacities of pulses mills in India vary from 10 to 20 tons/day.

Milling of pulses means removal of the outer husk and splitting the grain into two equal halves. Generally, the husk is much more tightly held by the kernel of some pulses than most cereals. Therefore, dehusking of some pulses poses a problem. The method of alternate wetting and drying is used to facilitate dehusking and splitting of pulses. In India, the dehusked split pulses are produced by traditional methods of milling. In traditional pulses milling methods, the loosening of husk by conditioning is insufficient. Therefore, a large amount of abrasive force is applied for the complete dehusking of the grains, which results in high losses in the form of brokens and powder. Consequently, the yield of split pulses in traditional mills is only 65%–70% in comparison to 82%–85% potential yield.

It is, therefore, necessary to improve the traditional methods of pulses milling to increase the total yield of dehusked and split pulses and reduce the losses.

Varieties, Composition, and Structure

Green gram, red gram, Bengal gram, horse gram, cluster bean, field pea, and *arhar* are some of the common types of pulses.

The botanical name of *arhar* is *Cajanus cajan*. Its chemical composition and structure are as follows:

Moisture	10.35%
Protein ($N \times 6.25$)	24.19%
Ether extract	1.89%
Ash	3.55%
Crude fiber	1.01%
Carbohydrate	59.21%

The average percentage of husk and endosperm in *arhar* is 15% and 85%, respectively. The structure is shown in Figure 1.4.

Traditional Dry Milling Method (Dhal's Milling)

There is no common processing method for all types of pulses. However, some general operations of dry milling method such as cleaning and grading, rolling or pitting, oiling, moistening, drying, and milling have been described in subsequent paragraphs.

Cleaning and Grading

Pulses are cleaned from dust, chaff, grits, etc., and then graded according to size by a reel-type or rotating sieve-type cleaner.

Pitting

The clean pulses are passed through an emery roller machine. In this unit, husk is cracked and scratched. This is to facilitate the subsequent oil penetration process for the loosening of husk. The clearance between the emery roller and cage (housing) gradually narrows from inlet to outlet. As the material is passed through the narrowing clearance, mainly cracking and scratching of husk take place by friction between pulses and emery. Some of the pulses are dehusked and split during this operation, which are then separated by sieving.

Pretreatment with Oil

The scratched or pitted pulses are passed through a screw conveyor and mixed with some edible oils like linseed oil (1.5–2.5 kg/ton of pulses). Then they are kept on the floor for about 12 h for diffusion of the oil.

Conditioning

Conditioning of pulses is done by alternate wetting and drying. After sundrying for a certain period, 3.5% moisture is added to the pulses and tempered for about 8 h and again dried in the sun. Addition of moisture to the pulses can be accomplished by allowing water to drop from an overhead tank on the pulses being passed through a screw conveyor. The whole process of alternate wetting and drying is continued for 2–4 days until all pulses are sufficiently conditioned. Pulses are finally dried to about 10%–12% moisture content.

Dehusking and Splitting

Emery rollers, known as Gota machine, are used for the dehusking of conditioned pulses. About 50% pulses are dehusked in a single operation (in one pass). Dehusked pulses are split into two parts also. The husk is aspirated off and dehusked split pulses are separated by sieving. The tail pulses and unsplit dehusked pulses are again conditioned and milled as mentioned earlier. The whole process is repeated two to three times until the remaining pulses are dehusked and split.

Polishing

Polish is given to the dehusked and split pulses by treating them with a small quantity of oil and/or water for the production of grade I dehusked and split pulses known as "dhal."

Commercial Milling of Pulses by Traditional Method

It is discussed earlier that the traditional milling of pulses are divided into two heads, namely, dry milling and wet milling. But both the processes involved two basic steps: (i) preconditioning of pulses by alternate wetting and sundrying for loosening husk and (ii) subsequent milling by dehusking and splitting of the grains into two cotyledons followed by aspiration and size separation using suitable machines. The 100% dehusking and splitting of pulses can seldom be achieved, particularly in cases of certain pulses like tur, black gram, and green gram. Of them, "tur" is the most difficult pulses to dehusk and split. Only about 40%–50% "tur" grains are dehusked and split in the first pass of preconditioning and milling. As sundrying is practiced, the traditional method is not only weather-dependent but also requires a large drying yard to match with the milling capacity. As a result, it takes 3–7 days for complete processing of a batch of 20–30 tons of pulses into "dhal." Moreover, milling losses are also quite high in the traditional method of milling of pulses.

In general, simple reciprocating or rotary sieve cleaners are used for cleaning while bucket elevators are used for elevating pulses.

Pitting or scratching of pulses is done in a roller machine. A worm mixer is used for oiling as well as watering of the pitted pulses.

The machines used for dehusking are either power-driven disk-type sheller or emery-coated roller machine, which is commonly known as "gota" machine. The emery roller is encaged in a perforated cylinder. The whole assembly is normally fixed at a horizontal position.

Sometimes, either a cone-type polisher or a buffing machine is employed for removal of the remaining last patches of husk and for giving a fine polisher to the finished "dhal."

Blowers are aspiration of husk and powder from the products of the disk sheller or roller machine. Split "dhals" are separated from the unhusked and husked whole pulses with the help of sieve-type separators.

Sieves also employed for grading of "dhals" mills vary from 68% to 75%. It may be noted that the average potential yields of common "dhals" vary from 85%

to 89% (Kurien, 1979). These milling losses in the commercial pulses mills can be attributed to small brokens and fine powders formed during scouring and simultaneous dehusking and splitting operations.

Modern CFTRI Method of Pulses Milling

CFTRI method of "dhal" milling is described as follows.

Cleaning

Cleaning is done in rotary reel cleaners to remove all impurities from pulses and separate them according to size.

Preconditioning

The cleaned pulses are conditioned in two passes in a dryer (LSU type) using hot air at about 120°C for a certain period of time. After each pass, the hot pulses are tempered in the tempering bins for about 6 h. The preconditioning of pulses helps in loosening husk significantly.

Dehusking

The preconditioned pulses are conveyed to the pearler or dehusker where almost all pulses are dehusked in a single operation. The dehusked whole pulses "gota" are separated from split pulses and mixture of husk, brokens, etc. (*chuni-bhusi*), and are received in a screw conveyor where water is added at a controlled rate. The moistened "*gota*" is then collected on the floor and allowed to remain as such for about an hour.

The flow diagram of the modern milling of pulses by CFTRI method is self-explanatory.

Lump Breaking

Some of the moistened "*gota*" form into lumps of varying sizes. These lumps are fed to the lump breaker to break them.

Conditioning and Splitting

After lump breaking, the *gota* is conveyed to LSU type of dryer where it is exposed to hot air for a few hours. The "*gota*" is thus dried to the proper moisture level for splitting. The hot conditioned and dried dehusked whole pulses are split in the emery roller. All of them are not split in one pass. The mixture is graded into grade I pulses, dehusked whole pulses, and small brokens. The unsplit dehusked pulses are again fed to the conditioner for subsequent splitting.

BY-PRODUCTS/ BIOMASS UTILIZATION

IV

Chapter 14

Rice Bran

Rice husk and rice bran are the two main by-products of the rice milling industry. Several valuable chemicals and other products have been produced from rice husk, but their commercial success has yet to be established. Commercially, rice bran is the most valuable by-product, obtained from the outer layers of the brown rice. Generally, rice bran consists of pericarp, aleurone layer, germ, and a part of endosperm. Bran removal amounts from 4% to 9% of the weight of paddy milled. True bran amounts from 4% to 5% only; rest is polish consisting of inner bran layers and portion of the starchy endosperm. Rice bran is characterized by its high fat and protein contents. It also contains vitamins, minerals, and many other useful chemicals.

Bran can be utilized in various ways. It is a potential source of vegetable oil. Because of its nutritional value, it is being used as feed for poultry and livestock. More stable defatted bran containing a higher percentage of protein, vitamins, and minerals than full fatted bran is an excellent ingredient for both food and feed. Crude bran oil of high free fatty acid (FFA) content is used for the manufacture of soap and fatty acids. Edible grade oil is produced by refining of the crude bran oil of low FFA content (about 5%). In addition to tocopherol, waxes of high melting point suitable for various industrial purposes are the by-products of the bran oil refining industry. Various uses of rice bran have been discussed at the end of this chapter.

Commercial bran is always contaminated with some amount of husk, which varies widely with the type of rice mill used. The huller bran contains a very high amount of husk, whereas sheller bran contains small amount of husk, and the bran, produced by the modern rubber roll type mill, is almost pure containing negligible amount of husk. The characteristics of rice bran obtained from different mills are given in Table 14.1. Therefore, complete exploitation of the usage of bran is largely dependent on its purity.

Rice bran can be classified into three groups: (1) full fatted raw bran (raw bran) obtained from milling of raw paddy; (2) full fatted parboiled bran (parboiled bran)

Table 14.1 Characteristics of Rice Bran from Hullers, Shellers, and Modern Rice Mills

S. No.	Variety	Moisture (%)	Oil (%)	Protein (%)	Ash Total (%)	Ash Insoluble (%)	Crude Fiber (%)
1	Raw						
	Huller	8.0–9.5	5.0–10.0	5.5–9.5	13.5–22.5	2.2–19.0	5.8–22.7
	Sheller	8.5–11.5	11.5–19.0	9.0–13.0	8.0–22.5	3.5–19.0	6.0–14.5
2.	Parboiled						
	Huller	4.0–9.5	6.5–12.5	4.5–9.5	17.5–20.0	14.5–18.5	20.5–30.0
	Sheller	3.0–11.5	12.0–27.5	10.0–15.5	8.5–22.5	3.0–21.0	6.5–32.0
3	Modern rice mill	8.0–9.5	15.0–18.5	13.0–14.0	9.0–11.5	2.5–4.5	8.5–10.5

Source: Oil Technological Research Institute (OTRI), Edible rice bran oil, Project Report, 1974.

obtained from milling of parboiled paddy; and (3) defatted/deoiled bran obtained after extraction of oil from either raw or parboiled bran. Raw bran contains 12%–18% oil, whereas parboiled bran contains 20%–28% oil. After extraction of oil from raw and parboiled bran, the deoiled bran contains about 1%–3% oil only. The chemical compositions of raw, parboiled, and deoiled bran are presented in Tables 14.2 and 14.3.

Raw bran is a light colored oily, unstable meal of various sized particle. Parboiled bran is relatively darker and oilier. The oiliness of parboiled bran causes clogging of screens during milling, particularly in rainy days and high humid atmosphere. Deoiled bran is lighter in color and dusty in nature. The physicothermal properties and particle sizes of rice bran are given in Tables 14.4 through 14.6.

Table 14.2 Chemical Composition of Raw and Parboiled Rice Bran

Constituents (%)	Raw Bran	Parboiled Bran
Moisture	11.9	89.1
Crude protein	12.5	12.8
Oil	14.6	20.3
Crude fiber	8.9	10.0
Ash	10.7	10.7
NFE	42.9	36.1

Table 14.3 Chemical Composition of Regular (Full Fatted) and Defatted Rice Bran

Constituents (%)	Regular Bran	Defatted Bran (Solvent Extracted)	Defatted Bran (Pressed)
Moisture	12.59	9.31	12.28
Crude protein	13.31	16.07	17.31
Crude fat	21.21	1.12	1.32
Crude fiber	9.05	10.79	11.59
Ash	9.40	11.41	12.01
NFE	33.44	41.13	45.49

Source: Yokochi, K., Rice bran processing for the production of rice bran oil and rice bran protein meal, UNIDO Publication, 1972.

Table 14.4 Physicothermal Properties of Rice Bran

Property	Moisture Content Range % (w.b.)	Temperature (°C)	Raw Bran	Parboiled Bran
Bulk density, kg/m³	8–14	—	276–291	267–314
	6–16.5[a]		285–239[a]	—
Specific gravity	8–14	—	1.4–1.48	1.45–
Porosity, %	8–14	—	81.3–79.3	82–76
Angle of repose	8–14	—	59.3–51.8	57–46.2
Static coefficient of friction on mild steel	8–14	—	0.593–0.508	0.61–563
Thermal conductivity (steady state), kcal/(h m °C)	8–14	43	0.035–0.049	0.028–0.049
Thermal conductivity (unsteady state), kcal/(h m °C)	8–14	36	—	0.14–0.26
	6–16.5[a]	30–43	0.065–0.102[a]	—
Specific heat cal/(g °C)	10.3–12.0	—	0.224–033	0.67–0.69
	6–16.5[a]		0.369–0.496[a]	
Thermal diffusivity, m²/h	8–14	—	0.0024–0.0021	0.0024–0.0021

[a] Devadattam and Chakaraverty (1986).

Table 14.5 Particle Size Distribution of Raw and Stabilized Bran

U.S. Sieve No.	Size (μm)	Raw Bran (% on Sieve)	Stabilized Bran by 15 min Steaming at 1.6 kg/cm² (% on Sieve)
12	—	0.7	2.3
20	—	7.5	7.7
30	—	9.0	9.4
40	420	8.8	11.8
50	—	9.2	59.0
70	—	16.0	9.8
100	149	21.2	—
200	—	26.2	—
Pan	—	1.5	—

Table 14.6 Particle Sizes of Raw and Defatted Rice Bran

U.S. Sieve No.	Size (μm)	Raw (Japonica) Weight% on Sieve[a]	Defatted Bran Weight, % Sieve
16	—	6	—
20	—	9	14.3
30	—	18	—
40	420	22	27.3
60	250	34	14.0
80	—	10	—
100	149	0.5	9.2
150	—	0.2	—
Through 100 mesh	—	—	11.9

[a] Yokochi (1972).

The most important and crucial property of rice bran is the instability of its oil caused by an oil-splitting enzyme, lipase, inherently present in it. The enzyme lipase acts as a catalyst. The fat and enzyme are spatially distributed in aleurone and testa layers, respectively, in intact rice grain. As long as the bran surface is uninjured and protected by the husk, the enzyme remains dormant and the enzymatic activity is not perceptible. As soon as the bran surface is ruptured and separated from the brown rice in milling operations, the lipase comes in contact with the oil-bearing layers and they are intimately mixed with each other, causing a very rapid rate of hydrolysis of fats into FFAs. As the reaction is hydrolytic in nature, it may be called hydrolytic type of rancidification. It is apparent that the rate of hydrolysis will be further enhanced with the increase of moisture in bran. The FFAs can then be more readily oxidized than the natural oils by the oxidative agents resulting in oxidative rancidity with the production of unpleasant odors and flavors. The composition of fatty acid esters is given as follows.

Composition of Fatty Acids in Rice Bran Oil (OTRI, 1974)		
1	Saturated acids	16%–20%
	(i) Palmitic acid	13%–18%
	(ii) Myristic acid	0.4%–1%
	(iii) Stearic acid	1%–3%
2	Unsaturated acids	80%–84%
	(i) Oleic acid	40%–50%
	(ii) Linoleic acid	20%–42%
	(iii) Linolenic acid	0%–1%

The hydrolysis of triglycerides of fatty acids (i.e., neutral fats) into FFAs in the presence of lipase enzyme (biocatalyst) is presented as follows:

$$
\begin{array}{l}
CH_2COOR_1 \\
\quad | \qquad\qquad +3H_2O \\
CHCOOR_2 \qquad\qquad \longrightarrow \\
\quad | \qquad\qquad Lipase \\
CH_2COOR_3
\end{array}
\qquad
\begin{array}{l}
R_1COOH \quad CH_2OH \\
\quad + \qquad\qquad | \\
R_2COOH \; + \; CHOH \\
\quad + \qquad\qquad | \\
R_3COOH \quad CH_2OH
\end{array}
$$

Immediately after milling, the FFA content of bran is normally below 3%. After milling, the rate of increase of FFA in bran may be as high as 1% per hour under favorable conditions. Alkali refining of crude oil for edible grade oil is considered uneconomical if its FFA content goes beyond the 10% level.

The problems in processing of rice bran for edible oil and defatted bran for feed, food, and other purposes are as follows:

1. Rapid hydrolysis of oil into FFA and glycerol
2. Oxidation of fat into fat peroxide in presence of peroxidase enzyme
3. The presence of excessive amount of fine particles in the bran
4. Abnormally high refining losses even in crude oil of low FFA content
5. Excessive color in the refined and bleached oil
6. The tendency of the finished oil to undergo flavor reversion
7. Use of traditional hullers leading to the production of bran of low oil content contaminated with high percentage of husk, earth, and other impurities
8. Lack of completely modernized rubber roll type rice mills for the production of pure bran free from any silica and other impurities
9. Scattered location of small- and medium-sized rice mills, which makes it difficult to transport fresh bran to the oil extraction plant within a short period
10. A complex cultural, socioeconomic problem

Problems 1 through 6 and 9 can be solved to a large extent by the effective stabilization of bran, whereas problems 7 and 8 can be solved by modernization of rice mills. But it is difficult to overcome problem 10.

Stabilization of bran extends the storage period of bran without any appreciable change in FFA content. Moreover, stabilization of rice bran offers the following advantages:

1. It imparts hardening effect to the bran for better extractability.
2. It increases the particle sizes and reduces the problem of fines and filtration.
3. It increases the bulk density and reduces the handling problem.

Factors affecting rate of formation of FFA in bran during storage are as follows:

1. Storage temperature
2. Moisture content of bran
3. Storage relative humidity
4. Variety, type, and particle size of bran
5. Contamination of bran with microflora or insects

Effect of Storage Temperature

Formation of FFA in raw bran was found to occur at a fairly rapid rate even at a storage temperature of 3°C, whereas parboiled rice bran was quite stable when stored at a temperature of 25°C (Loeb et al., 1949).

Effect of Storage Moisture Content of Bran

The rate of hydrolysis of fats into FFA decreases as the storage moisture content of bran decreases. Even the stabilized bran of low moisture content should be stored in moisture-proof polythene-lined or polythene bags at room temperature for a long and safe storage.

Effect of Relative Humidity

The rate of FFA formation increases with the increase of the relative humidity of storage and the final moisture content of bran as well (Loeb et al., 1949). The equilibrium moisture content of bran depends on the relative humidity and temperature of the environment.

Particle Size

The finer the particle size of the bran, the higher the rate of formation of FFA.

Insect Infestation

Some of the insects and microflora can contribute lipase to the bran. Therefore, for the control of insect infestation of stabilized bran during storage, fumigation of the stabilized bran and its container is necessary. The bran can also be stored safely in sealed thicker polythene bags without any insect infestation.

Chapter 15

Utilization of Rice Bran

So far, six basic approaches have been tried. They are as follows:

1. Low temperature storage tends to reduce the free fatty acid (FFA) rise but unacceptable high levels of FFA are reached in a comparatively short time.
2. Storage of bran at low relative humidity retards the rise of FFA in bran but it should be preceded by drying of bran to a low moisture content (by heat treatment). Otherwise, it is not very effective.
3. Simultaneous milling and extraction with the hexane solvent (X–M process) shows considerable promise, but it is a very expensive process and its high technical know-how is not available.
4. Various chemical treatments (except HCl treatment) and exposure to inert atmosphere have been proved to be ineffective.
5. Treatment of bran with microwaves, etc., is effective, but the cost of processing becomes prohibitive.
6. Heat treatment of bran has been found to be most effective, feasible, and economic and studied extensively.

The results of the studies on heat treatment of bran are summarized later.

The entire heat treatment process can be divided into two major heads: (i) dry heat treatment and (ii) wet heat treatment.

Dry Heat Treatment

Dry heat treatment can be further divided into (1) convection heat treatment; (2) conduction heat treatment; (3) radiation heat treatment including infrared heating; (4) dielectric heat treatment; and (5) fractional heat treatment.

Convection Heating

When bran was heated by convection for 1 h at 110°C, the FFA contents of bran were 4% and 7% after 25 and 50 days of storage at 25°C. Therefore, convection heating requires a long time for stabilization.

Conduction Heating

If bran in a thin layer is dried uniformly by conduction heating for 20 min at a bran temperature of 90°C, then it is stabilized and can be stored safely for 2 months at 25°C–30°C provided there is no insect infestation. Therefore, conduction drying is very efficient provided a thorough mixing arrangement for uniform heating of bran is ensured.

Infrared Heating

When bran was heated for 10–15 min at 110°C by infrared lamps, lipase was inactivated and the treated bran could be stored safely in an airtight container for 2 months keeping FFA level below 7%.

Frictional Heating

There is possibility of using some form of frictional heat for developing the required temperature for inactivation of lipase. Particularly, Handler type of oil expeller can be used for the aforementioned purpose (Viraktamath and Desikachar, 1971).

Wet Heat Treatment

Yokochi (1972) reported that for the stabilization of rice bran, it should be steamed at 95°C and then dried to about 3% moisture content followed by an air cooling. The stabilized bran can be stored for 3–4 months.

It is recommended that the bran should be steamed with live steam for 3–5 min and then dried to the desired low moisture content to bring down the residual enzymatic activity to zero.

It is reported that steaming of raw bran for about 15 min and subsequent drying could inactivate the enzyme. They also reported that steaming of raw paddy or steaming of soaked paddy during parboiling could inactivate the lipase enzyme.

Rice Bran Stabilizers under Development in India

1. *On the basis of wet heat treatment, the following rice bran stabilizer of 0.5 ton/day capacity has been developed by Jadavpur University, Calcutta.* It consisted of a screw conveyor (s.c.) for direct steaming of bran and a pneumatic type of bran dryer fitted with an air suction unit and cyclone separator. Bran is exposed to live steam for a period of 5 min in a semicircular trough fitted inside with s.c. The speed of the s.c. is so adjusted that the retention time of bran is 5 min. Then the steamed bran is dried with hot air in a heat-insulated 4 m tall vertical column under fluidized condition, and then the treated bran is separated from the hot air by a cyclone separator and finally collected in a close container. Though the capacity of the plant was originally designed for 0.5 ton/day only, its capacity can be increased by changing the design of the cyclone separator, increasing the speed of the screw feeder, and increasing the suction capacity of the blower.

2. *A close conduction heating–type rice bran stabilizer has been developed by CFTRI, Mysore.* It consists of an electrically heat-jacketed revolving drum provided with a tight-fitting lid. The lid is also fitted with a vent valve and temperature-cum-pressure gauge. The stabilizer is operated in the following manner. After charging the bran into the drum and closing the lid with vent valve open, the bran is heated until steam begins to emerge from the valve. After allowing the air to escape for about 2 min, the valve is closed and heating is continued until dial thermometer records a temperature of 110°C–115°C. The temperature is maintained for about 5 min, after which the vent valve is opened and steam is allowed to escape. The material is discharged from the toaster by tilting and is allowed to cool. Bran thus treated could be stored for 3 months at 37°C and 70% RH with FFA below 10%. This revolving toaster unit has been demonstrated in some rice mills. But its use has been restricted as most of the remote villages are not electrified and the capacity of the unit is also limited.

3. *A continuous rice bran stabilizer of 0.5 ton/day capacity developed and patented by Chakraverty at PHTC.* It consists of a steam-jacketed s.c. with mixing device and a cylindrical conduction-type dryer (Figure 15.1). The s.c. is placed above the dryer. The outlet of the s.c. and the inlet of the dryer are connected by a chute. The s.c. connected with a steam pipe is used for direct steaming of bran, whereas the steam-jacketed cylindrical dryer fitted inside with paddle-type agitator is used for conduction heating as well as drying of bran. The dryer with the internal rotary paddle-type bran mixing system is inclined at an angle of about 3° horizontally. Fresh raw bran (with low FFA content) is fed at the hopper and passed through the s.c. where it is steamed at 100°C for about 5 min. The steam-treated bran is then allowed to pass through the steam-jacketed and internal agitated dryer where it is thoroughly mixed and heated to 100°C for 10 min by conduction and uniformly dried to 4%–5% moisture content. The stabilized bran

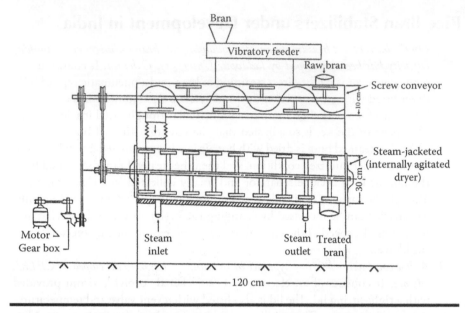

Figure 15.1 Schematic diagram of rice bran stabilizer using wet and dry heat treatment methods.

can then be stored for 2 months at 25°C–30°C and 70% RH with FFA content below 5%. The steam-jacketed dryer unit alone can be used for the stabilization of bran by dry conduction heating. The fresh raw bran in thin layer is to be uniformly heated at 90°C for 20 min in the dryer and dried to 4%–5% moisture content. Then the bran is to be cooled. The dry heat-treated bran can be kept in the sealed polythene bag and stored at 25°C–30°C for 45 days with its FFA content below 5%.

Examples of Stabilizer Design

Design a continuous rice bran stabilizer equipped with an overhead steam-jacketed screw conveyer (s.c.) for direct steaming as well as conduction heating of raw rice bran at a 25 kg/h feed rate to feed the steam-jacketed stationary drum dryer fitted inside with an internally rotated paddle-type agitator to heat and stabilize the steam-treated rice bran at 100°C the same rate by conduction heating with the steam-heated surface of the half-jacketed dryer. Estimate: 1(a) rpm of s.c.; 1(b) power needed to drive the s.c.; 2(a) dimensions of the jacketed dryer-steel shell; and 2(b) heat required to increase the treated bran from 30°C to 100°C using the data furnished hereafter and finding other relevant data from various reliable sources/references or otherwise assuming properly.

Screw conveyor. The capacity of s.c. is given by

$$Q = 15\pi D^2 Sn\psi\rho C$$

where
  Q is the s.c. capacity, 0.25 ton/h
  D is the screw diameter, 0.98 m
  S is the screw pitch, 0.07 m
  n is the speed of screw shaft, rpm
  ψ is the loading efficiency, 0.17
  ρ is the bulk density of bran, 0.28 ton/m^3
  C is the factor of inclination for horizontal conveyor

The dimensions of a common steam-jacketed s.c. are used to design its speed.
  Putting these values in the previous equation,

$$n = \frac{0.025}{15 \times \lambda(0.098)^2 \times (0.074) \times 0.17 \times 0.28 \times 1} = 15.68 \approx 16 \text{ rpm (approximately)}$$

Power required: The power required, P, for driving the s.c. shaft is given by

$$P = \frac{QL}{367}[w_0 \pm \sin\beta]\frac{1}{\eta}$$

where
  P is the power, kW
  L is the length of s.c., m
  β is the angle of inclination of screw shaft to horizontal, degree
  w_0 is the material factor
  η is the efficiency of gear reducer

Data available or assumed: L = 1.21 m, Q = 0.025 ton/h, β = 0°, w_0 = 4.0, and η = 0.90 (assumed):

$$P = \frac{0.025 \times 1.21}{367}(4.0 \pm 0)\frac{1}{0.90} = 3.66 \times 10^{-4} \text{ kW}$$

Design of a *steam-jacketed dryer shell* is based on mass flow rate. Let R and L be the radius and effective length. The cross-sectional area of the cord of 3 cm bran layer (assumed) is given by

$$a = \left[\frac{\pi}{360} \times 2\cos^{-1}\left(\frac{R-3}{R}\right)R^2 - \frac{1}{2}(R-3)2 \times \sqrt{R^2 - (R-3)^2}\right] \times 10^{-4} \text{ m}^2$$

The volume of bran in the dryer, $V = a \times L$

$$V = \frac{0.025}{60} \times 2.5 \times \frac{1}{0.280} = 3.72 \times 10^{-3} \text{ m}^3$$

Assuming the value of R = 19 cm and putting it in the previous equation,

$$a = 41.68 \times 10^{-4} \text{ m}^2$$

$$V = 41.68 \times 10^{-4} \times L$$

$$\text{or} \quad L = \frac{3.72 \times 10^{-3}}{41.68 \times 10^{-4}} = 0.89 \text{ m} = 89 \text{ cm}$$

Keeping the allowance of about 6 cm, the dryer effective L = (89 + 6) cm = 95 cm. The overall length of the dryer having 38 cm diameter is kept suitably equal to 120 cm.

Heat requirement and dimensions of the steam jacket: The amount of total heat required H to increase the bran temperature from 30°C to 100°C for stabilization is given by the equation:

$$H = h_1 + h_2 = mc_p(\Delta t) + m\lambda = 25 \text{ kg} \times (100 - 30) + 2.5 \times 539$$

$$= 392 + 13{,}349 = 1741 \text{ kcal/h}$$

where
h_1 is the sensible heat required, kcal/h
λ is the latent heat of water vaporization, 539 kcal/kg
h_2 the is total amount of latent heat required to evaporate, kcal/kg °C
Δt is the increase of bran temperature from initial 30°C to final 100°C

Adding 10% heat losses (assumed) to the value of H,

$$\text{Net heat required, } H = 1741 \times 1.10 = 1915.1 \text{ kcal/h}$$

This heat is to be supplied by the steam of the jacket calculated earlier, U = overall heat transfer coefficient, kcal/m² h °C, ΔT_m = log mean temperature difference °C and heating area, m². The overall heat transfer coefficient

$$U = \frac{1}{\dfrac{1}{h_s} + \dfrac{x_m}{k_m} + \dfrac{x_b}{k_b} + \dfrac{1}{h_a}} \quad \text{or} \quad U = \frac{1}{\dfrac{1}{9760} + \dfrac{1.6 \times 10^{-3}}{43.2} + \dfrac{0.005}{0.19}}$$

$$= 37.8 \text{ kcal/h m}^2 \text{ °C}$$

where

Steam side heat transfer coefficient h_s = 9760 kcal/h m^2 °C
Steel sheet thickness x_m = 1.6 × 10^{-10} m
Bran layer thickness in the dryer x_b = 0.005 m
Thermal conductivity of bran k_b = 0.19 kcal/h m °C
Negligible heat transfer coefficient of air inside the dryer k_m = 0

If length of steam jacket is 1 m and diameter of dryer D_o = 0.38 m, the heat transfer area A m^2 for partially jacketed dryer is given by A = (π/2) × 0.38 × l. Steam temperature at 2.0 kg/cm^2 (abs) found from steam table = 119.62°C. The log mean temperature difference ΔT_m is calculated as follows:

$$\Delta T_m = \frac{(119.62 - 30) - (119.62 - 100)}{\ln\left(\dfrac{119.62 - 30}{119.62 - 100}\right)} = 46.08°C$$

Substituting the values of U, A, and ΔT_m in the equation Q = $UA\Delta T_m$, H = 1919.1 = 37.8 × π/2 × 0.38 × l × 46.08. Therefore, length of the jacket, l = 1.84 m. Since the maximum possible length of jacket can be 87 cm with 4 cm annular space for steam. Therefore, the bran has to be recirculated for about 2 times in order to raise its temperature up to the desired level.

Steam requirement: Assuming the dry saturated steam supplied at 2.0 kg/cm^2, the enthalpy of dry saturated steam and water at the pressure of 2 kg/cm^2 are 646.3 and 119.94 kcal/kg (steam table). Total heat to be supplied = 1915.1 kcal/h

$$\text{Steam required} = \frac{1915.1}{646.3 - 119.94} = 3.63 \text{ kg/h}$$

Exercise

Following the stabilizer problem 1, calculate i(a) the thickness of jacketed dryer and s.c. steel shell for the working steam pressure inside the jacket, p_i; i(b) external critical steam pressure, p_c kg/cm^2, for 1.6 mm thick steel shell of dryer and s.c.; and (ii) the torque and the power required, kW, to drive the dryer's inner rotary paddle-type agitator shaft using the related data furnished in Example 1 and finding or assuming other related data.

Additional useful equations and data:

Thickness of the jacket shell, t_j mm, is given by $t_j = p_i D_i/(2fJ - p_i)$, where p_i internal design pressure, kg/cm^2 (g), D_i is internal diameter of jacket, mm, f is allowable stress, kg/cm^2 = 980 kg/cm^2 for steel plates, J = 0.50 (lowest value of class 3 weld joint).

The external critical steam pressure, p_c kg/cm^2, for 1.6 mm thick shell is expressed by

$$p_c = \frac{2.42E}{\left(1-\mu^2\right)^{3/4}} \cdot \frac{\left(t/D_o\right)_o^{5/2}}{\left(L/D_o\right) - 0.45\left(t/D_o\right)^{1/2}}$$

where

E is the modus of elasticity, kg/mm^2 = 2070 × 10^3 kg/cm^2
μ is the Poisson's ratio = 0.3
t is the thickness, 1.6 mm
D_o is the outside diameter of dryer shell = 383.2 mm
L is the unsupported length of vessel = 870 mm

Extraction of Rice Bran Oil

Oil extraction methods can be divided into two major groups: mechanical method and solvent extraction method.

In mechanical method, either hydraulic press or oil expeller is used to press oil out of the bran, which has already been preprocessed by steaming and drying. The high pressure employed is in the range of 70–280 kg/cm^2. But the output of oil is low compared to the solvent extraction method. The mechanical method had been in use in Japan for many years in the past. Nowadays, the mechanical method has been abandoned but it may be popular once again in near future because of the shortage of petroleum products.

In solvent extraction method, usually, *n*-hexane (b.p. 66°C–70°C) is used. This method is now universally used mainly due to high oil output.

Solvent Extraction Method

Three systems of solvent extraction operations are in use, namely, batch, semicontinuous, and continuous systems. The continuous system has certain advantages over the batch system, as it can be used for the production of oil from various oilseeds and meal and rice bran as well. But the capacity of the plant is higher and the initial investment is high.

The batch type is believed to be exclusively suitable for rice bran oil extraction. It is popular due to its simplicity in operation and low cost of installation. The solvent extraction method is described as follows.

Pretreatment of bran prior to extraction is an essential step for either batch or continuous system. Pretreatment consists of either direct steaming and drying of bran or drying of bran alone at 90°C–100°C to 6%–8% moisture content.

Pretreatment of bran reduces amount of fines and moisture content and thereby increases the particle sizes, aids the release of oil from bran, imparts hardening effect to bran particles for better extractability lower filtration time, and eliminates the problem of fines. Without pretreatment, the fines would create problems like resistance to percolation of oil, channeling resulting in longer steaming time for the desolventization of meal, and low rate of extractability.

In the continuous system, the steam-treated bran is pelletized in the pelletizing equipment to enlarge the particle size of the bran to 6–8 mm pellets. It not only eliminates fines but also reduces the moisture content to some extent. But it is a very costly method because of the requirement of high electrical power per unit mass of pellets. Due to limited scope, only batch extraction method has been described here.

Batch Extraction Method

The treated bran containing 6%–8% moisture is charged to each of the five station-ary batch extraction vessels that were fitted with a sugar bag at the false bottom, each holding 0.5 ton of raw bran. The thick sugar bag serves as a filter. Steam, at 2 atm pressure, is applied to force out miscella. The miscella from the extractor passes through a strainer. Hot vapor is effectively used for heating the miscella in the heat exchanger. Last traces of solvent in the oil are removed in the stripper. The bottom steam is used to drive the solvent out of the meal. Meal with a residual oil content of 1%–3% and moisture content of 8%–12% is discharged from the extractor by opening the door of the extraction vessel and by raking it out.

It takes 2½ h for each cycle of batch. Typical capacity of a plant is 24–25 tons/day of 24 h. A simplified flow diagram of the batch extraction method is shown later.

Refining of Crude Rice Bran Oil into Edible Grade Oil

Because of the presence of fatty acids, gum, wax, coloring and odoring matters, etc., the rice bran oil is the most difficult oil among all vegetable oils to refine.

A simplified flow diagram of the refining process is shown later.

Several patented methods are available for the refining process. Generally, the following steps are adopted:

1. Preliminary dewaxing and degumming process to remove hard wax, gums, mucilages, and some other impurities
2. Neutralization process for the removal of FFA
3. Decolorization process for the removal of coloring matters
4. Deodorization process for the removal of odorous matters and unsaponifi-able matters
5. Winterization operation for the removal of soft wax

Uses of Bran, Bran Oil, and Various Constituents

Edible Grade Oil

Because of the very low content of linolenic acid and high content of tocopherol, bran oil has distinct advantages over other vegetable oils. Different grades of bran oil, such as salad oil and cooking oil, and also shortenings can be produced by refining and suitable hydrogenation of bran oil.

Industrial Grade Crude Oil

Soap Manufacture

Rice bran oil with high FFA content is highly suitable for the manufacture of soft soap and liquid soap. In addition to alkali soaps, other metallic soaps obtained from bran oil, namely, aluminum, barium, and calcium soaps, find market as components of lubricants.

Free Fatty Acid Manufacture

The process consists of hydrolysis of the triglycerides of fatty acids into fatty acids and glycerol, separation of glycerol, and purification of mixed fatty acids. The use of hydrogenation in combination with fractional distillation enables the manufacture of pure stearic and oleic acid of desired quality.

Protective Coatings

The manufacture of surface coatings like alkyl- and resin-based paints, enamels, varnishes, and lacquers also represents the usage of rice bran oil.

Plasticizers

Recently, fatty acids and fatty oil–based plasticizers are being used in the plastic and rubber industries.

Tocopherol

Tocopherol has nutritional and antacid effects. Crude oil contains 2%–4% tocopherol. During deodorizing process, a significant amount of it is lost. Only 1%–2% of tocopherol remains in edible oil.

Rice Bran Wax

Rice bran wax is a proper substitute of carnauba wax due to high milling point (72°C–84°C), hardness, and nontackiness. It is being used for coatings of candy, fruits, and vegetables as it prevents moisture loss and shrinkage of the said products. It can also be used as component in formulations like carbon paper base, stencils, candles, etc.

Uses of Defatted Bran

Feed

Defatted bran can be best utilized as an ingredient of cattle and poultry feed. In regard to this aspect, defatted bran is more suitable than raw bran due to its higher protein and fat contents, higher digestibility, and more stability toward storage quality.

Food

Defatted bran can be successfully used as an ingredient in bakery products such as bread, cake, biscuits, etc. After finer grinding, it can be added to baking flour up to 20%.

Fertilizer

On account of the presence of high fat and wax contents, regular raw bran is not only unsuitable for plants but also harmful for their roots. But defatted bran contains all three manurial factors (NPK valucs) in the right proportion.

Medicinal Use

As rice bran contains valuable vitamin B complexes, amino acids, phosphoric acid compound, etc., it is useful in the field of medicine and dietrics. Protein can also be easily extracted from rice bran.

Chapter 16

Biomass Conversion Technologies

Biomass

Biomass is the renewable and biodegradable organic matter generated through life processes. The enormous quantity of biomass produced annually by land and aquatic plants must be utilized effectively for the future need of the world. It is in this context that the biomass conversion technologies assume great importance. Biomass is the main noncommercial source of energy. In the future, with the inherent scarcity of fossil fuels, biomass is supposed to play an important role in the commercial sector as well.

The major problems involved in the utilization and conversion of biomass are as follows:

- The cost of collection, handling, pretreatment, and enrichment
- The selection of an alternative and economically viable technology
- The development of integrated systems of the biomass conversion processes
- Ecological balance

In many countries, a major part of the agricultural residues is used as cattle feed, fiber, fuel, organic fertilizer, and housing materials. The huge quantity of cattle dung produced annually is not utilized properly, which can also be easily converted into biogas and organic manure by its biogasification.

Biomass and other solid waste resources may be classified under the following categories: (i) crop residues, (ii) agro-industrial wastes, (iii) forest products,

(iv) silvicultural products, (v) aquatic biomass, (vi) marine products, (vii) animal wastes, and (viii) municipal solid wastes and (DC) municipal sewage sludges.

Biomass Conversion Technologies

The biomass conversion technologies involve the following: (i) chemical (ii) biochemical, (iii) thermochemical, and (iv) thermal processes.

These conversion technologies offer alternative options for the production of fuels, chemicals, food, and feed from biomass to meet the growing needs of the world.

Chemical and Biochemical Processing

Silica and Silicon from Rice Husk

Rice Husk

Paddy is the major crop cultivated in India as well as in many other countries of the world. One-fifth of the weight of paddy is usually considered as its husk. In India, about 20 million tons of rice husk is produced annually. Disposal of this low value by-product poses a global problem because of its abrasive characteristics, low nutritive value, low bulk density, and high ash content.

Some of the common uses of the rice husk are given in Table 16.1.

The chemical compositions and physicothermal properties of rice husks and their ash are given later (Tables 16.2 through 16.4).

Table 16.1   Common Uses of Rice Husk

Uses of Rice Husk	% Used in the United States, 1972
Feeds including chemically treated husk	39.02
Litter and bedding	11.46
Pressing and filtering aids	2.20
Furfural and oilier chemicals	2.07
Silica and carbon source (husk equivalent)	4.71
Fuels	—
Panel boards, absorbents, and other uses	3.90
Excess	36.59

Source: Houston, D.J. (Ed.), *Rice Chemistry and Technology*, AACC, St. Paul, MN, 1972.

Table 16.2 Average Organic and Other Components (%, d.b.) of Rice Husk

Ash	Lignin	Cellulose	Pentosans	Other Organics
20.0	22.0	38.0	18.0	2.0

Table 16.3 Rice Husk Ash Composition (%, d.b.)

Constituent	Percentage	Constituent	Percentage
SiO_2	94.50	Fe_2O_3	Trace
K_2O	1.10	P_2O_5	0.50
Na_2O	0.80	SO_3	1.10
CaO	0.25	Cl	Trace
MgO	0.20		

Source: Chakraverty, A., *Biotechnology and Other Alternative Technologies for Utilization of Biomass/Agricultural Wastes*, Oxford and IBH Publication Co., New Delhi, India, 1996.

Table 16.4 Physical and Thermal Properties of Rice Husk Containing 11%–21% (w.b.) Moisture (M)

	Properties	Equation
1	Bulk density [ρ_B], kg/m^3	$\rho_B = 0.37M + 97.44$
2	True density [ρ_t], kg/m^3	$\rho_t = 3.05 + 985.25$
3	Angle of repose [θ], °	$\theta = 0.39M + 46.31$
4	Thermal conductivity [K], kcal (h m °C)$^{-1}$	$K = 0.00339M + 0.003$
5	Specific heat [C_p], kcal (kg °C)$^{-1}$	$C_p = 0.024M + 0.02426$

Source: Mishra, P., Investigation on physico-thermal properties and thermal decomposition of rice husk, production of pure amorphous white ash and its conversion to pure silicon, PhD thesis, HT, Kharagpur, India, 1986.

Thermal Analysis of Rice Husk

The simultaneous TG, DTG, and DTA curves for rice straw and husk samples are shown in Figures 16.1 and 16.2, respectively. It can be concluded from these figures that the TG analysis of rice husk and rice straw reveals three distinct stages of mass loss, namely, removal of moisture, devolatilization with fast-flaming combustion of volatile matter, and slow smoldering combustion of fixed carbon. The removal of moisture from the samples takes place up to a temperature of about 150°C. The volatile matter of the samples was released at temperatures ranging from 220°C to 375°C. The combustion of fixed carbon of the samples takes place at temperatures above 375°C. The DTA records exhibit an overall exothermic reaction during the course of thermal decomposition and endothermic peaks during the removal of moisture and volatile matters.

Pure Silica and Silicon

Amorphous silica finds varied uses depending on its form and purity. It is used as a reinforcing material in rubber and industries. The amorphous silica, being chemically active, can be easily utilized in different silicate industries. It can also be

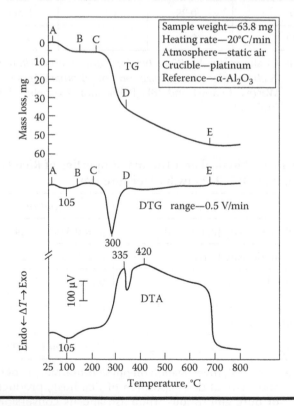

Figure 16.1   TGA, DTG, and DTA curves for rice husk (40 mesh). (From Chakraverty, A. et al., *Thermochim. Acta*, 120, 241, 1987.)

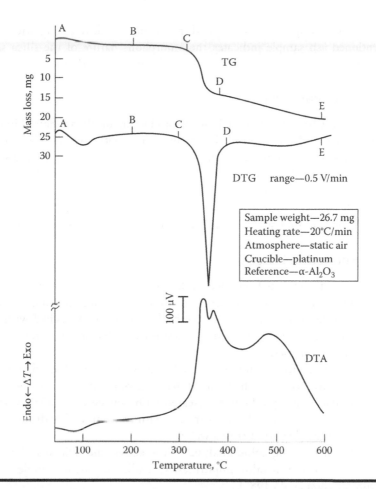

Figure 16.2 **TGA, DTG, and DTA for 1 N HCl-treated rice straw at 90°C for 50 min.** (From Chakraverty, A. and Kaleemullah, S., *Energy Convers. Manage.*, 32(6), 565, 1991a; Chakraverty, A. and Kaleemullah, S., *J. Mater. Sci.*, 26, 4554, 1991b.)

reduced to silicon at a comparatively lower temperature by metallothermic reduction processes. The semiconductor-grade and solar-grade silicon need to be very pure. These can be produced from pure amorphous rice husk silica.

Pure Amorphous Silica from Rice Husk

Clean and dry rice husk is ground to 40 mesh size. These are then treated with normal HCl solution at an elevated temperature for about 2 h. After acid leaching, the husk is washed thoroughly with water and dried. The acid-treated husk is subjected to a combustion temperature of 700°C for about 1½ h in an electric furnace for complete combustion. The ash containing mainly silica obtained from the method mentioned earlier is completely white in color due to prior removal of most of the metallic and other impurities during acid treatment. The diffraction pattern of the aforementioned ash sample indicates the amorphous nature of the silica sample (Chakraverty et al., 1987; Chakraverty and Kaleemullah, 1991a,b).

Production of Pure Silicon

Silica (SiO_2) is the basic raw material for the production of silicon. Found in nature as quartzite rock and quartz sand, its quality varies with sources. On the other hand, volatile silicon compounds such as $SiCl_4$, $SiHCl_3$, SiF_4, etc., are the secondary sources for the production of silicon. All these volatile compounds have a significant advantage as they can be purified by fractional distillation in the vapor phase.

Rice husk has been identified as an attractive source of high-grade amorphous silica and silicon as its undesirable metallic and other impurities can be removed easily.

Calcium Reduction Process

Amorphous silica prepared from rice husk can be reduced to pure silicon by a metallothermic reduction process using calcium as a strong reducing agent. The white ash produced from acid-treated rice husk is once again leached with warm 1 N hydrochloric acid for about 2 h and then washed with distilled water to remove the acid-soluble impurities.

The silica mixed with calcium is then reduced in an electrical muffle furnace at around 720°C. After cooling, the reduced mass is ground to fine particles.

The ground reaction products are treated with concentrated nitric acid. The unreacted silica in the material is then treated with concentrated hydrofluoric acid and washed thoroughly with distilled water.

The emission spectrographic analysis of the above said silicon indicates the absence of most of the metallic impurities. Overall, the silicon sample is about 99.9% pure (Mishra et al., 1985).

It is reported that about 99% pure silicon can be produced by magnesium reduction of rice husk ash at a temperature of 550°C. In an alternative process, $SiCl_4$ is synthesized by chlorinating rice husk ash. The silicon is produced by the reduction of $SiCl_4$ with Zn at a temperature of 500°C in a vapor phase reaction (Acharya et al., 1982).

Ceramics from Rice Husk Ash

The ceramic industries are also called silicate industries. The common ceramic products are made from three basic raw materials, namely, (1) clay (Kaolinite—Al_2O_3, $2SiO_3$, $2H_2O$), (2) feldspar (Spar—K_2O, Al_2O_3, $6SiO_2$), and (3) sand/quartz (SiO_2). Besides these three materials, some flushing agents such as borax ($Na_2B_4O_7$, $10H_2O$), boric acid (H_3BO_3), and fluorspar (CaF_2) are also used. All ceramic materials are produced by mixing various amounts of the aforementioned raw materials, shaping, and firing at temperatures ranging from 900°C to 1400°C, depending on the type of the ceramic products.

The white ash produced by burning rice husk containing about 95% silica can be used as one of the basic raw materials in glass and ceramic industries.

Raw materials: The following three compositions of the raw materials for different ceramic materials have been found to be feasible (Goyal, 1977). However, in place of sodium oxide, boric oxide, and calcium oxide, sodium carbonate, boric acid, and calcium carbonate, respectively, can be used.

Production of ceramics: The process flowchart for making ceramic materials is shown in Figure 16.3.

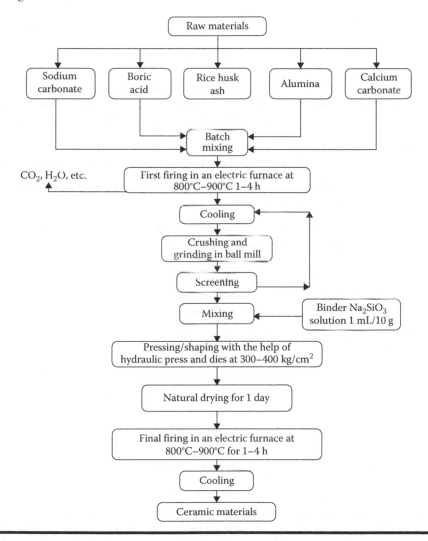

Figure 16.3 Flowchart for making ceramics. (From Goyal, Ceramic materials from rice husk, MTech thesis, ITT, Kharagpur, India, 1977.)

Table 16.5 Various Compositions of Raw Materials for Ceramics

Composition	White Rice Husk Ash%	Na_2O%	B_2O_3%	Al_2O_3%	CaO%
1	75.00	—	15.00	5.00	5.00
2	80.00	10.00	—	5.00	5.00

Source: Goyal, Ceramic materials from rice husk, MTech thesis, ITT, Kharagpur, India, 1977.

Mixing of raw materials: White ash obtained from rice husk is mixed with other raw materials, namely, Na_2CO_3 or H_3BO_3, $CaCO_3$, and Al_2O_3, in powder form according to the required quality of the ceramics (Table 16.5).

First firing: The mixture composition is heated in an electric furnace at temperatures ranging from 800°C to 900°C for 1–4 h so that CO_2, H_2O, and volatile matters (V_m), etc., are released and the shrinkage during final firing is also reduced.

Crushing, grinding, and sieving: After cooling, the partially fused material is subjected to grinding in a ball mill; ground materials are sieved through 212 μm sieve.

Mixing of binder: Sodium silicate solution is added to the material already sieved. It is mixed thoroughly to make it homogeneous.

Shaping: The mixture with the binder is pressed and given shapes with the help of a hydraulic press and dies at a pressure of 300–400 kg/cm².

Drying: After being shaped, the specimens are then dried.

Firing of body: The bodies are finally fired in a closed electrically heated chamber at temperatures ranging from 800°C to 900°C for 1–4 h.

Cooling of fired body: After firing, the ceramic bodies are left in the furnace to cool down gradually.

Properties of the aforementioned ceramics: (a) The modulus of rupture of the earlier samples varies from 215 to 360 kg/cm². (b) The electrical resistivity of the samples containing no Na_2O is of the order of 7×10^{15} Ω cm. (c) The material also had high chemical resistance toward water as well as mineral acids.

Paper Production from Cellulosic Biomass

Wood cellulose is one of the most versatile raw materials used for production of paper. There are two distinct phases in the conversion of raw wood to finished paper. These include (a) the production of pulp from the raw wood and (b) the conversion of pulp to paper. There are four different kinds of wood pulp, namely, mechanical pulp, sulfite pulp, sulfate pulp, and soda pulp. Chemical pulps are much superior to mechanical pulp for fine paper making. The pulp is converted to paper by two common methods, namely, beating and refining. Paper mills use either of the two or both together, depending on the desired quality of the paper. The most commonly used type of beater is known as Hollander.

Raw materials: Wood is the basic raw material used in all parts of the world for pulp and papermaking. At present, non-wood pulps have also been in use and these account for about 5% of the world production. The common agricultural fibers can be divided into the following groups: (a) straw and grasses other than bamboo; (b) canes and reeds; (c) woody stalks with blast fiber; (d) bamboos; and (e) others. The characteristics of some agricultural wastes and their percentage recovery of pulp are shown in Tables 16.6 and 16.7.

Table 16.6 Cellulose and Lignin Contents of Some Agricultural Wastes

Material	Cellulose (%)	Lignin (%)
Corn cobs	30.0	22.2
Sorghum stalk	28.7	20.4
Groundnut shells	26.2	30.5
Cotton stalk	31.5	26.0

Source: Swaminathan, K.R., Project report for using agricultural waste for paper-board making, College of Agricultural Engineering, TNAU, Coimbatore, India, 1980.

Table 16.7 Percentage Recovery of Pulp from Various Raw Materials by Caustic Soda Treatment

Material	Pulp Production (%)
Waste paper	66.7
Napier grass	34.0
Sunflower stalk	31.3
Sugarcane bagasse	33.0
Cotton waste	52.0

Source: Swaminathan, K.R., Project report for using agricultural waste for paper-board making, College of Agricultural Engineering, TNAU, Coimbatore, India, 1980.

Paper Production

The fibrous cellulosic wastes can be converted to pulp and paper following the method of production of soda pulp (Swaminathan, 1980; Balasubramanian et al., 1982) as shown in the flow diagram (Figure 16.4).

The different operations involved in the conversion of cellulosic wastes to paper card are described in the subsequent paragraphs.

Digestion: The chipped cellulosic wastes are cooked in the digester with a solution of caustic soda and other chemicals for 3–8 h, depending on the raw material. The digester consists of a cylindrical vessel equipped with false bottom, vomiting pipe, safety valve, etc.

Beating: The fibers are beaten to make the paper stronger, more uniform, more dense, and more opaque. A Hollander beater, mainly used for beating digested pulp, consists of a tank with rounded ends and a partition around which the pulp circulates continuously. A heavy rotating roll protruding with bars is directly placed above a stationary bed plate equipped with more bars. The circulating pulp is forced between these two. The entire roll can be raised or lowered to achieve

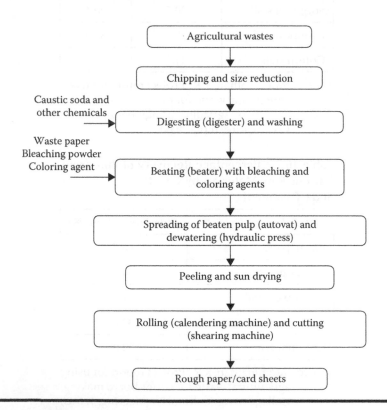

Figure 16.4 Paper production from agricultural wastes.

the desired product. In the beater, digested material and waste paper are mixed in proper proportion for the desired quality of pulp. Bleaching and coloring agents are also added. All papers, except blotting, tissue, and other rough papers, should have a filler-like talc and a sizing agent like wax emulsion. Only after sufficient beating is the desired consistency of the pulp achieved.

Spreading, lifting, and couching: An autovat consisting of a wooden tank is filled with water. A wooden-framed wire mesh mould is suspended in water. A measured quantity of the pulp is poured into the wire mesh mould and stirred. Then it is lifted after being shaken. The sheet is formed and couched.

Pressing: Most of the water is squeezed out from the wet paper sheet with help of a two-rammed hydraulic press. After peeling the cloths, the wet paperboards are then piled up.

Drying: The wet paperboards are then dried in the sun for a day or two depending on the weather.

Calendering and sizing: The hydraulically loaded calendering machine irons out the wrinkles on the surfaces of the paper. It consists of two rotating mirrors—polished cast iron rolls. The smooth paper sheets are cut to sizes by the shearing machines.

Production of Biodegradable Plastic Films

Utilization of starch, a naturally occurring polymer, in various polymers has created great interest. A process of extrusion blowing of plastic films from low-density polyethylene (LDPE) and starch and other biomaterials in the presence of a suitable coupling agent has been developed (Nanda et al., 1990).

Materials: LDPE, corn starch of 9.8% moisture content, urea, and a coupling agent are used.

Film preparation: A mixture of corn starch, LDPE, the coupling agent, and urea is prepared. A small amount of water is added to moisten the starch enough to smear itself uniformly around the granules of LDPE and the coupling agent. The mixture is compounded in a kneader at 122°C–124°C and crosshead extruder at 122°C–126°C. The compounded material is discharged in the form of a continuous strand and cut with rotary knives in a granulator. The granules are blown into a film on film-blowing extruder.

Products: Smooth, clear, and good quality films containing 10%–40% starch are obtained. The correct amount of water and urea is critical for the quality of the film. The amount of urea present in the composition is also important as it helps gelatinization and prevents retrogradation of starch. With the help of the coupling agent, the nonpolar polyethylene polymer can intimately mix with the polar starch polymer to make the mass almost a single-phase system. These films are suitable for various uses where biodegradation of films in the field condition or soil environments is desired.

Alcoholic Fermentation

Ethyl alcohol/ethanol, commonly known as alcohol, is one of the most important and popular fermented industrial products. Ethanol is a liquid fuel alternative to automotive fuel and nonpetroleum fuel option, which can be produced from renewable resources. The production of ethanol by fermentation is proven technology. It is the base chemical for the production of ethylene, ether, esters, and many other chemical products. Sugars, starch, and cellulose are the basic organic compounds used for the production of ethanol.

Production of Ethyl Alcohol from Rice Straw, Rice Husk, and Bagasse

Ethyl alcohol can be produced by simultaneous saccharification and fermentation (SSF) of lignocellulosic residues. Ethanol fermentation can be carried out by simultaneous hydrolysis and fermentation in the same reactor (Roy Chowdhury et al., 1980). This SSF process is described as follows.

Bagasse and rice straw are ground separately and then delignified by autoclaving them in 1% solution of caustic soda at 120°C for 1 h. The yeast *Pichia etchellsii* is used for SSF. The enzyme cellulase culture filtrate is obtained from *Trichoderma reeseif* QM 9414. The concentration of the cellulosic substrate may be kept at 140 g/L. The pH and temperature of the medium are maintained at 4.8°C and 40°C, respectively. The medium is fortified with cheap nitrogen sources to meet the growth requirements of yeast during the SSF. The amount of ethyl alcohol produced in the medium is 32 g/L in a period of 58 h with the aforementioned initial substrate concentration.

By-Products of Fruit and Vegetable Processing

The world production of fruits and vegetables was about 429 and 596 million tons, respectively. Among the fruits and vegetables, citrus, apple, pineapple, grape, tomato, and potato are mainly processed.

The fruit and vegetable processing industries generate 10%–50% of the raw materials as solid waste. Their compositions suggest potential for producing value-added products, as for examples pineapple core, peel and citrus peel, pulp wash generated from their juice, and juice concentrate processing industries, can be further processed to produce value-added by-products. In fact, utilization of these wastes rather than their disposal should be the aim of the food industry.

The future of the food processing industry as a whole and the fruit and vegetable processing industry in particular rests in achieving zero-waste processing systems.

The main fruit and vegetable producing countries and their production in 1996 are given as follows.

Major potato-producing countries with their production, 1000 MT in 1997, and its processing wastes are presented as follows.

Country	Production	Processing Waste	By-Products	References
China	45,534	Potato pulp and liquor	Starch, ethanol, and enzymes	Klingspohn et al. (1993)
Russian Federation	40,000			
India	19,240			
World	295,407			

Pineapple: The total waste generated by the pineapple industry itself is about 30%–60% of the fruit. The important waste components include leaves, crown, stem, peel, core, and mill juice. Potential by-products that could be produced from the wastes of pineapple and apple are presented as follows.

Pineapple Waste	By-Products	References	Apple Waste	By-Products	References
Mill juice	Sugar			Alcohol	
	Syrup			Flavor	Almosnino et al. (1996)
	Wine		Pomace	Pectin	Sulc et al. (1982)
Core	Syrup			Fiber	
	Vinegar			Biogas	Kranzier and Davis (1981)
Peels	Tit bit, jam				
Crown and leaves	Fiber	Ghosh and Sinha (1977)			

Potato: Major waste material in the potato-processing plant is in the form of peel and juice-containing solids. Potatoes are processed for different products including starch and starch granules. The waste effluent from the potato starch manufacturing is similar to that of other effluents from cutting and chipping operations of other processes. The other sources of potato waste are damaged produce from harvest; sprouted, green, and damaged potato from storage; and rejected potato. A number of products can be prepared from potato wastes.

Tomato: Tomato pomace is one of the major wastes that is mainly composed of seeds and skin.

Seed oil: Tomato seed oil is characterized by its high content of monounsaturated fatty acids, namely, oleic acid. Although this edible fatty acid is of low nutritional value compared to polyunsaturated fatty acids, especially linoleic acid, which is predominant in such crops as safflower, sunflower, and corn, the oleic acid can be widely used as a soap base in the manufacture of value-added oleate ointments, cosmetics, food-grade additives, and other products.

Coloring matter: The major attractive pigment in tomato skin is red-colored lycopene (about 12 mg/100 g), which corresponds to about 71% of the lycopene found in tomato paste. The chemical structure, nutritional value, and other utilization of lycopene were discussed in Chapter 19.

Biogas: A promising method for economic utilization of fruit and vegetable processing wastes is biomethanation. This method has certain advantages because its products include methane, as a gas and digested slurry that can be further used as manure. Solid-state fermentation of wastes using selected strains of *Sporotrichum*, *Aspergillus*, *Fusarium*, and *Penicillium* spp. can reduce the level of antimicrobial substances, allow the use of a loading rate of 8%–10% (dry weight), and improve the overall productivity of biogas methane. Carbon, nitrogen, and phosphorus (C:N:P) ratio is an important parameter for anaerobic digestion. Mango peel waste supplemented with rice bran could generate biogas of 0.6 m³/kg volatile solid (VS) added (Krishna Nand, 1991).

Composting: The organic composting of fruit and vegetable processing waste is an environment-friendly process. Their end products obtained by composting can be added back into the soil as organic fertilizer. During the process of composting, the volume of the waste is reduced by 25%–40%. For an efficient composting process, the prerequisites for the microorganisms to survive and grow are specific. The major controlling parameters are moisture range (40%–60%), carbon-to-nitrogen ratio (2:5), and pH (6–8) of the waste material; temperature; and aeration (oxygen concentration). Therefore, in the preparation of the waste material for composting one should take care of these parameters. Proper mixing of various waste materials to achieve the desired C:N ratio is very important prior to composting. The main problems of *composting of fruit and vegetable processing wastes* are their high moisture content (80%–90%) and acidic nature. Moisture content can be partially reduced by draining and pressing. Bulky agents used to reduce moisture content and increase the porosity are sawdust, paper, and coffee ground residues. However, care should be taken to maintain the C:N ratio. To reduce high C:N ratio, nitrogen in the form of urea may be mixed and lime can be added to neutralize excess acidity. Aerobic and thermophilus bacteria, namely, *Bacillus sterothermophilus* and *Thermus* species obtained from sewage, can be used as starting material for the treatment. Composting can be completed in ~2 weeks. Rotary device can be introduced to produce a batch per day. The major benefits of composting are environmental friendliness, waste volume reduction, and value additions.

Thermal and Thermochemical Processing

Pyrolysis

Before World War I, the wood pyrolysis/wood distillation industry was the only source of acetic acid, methanol, acetone, and wood oil. Even now, there is no synthetic substitute for charcoal. The pyrolysis of solid wastes refers to its thermal decomposition in an inert atmosphere. In this process, a mixture of gases, tars, oils, acetic acid, methanol, etc., and a solid residue containing inerts and char is produced. The yield of char is about 15%–25% by weight of the refuse on dry ash-free basis (dafb).

Pyrolysis, or charring of biomass fuel, has the main objective: production of a less smoky, more reactive fuel with a higher calorific value. However, the disadvantages of pyrolysis are (i) considerable fraction of the energy in the original feedstock is lost in the gas and liquid products and (ii) the charred products are more brittle such as rice husk char, soft wood charcoal, etc.

Principles

The different stages of the pyrolysis of wood and their products are presented later (Table 16.8).

Table 16.8 Different Stages of Wood Pyrolysis and Their Corresponding Products

Temperature Range (°C)	Product
0–170	Evaporation of moisture and tar
170–270	Evolution of CO, CO_2, and other gases
270–400	Evolution of methanol, acetone, and acetic acid
400–500	Optimum carbon in charcoal

The thermogravimetric analysis (TGA), differential thermogravimetric analysis (DTG), and differential thermal analysis (DTA) of rice husk as well as rice straw are presented in Figures 16.1 and 16.2.

Production of Wood Chemicals and Charcoal

About 4–5 tons of dry wood of around 25% moisture content is loaded into 20 m long steel retorts and heated for about 20 h at 350°C–370°C. Charcoal is removed from the retorts and cooled in absence of air for about 2 days, placed in open sheds for another 2 days, and is then ready for supply to the consumers. The wood gas of calorific value 4–7 MJ/kg produced is led to a scrubber and sent to the gas pipe connected to the burners. The liquor from the far-soil is pumped to a continuous demethanolizing distillation column where crude methanol is distilled. The demethanolized from the bottom of the column is sent to a pre-evaporator. The remaining volatile acetic acid and water are distilled in the evaporator. The water is then removed in the azeotropic column by adding a water-insoluble component such as butyl acetate, a water-withdrawing liquid. This method of removing water from acetic acid is known as Othmer process. The butyl acetate and water vapors are condensed, cooled, and separated into two layers. The crude acetic acid is further rectified in other columns to separate glacial acetic acid from formic acids and higher acids.

Gasification

Man has been aware of wood pyrolysis and wood gasification technologies for a long time. Nowadays, of course, various raw materials from crop residues to municipal solid wastes have been suitably used for gasification to generate low calorie gas. However, the following points should be made clear:

- A gasifier either generates fuel gas for direct combustion or produces fuel gas for an internal combustion (IC) engine.
- If the gas is produced for IC engine, it must be thoroughly cleaned to make it free from tar, particulate, etc., to conform the specification of 5–15 mg/kg gas.
- Owing to the presence of CO in the fuel gas, it cannot be used as a cooking gas.

Principles

Gasification is the process of thermochemical conversion of an organic matter to the chemical energy in the form of fuel gas (Kaupp, 1984). In gasification always a sub-stoichiometric quantity of air is used.

With reference to gasification, the equivalence ratio, ψ, can be defined as

$$\psi = \frac{\text{Actual air supplied for gasification}}{\text{Theoretical (stoichiometric) air required for complete combustion}}$$

In the course of gasification, both total energy and the sensible heat energy in the fuel gas increase with the increase of equivalence ratio, Ψ. But the chemical energy in the gas increases up to a certain value of Ψ (say, 0.2–0.3 approximately) and then it decreases sharply with a further increase in Ψ. Hence, invariably a sub-stoichiometric quantity of air is supplied for gasification.

Design Parameters

The following design parameters should be taken into consideration for a rice husk gasification system (Kaupp, 1984): (i) air to fuel ratio; (ii) equivalence ratio; (iii) diameter of the gasifier; (iv) rice husk feed rate; (v) specific gasification rate; (vi) time of operation; (vii) gas composition; (viii) ultimate analysis of rice husk; (ix) degree of rice husk conversion; (x) efficiency of the gasification process; (xi) carbon conversion percentage; (xii) higher heating value of the dry gas produced; (xiii) gas composition; (xiv) volume reduction of rice husk; and (xv) ash removal rate for a continuous system.

Reaction Zones

In a biomass gasification process, CO, H_2, CH_4, CO_2, N_2, H_2O, tars, and some organic liquids are generally produced. In different stages of gasification, many thermochemical reactions take place. After ignition, four reaction zones are set up in the gasifier unit, namely, drying, pyrolysis, gasification, and oxidation zones. Few important thermochemical reactions corresponding to gasification and oxidation reaction zones are shown as follows.

Zone	Reactions
	$C + H_2O + Heat \rightarrow CO + H_2$
Gasification	$C + 2H_2O + Heat \rightarrow CO_2 + 2H_2$
	$C + CO_2 + Heat \rightarrow 2CO$
Oxidation	$C + O_2 \rightarrow CO_2 + Heat$

Heat energy is necessary for the endothermic reactions of the drying, pyrolysis, and reduction zones. These heat energies are supplied by the heat of the exothermic reaction between C and O_2 to produce CO_2 in the oxidation zone.

Gasifiers

Either fixed bed or fluidized bed gasifiers can be employed for gasification. The various designs of fixed bed gasifiers include updraft, downdraft, and cross-draft gasifiers. However, other gasifiers, namely, (a) stirred bed, (b) tumbling bed, and

(c) entrained bed gasifiers can also be designed. An air-blown gasifier produces a low calorie gas, whereas a medium calorie gas is generated by oxygen-blown gasifier. Generally, fixed bed gasifiers operate under atmospheric pressure.

Updraft (countercurrent flow) gasifier: In this gasifier, the air moves upward from the bottom, whereas the feed moves downward. The oxidation zone lies at the bottom (Figure 16.5). The fuel gas flows upward and encounters the relatively cold feed in the opposite direction and leaves the gasifier at a lower temperature. Thus, the fuel gas in an updraft gasifier contains a high quantity of uncracked tar. Therefore, an effective cleaning system is required for a thorough cleaning of the gas before using it for IC engines.

Downdraft (co-current flow) gasifier: In this gasifier, the feed and air move in the same direction toward its bottom. As the gas passes through the oxidation zone of

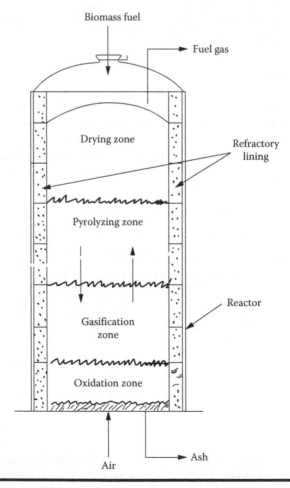

Figure 16.5   Schematic representation of an updraft gasifier.

higher temperature at the bottom, some amount of tar is thermally cracked. It is expected that this system should produce relatively clean gas with less amount of tar, etc.

Cross-draft (cross-flow) gasifier: Here, air is forced through the horizontal nozzles and the fuel gas generated is discharged through a vertical grate on the opposite side of the air nozzles.

Fluidized bed gasifier: The desired fluidization is achieved by blowing air or oxygen at a fluidization velocity. In consequence, the entire bed remains under a state of suspension and the bed temperature is uniformly maintained due to a thorough mixing of the fluidized bed. Both oxidation and gasification occur simultaneously at a rapid rate. The bed temperature may go up at a high level. For a fluidized bed reactor, the height-to-diameter ratio may be 10:1.

Combustion

The process of liberation of heat energy from a fuel is known as combustion. The heat energy is liberated by the exothermic reaction due to oxidation of the combustible constituents of the fuel. Normally, fuels are compounds of carbon and hydrogen; in addition, variable percentages of oxygen, nitrogen, and small percentages of sulfur, etc., are also present. Combustion and incineration are synonymous. However, incineration is commonly used for disposing municipal and other solid wastes by burning at a high temperature.

Principles

The plant fuels mainly contain moisture, organic, and inorganic compounds. If biomass fuels contain high moisture, they should be dried to a lower moisture level. The high moisture content not only acts as a heat sink and lowers the combustion efficiency, but it also affects the economics of the fuel utilization.

Generally, biomass is composed of cellulose, hemicellulose, and lignin and at times, also some lipids, carbohydrates, and proteins. Combustion is influenced with the variation in composition of these organics.

Biomass fuels also contain variable amounts of minerals (ash) depending upon the species, locality, and soil. The presence of high silica in any fuel acts as a heat insulator, whereas some soluble electrovalent compounds may act as catalysts in the combustion process.

Some important terms used in combustion are defined as follows.

The gross calorific value or higher heating value is defined as the number of heat units released when a unit mass of solid or liquid fuel or unit volume of gaseous fuel is completely burnt and the products of combustion are cooled to 15°C, thereby condensing the water vapor present therein.

The net calorific value refers to the gross calorific value minus the latent heat of condensation (at 15°C) of the water vapor present in the products of combustion.

The thermochemical equations involved in the combustion of a fuel containing carbon, hydrogen, sulfur, etc., are presented as

$$C + O_2 = CO_2 + 97,644 \text{ kcal/kgmol}$$

$$H_2 + 1/2O_2 = H_2O + 69,000 \text{ kcal/kgmol}$$

$$S + O_2 = SO_2 + 71,040 \text{ kcal/kgmol}$$

The gross calorific value, C_G, kcal/kg of a fuel is mathematically expressed as follows:

$$C_G = \frac{C \times 8137 + \left(H - \dfrac{O + N - 1}{8}\right) \times 34,500 + S \times 2220 - H_2O \times 600}{100}$$

where C, H, O, S, N, and H_2O are present in the fuel as percentages of carbon, hydrogen, oxygen, sulfur, nitrogen, and water, respectively.

Conditions for Efficient Combustion

Normally, combustion reactions occur at high temperature and high speed. But the same amount of heat will be released independent of the conditions of the reactions, provided the end products are the same and the combustion reactions are complete.

In order to make the heat of combustion of the fuel fully available, the thermochemical reactions should be as complete as possible. Therefore, at least stoichiometric (theoretical) amount of air must be supplied in proper manner. In practice, however, the best results are achieved by using small proportions of excess air as operation with theoretical air may lead to the escape of some unreacted fuel.

The basic conditions for efficient combustion of any fuel in a furnace are as follows:

■ More than theoretical amount of air must be supplied to ensure complete combustion.
■ The fuel and oxygen of the air must be in free and intimate contact.
■ The secondary air must be intimately mixed with the combustible volatiles leaving the fuel bed for flaming combustion.

- The gases and vapor cannot be allowed to cool below the ignition point until the combustion reactions are complete.
- While designing furnace volume, provision for expansion of the gases during combustion at high temperature and pressure should be made.

Combustion Air, Flue Gas, and Enthalpy

The quantity of theoretical (stoichiometric) air required and the volume of flue gas produced for complete combustion of a fuel can be calculated stoichiometrically if the ultimate analysis of the fuel is available. The air actually required for different excess air factors and the corresponding flue gas produced and the enthalpy content of the flue gas can also be conveniently determined from the empirical equations and statistical tables available for combustion calculation where only information about net calorific values of the fuels are needed. The empirical equations are developed on the basis of the following facts:

1. There exists statistical relationship between the net calorific values of all industrial fuels and (a) the air required for their combustion and (b) the volume of the flue gas produced by combustion.
2. The enthalpy contents at the same temperature per unit volume of the theoretical wet flue gas are the same for different industrial fuels.

The net calorific value (C_N), kcal/kg of a fuel, namely, bituminous coal, can be derived from the gross calorific value (C_G), kcal/kg, with the help of the relationships presented as follows:

$$C_N = 0.954\ C_G + 110$$

The theoretical air requirement, V_{ao}, and the corresponding volume of flue gas produced, V_{fo}, can be calculated using the empirical equations presented in Table 16.9.

Here, the excess air factor n' is expressed as $n' = V_a/V_{ao}$.
The volume of excess air is given by $V_{ex} = (n' - 1)V_{ao}$.

Table 16.9 Empirical Equations for V_{ao} and V_{fo} for Industrial Fuels

Fuels	V_{ao} (nm³/kg)	V_{fo} (nm³/kg)
Solid fuels		
C_N < 5500 kcal/kg	1.01 (C_N/1000) + 0.5	0.89 (C_N/1000) + 1.65
C_N > 5500 kcal/kg	1.01 (C_N/1000) + 0.55	(C_N/1000) + 0.9

Corresponding to any excess air factor n', the volume of flue gas produced is calculated as

$$V_f = V_{fo} + (n' - 1)V_{ao}$$

For complete combustion of a fuel without any heat loss, the enthalpy of the flue gas per unit volume is given by

$$h' = \frac{C_N}{V_f}$$

where h' is enthalpy of flue gas, kcal/nm³.

The values of V_a, V_f, and h' for solid fuels corresponding to different excess air factors for combustion are given in Table 16.10.

C_N—Net calorific value, kJ/kg fuel; V_o and V_{fo}—theoretical air for complete combustion and theoretical flue gas produced, respectively, nm³/kg fuel; V_a and V_f—actual air used for combustion and actual flue gas produced, respectively, nm³/kg fuel; n'—excess air factor (V_a/V_{ao}); h'—enthalpy of the flue gas (C_N/V_f), kJ/nm³ flue gas.

Furnaces

Both fixed grate furnaces and cyclone furnaces can be used for combustion of biomass fuels. However, various other furnaces are in vogue in different industries (Shvete et al., 1975).

Horizontal fixed grate furnace: It is a very simple furnace, which comprises a furnace chamber, a precipitation chamber, and an air-mixing chamber. The furnace walls and roof are made of refractory bricks to minimize heat losses and stand high temperature (Chakraverty, 1996). Cast iron bars are used for the fire grates due to their stability under thermal shock and better resistance to corrosion.

Inclined fixed grate furnace: A box-type inclined step grate furnace consists of a hopper, inclined grates set at an angle of 45°. The cast iron grates are arranged in a staircase fashion. The furnace walls are made of refractory bricks to reduce heat losses and stand high temperatures (Chakraverty, 1996).

Cyclone furnaces: The cyclone furnaces have been found to be highly efficient for the combustion of pulverized coal. These may also be suitable for firing biomass fuels. The cyclone furnaces may be horizontal or vertical. A horizontal cyclone furnace (Figure 16.6) consists of a slightly inclined cylinder, lined with fire bricks, into which air is ejected tangentially at a very high speed. The fuel introduced at one end is entrained by the revolving mass and is thrown against the cyclone walls, where it

Table 16.10 Values of V_a, V_f, and h' for Solid Fuels for Excess Air Factors from 1.0 to 1.5

C_N	$n' = 1.0$			$n' = 1.1$			$n' = 1.2$		
	V_{ao}	V_{fo}	h'	V_a	V_f	h'	V_a	V_f	h'
8,372	2.51	3.42	2440	2.78	3.67	2273	3.03	3.94	2126
10,464	3.02	3.87	2704	3.33	4.16	2512	3.61	4.47	2340
12,558	3.54	4.32	2905	3.88	4.68	2687	4.22	5.04	2495
14,650	4.04	4.76	3077	4.44	5.17	2938	4.82	5.58	2629
16,743	4.54	5.22	3210	5.01	5.66	2955	5.45	6.13	2733
18,836	5.04	5.65	3332	5.56	6.16	3060	6.08	6.65	2830
20,929	5.56	6.11	3432	6.11	6.66	3148	6.67	7.22	2901
23,022	6.11	6.42	3595	6.71	7.02	3281	7.33	7.62	3018
25,115	6.61	6.90	3637	7.27	7.56	3319	7.93	8.23	3055
27,208	7.12	7.40	3675	7.83	8.12	3353	8.54	8.83	3085
29,301	7.62	7.90	3709	8.38	8.66	3382	9.14	9.42	3110
31,394	8.12	8.40	3738	8.94	9.22	2407	9.76	10.01	3135
33,487	8.62	8.90	3768	9.50	9.77	3391	10.36	10.61	3152

C_N	$n' = 1.3$			$n' = 1.4$			$n' = 1.5$		
	V_a	V_f	h'	V_a	V_f	h'	V_a	V_f	h'
8,372	3.28	4.20	1997	3.53	4.45	1884	3.79	4.68	1784
10,464	3.93	4.78	2190	4.23	5.08	2060	4.52	5.38	1960
12,558	4.58	5.39	2332	4.94	5.74	2190	5.28	6.08	2064
14,650	5.24	5.97	2453	5.64	6.37	2298	6.04	6.78	2160
16,743	5.91	6.56	2543	6.36	7.04	2382	6.82	7.48	2236
18,836	6.56	7.16	2629	7.07	7.66	2453	7.57	8.18	2307
20,929	7.22	7.77	2696	7.77	8.33	2516	8.34	8.86	2361
23,022	7.93	8.24	2796	8.54	8.84	2604	9.14	9.46	2437
25,115	8.59	8.88	2826	9.25	9.54	2629	9.92	10.20	2462
27,208	9.26	9.52	2851	9.96	10.24	2654	10.67	10.96	2483
29,301	9.90	10.18	2880	10.67	10.96	2675	11.44	11.72	2503
31,394	10.56	10.83	2897	11.37	11.65	2696	12.18	12.46	2520
33,487	11.21	11.50	2913	12.08	12.36	2713	12.95	13.22	2533

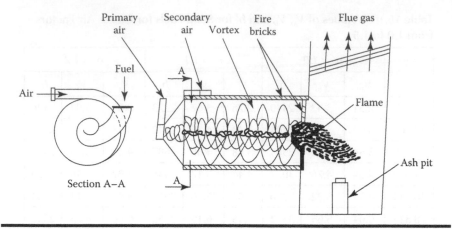

Figure 16.6 Schematic representation of a horizontal cyclone furnace. (From Chakraverty, A., *Biotechnology and Other Alternative Technologies for Utilization of Biomass/Agricultural Wastes*, Oxford and IBH Publication Co., New Delhi, India, 1996.)

intimately mixes with air and burns. The flue gases leave through the aperture at the other end of the cyclone furnace.

Cyclones are designed for positive pressure operation. Generally, cyclones require an air blower with high static pressure. The schematic representation of a cyclone furnace as shown in Figure 16.6 is self-explanatory.

Furnace Design

The two important terms used for the design of a furnace are defined as follows.

The thermal load of a fire grate is the amount of heat generated in kilocalories by complete combustion of a solid fuel on 1 m² area of a fire grate in 1 h.

The thermal load of the furnace volume is the amount of solid fuel in 1 m³ of the furnace volume in 1 h.

Mathematical expressions for these are as follows:

$$Q_a = \frac{w_f \times C_N}{a} \quad \text{and} \quad Q_v = \frac{w_f \times C_N}{v}$$

where
Q_a is the thermal load of furnace grate area, kcal/(m² h)
Q_v is the thermal load of furnace volume, kcal/(m³ h)
w_f is the fuel burnt, kg/h
C_N is the net calorific value of the fuel, kcal/kg
a is the furnace grate area, m²
v is the volume of furnace space, m³

The thermal grate load and thermal load of furnace volume for rice husk are about 2×10^5 kcal/(m² h) and 2×10^5 kcal/(m³ h), respectively.

The efficiency of a furnace (η_f) can be expressed as follows:

$$\eta_f = \frac{G \times h'}{w_f \times C_N}$$

where

G is the flow rate of the flue gas, nm³/h

h' is the enthalpy of flue gas, kcal/nm³

Problem on Combustion of Rice Husk

The ultimate analysis of rice husk is as follows: C, 39%; H_2, 5%; O, 32.7%; S, 0.1%; N_2, 2.0%; H_2O, 3.6%; and ash, 17.6%.

Assuming molecular weights of air and flue gas as 29, compute the actual air required and the flue gas produced per kilogram of rice husk, if 20% excess air is supplied for complete combustion of the rice husk.

Solution: Stoichiometric air required and flue gas produced for combustion of 100 kg rice husk.

Basis: 100 kg rice husk

Risk Husk Constituents	Constituents (kg)	Constituent (kg mol)	O_2 Required (kg mol)	Combustion Products (kg mol)
C	39.0	3.250	3.250	3.250 (CO_2)
H_2	5.0	2.500	1.250	2.500 (H_2O)
O_2	32.7	1.022	−1.022	—
S	0.1	0.003	0.003	0.003 (SO_2)
N_2	2.0	0.071	—	0.071 (N_2)
H_2O	3.6	0.200	—	0.200 (H_2O)
Total			3.481	6.624

Theoretical air required for combustion is as follows:

$$\text{Theoretical air required per kilogram husk} = \frac{100 \times 3.481 \times 29}{21 \times 100}$$

$$= 4.8 \text{ kg}$$

Actual air supplied and actual flue gas produced is as follows:

$$\text{Actual } O_2 \text{ supplied} = 3.481 \times 1.2 = 4.18 \text{ kg mol}$$

Therefore, excess O_2 in the flue gas = $3.481 \times 0.2 = 0.69$ kg mol

$$\text{Actual air supplied} = \frac{100}{21} \times 4.18 = 19.9 \text{ kg mol}$$

$$\text{Actual air supplied per kilogram husk} = \frac{100 \times 4.18 \times 29}{21 \times 100} = 5.77 \text{ kg}$$

$$N_2 \text{ present in the actual air} = \frac{79}{100} \times 19.9 = 15.72 \text{ kg mol}$$

Therefore, total flue gas produced

= Stoichiometric combustion products + inert nitrogen (N_2) present in the actual air + excess oxygen (O_2) present in the air = 6.024 + 15.72 + 0.69
= 22.434 ≅ 22.43 kg mol

Therefore, flue gas produced per kilogram husk $= \dfrac{22.43 \times 29}{100} = 6.50$ kg

Problem on the Design of an Inclined Grate Furnace

Design an air inclined step grate husk-fired furnace for a boiler to generate 0.5 ton of steam at an absolute pressure of 8.5 kg/cm² with the following assumptions:

Average net calorific value of rice husk = 3000 kcal/kg
Efficiency of the furnace = 50%
Thermal grate load of the furnace for husk = 2.1 × 10⁵ kcal/h m²
Thermal load on furnace volume for husk = 1.5 × 10⁵ kcal/h m²
Temperature of the inlet water to the boiler = 25°C
Enthalpy of 1 kg saturated steam 8.5 kg/cm² = 660 kcal/kg
Enthalpy of 1 kg water at 25°C = 25 kcal/kg
Temperature of the exhaust flue gas from the boiler = 325°C
Specific heat of the flue gas = 0.25 kcal/kg °C

Heat required to generate 0.5 ton of steam from water is as follows:

500 × (Enthalpy of 1 kg saturated steam at 8.5 kg/cm² pressure – Enthalpy of 1 kg of water at 25°C and 1 atm pressure) = 500 × (660 – 25) = 317,500 kcal

Heat available from 1 kg rice husk is as follows:
From Table 16.10, the actual volume of flue gas produced in combustion of 1 kg rice husk (with 40% excess air)

$$= 5.74 \text{ nm}^3/\text{kg}$$

Weight of flue gas produced = 5.74 × (29)/(22.4) = 7.43 kg
The heat available from 1 kg husk

$$= 3000 \times 0.5 - 7.43 \times (325 - 25) \times 0.25$$

$$= 1500 - 555 = 945 \text{ kcal}$$

Furnace grate area,
Therefore, husk required for the 0.5 ton steam generation

$$= \frac{317,500}{945} = 336 \text{ kg/h}$$

Hence, the required furnace grate area

$$= \frac{336 \times 3,000}{210,000} = 4.8 \text{ m}^2$$

Area of the inclined grate = $\sqrt{2} \times 4.8 = 6.8 \text{ m}^2$
Furnace volume is calculated as follows:

$$\text{Furnace volume} = \frac{336 \times 3000}{2 \times 10^5} = \frac{336 \times 300}{200,000} = 5.2 \text{ m}^3$$

FOOD PROCESS ENGINEERING

Chapter 17

Postharvest Management of Fruits and Vegetables

Fruits and vegetables are highly perishable and require proper postharvest management, using appropriate techniques for handling and storage to minimize loss. The range of postharvest losses varies anywhere from 5% to 50% and higher. In developing countries, the postharvest losses of fruits and vegetables are enormous due to lack of adequate infrastructure and poor postharvest handling practices. As a consequence, growers and those involved with the food handling chain suffer major financial losses. The shelf life of these products is severely reduced and the consumer receives a product of poor quality and lower nutritional value. It is not uncommon that the gains made in increasing the production yields of fruits and vegetables are compromised by increased postharvest losses due to inadequate practices.

In countries with well-developed agricultural systems, significant progress has been made in developing proper systems for handling of fruits and vegetables. These technical and commercial developments have significantly reduced postharvest losses and improved the quality of the product for the consumer. Moreover, improved postharvest practices have played a major role in increasing the competitiveness of fruits and vegetables in distant international markets.

In this chapter, we will first examine the major causes of changes in fruits and vegetables that precipitate their deterioration after harvest and then consider systems that are useful for controlling the rates of deteriorative changes, such as cooling systems. Controlling the ripening process is important to extend periods during which a product may be successfully marketed. The use of proper packaging systems is also vital to ensure proper handling of fruits and vegetables from the farm gate to the consumer. Finally, the role of polymeric materials in packaging to extend the shelf life by modifying the internal package atmosphere will be discussed in the concluding section of this chapter.

Respiration Process

Physiological changes in a fruit or vegetable continue after harvest sometimes at an accelerated rate. These changes are largely due to the respiration process, since fruits or vegetables continue to respire even after harvest.

The metabolic pathways active in a respiration process are complex. Overall, starch and sugars present in the plant tissue are converted into carbon dioxide and water. The process is influenced by the availability of oxygen. The oxygen concentration within produce is very similar to that in the normal atmosphere. When oxygen is readily available, the respiration process is called *aerobic*. However, if the surrounding atmosphere becomes deficient in oxygen, then the metabolic pathways shift and anaerobic respiration occurs. The products of anaerobic respiration include ketones, aldehydes, and alcohols. These products are often toxic to plant tissue and hasten its death and decay. Therefore, while it is important to prevent anaerobic respiration, the oxygen concentration may be controlled in such a manner as to permit aerobic respiration at a reduced rate. Control of the respiration process has become one of the most important methods to control the shelf life of fruits and vegetables in commercial practice.

In addition to the final products of aerobic respiration, carbon dioxide, and water, there is also the evolution of heat. The amount of heat generated due to respiration varies with different commodities, as shown in Table 17.1. The rate of heat respiration is associated with the functional aspect of the vegetable tissue. For example, a growing part of a plant such as a leaf (in leafy vegetables) has a higher rate of heat generation than plant tissue where growth has ceased, such as in a tuber crop.

Furthermore, the respiration process produces a decrease in the product mass as some of the food components are oxidized as well as a loss of sweetness in many commodities. However, the rate of respiration may be controlled by reducing the storage temperature. The respiration rate is expressed in terms of the rate of carbon dioxide production per unit mass. For example, in Table 17.2, a classification of commodities based on their respiration rates is given.

Related to the respiration process, another important physiological change is the production of ethylene gas. Based on their ethylene production, fruits are classified as either climacteric or non-climacteric. Climacteric fruits exhibit a high production of ethylene and carbon dioxide at the ripening stage, as shown in Figure 17.1. In non-climacteric fruits, the general production of carbon dioxide and ethylene gas remains quite low and there is no increased evolution rate of these gases at the ripening stage. Table 17.3 lists some of the fruits according to this classification.

Ethylene gas, a product of natural metabolism, largely affects the physiological processes of a plant tissue. The rates of ethylene production vary for different fruits

Table 17.1   Heat of Respiration of Selected Fruits and Vegetables

Commodity	Watts per Megagram (W/Mg)			
	0°C	5°C	10°C	15°C
Apples	10–12	15–21	41–61	41–92
Apricots	15–17	19–27	33–56	63–101
Beans, green, or snap	—	101–103	161–172	251–276
Broccoli sprouting	55–63	102–474	—	514–1000
Cabbage	12–40	28–63	36–86	66–169
Carrots, topped	46	58	93	117
Garlic	9–32	17–29	27–29	32–81
Peas, green (in pod)	90–138	163–226	—	529–599
Potatoes, mature	—	17–20	20–30	20–35
Radishes, topped	16–17	23–24	45–47	82–97
Spinach	—	136	327	529
Strawberries	36–52	48–98	145–280	210–273
Turnips, roots	26	28–30	—	63–71

(Table 17.4) and may be controlled by storage temperature, atmospheric oxygen, and carbon dioxide concentration.

The consequences of ethylene gas on the maturation process of fruits and vegetables include a change in the green color due to loss of chlorophyll, browning of tissues due to changes in anthocyanin and phenolic compounds, and development of yellow and red colors due to the development of anthocyanin and carotenoids, respectively.

In certain cases, the conversion of certain compounds may be undesirable. For example, if potatoes are to be processed as French fries, the conversion of starch to reducing sugars causes browning in the final product. Similarly, adverse flavor changes due to the formation of organic acids may influence the edible quality of fruits.

Asparagus becomes tough due to continuing elongation even after they are harvested. Other changes, such as sprouting of potatoes and root corps (e.g., onions, garlic, and potatoes), reduce their utilization and increase their rate of deterioration. Furthermore, there are nutritional losses.

The loss of water from a commodity causes major deteriorative changes. Not only is there a loss of weight, but the textural quality is altered, causing a commodity to lose its crispness and juiciness.

Poor management of storage temperature accelerates physiological breakdown in fruits and vegetables. For example, the chilling injury occurs mostly in commodities from tropical and subtropical regions when they are stored at temperatures

Table 17.2 Classification of Fruits and Vegetables Based on Their Respiration Rates

Respiration Rates	Range of CO_2 Production at 5°C (mg CO_2/kg h)	Commodities
Very low	<5	Nuts, dates, dried fruits, and vegetables
Low	5–10	Apple, citrus, grape, kiwifruit, garlic, onion, potato (mature), sweet potato
Moderate	10–20	Apricot, banana, cherry, peach, nectarine, pear, plum, fig (fresh), cabbage, carrot, lettuce, pepper, tomato, potato (immature)
High	20–40	Strawberry, blackberry, raspberry, cauliflower, lima bean, avocado
Very high	40–60	Artichoke, snap beans, green onion, Brussels sprouts
Extremely high	>60	*Asparagus*, broccoli, mushroom, pea, spinach, sweet corn

Source: Adapted from Thompson, F.F. et al., *Commercial Cooling of Fruits, Vegetables and Flowers*, DANR Publication 21567, University of California, Davis, CA, 1998.

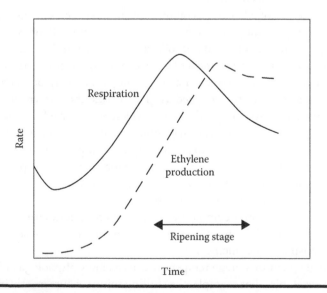

Figure 17.1 The rate of respiration and ethylene production in a climacteric fruit.

Table 17.3   Classification of Some Fruits according to Their Respiratory Behavior during Ripening

Climacteric Fruits		Non-Climacteric Fruits	
Apple	Muskmelon	Blackberry	Olive
Apricot	Nectarine	Cacao	Orange
Avocado	Papaya	Cashew apple	Pepper
Banana	Passion fruit	Cherry	Pineapple
Blueberry	Peach	Cucumber	Pomegranate
Breadfruit	Pear	Eggplant	Raspberry
Cherimoya	Persimmon	Grape	Satsuma
Feijoa	Plantain	Grapefruit	Mandarin
Fig	Plum	Jujube	Strawberry
Guava	Capote	Lemon	Summer squash
Jackfruit	Soursop	Lime	Tamarillo
Kiwifruit	Tomato	Loquat	Tangerine
Mango	Watermelon	Lychee	

Source: Adapted from Kader, A.A., *Postharvest Technology of Horticultural Crops*, Publication 3311, Agriculture and Natural Resources, University of California, Davis, CA, 2002.

Table 17.4   Ethylene Production of Selected Fruits and Vegetables

Commodity	Range (µL/kg h)
Cherry, citrus, grape, strawberry	0.01–0.1
Blueberry, cucumber, okra, pineapple	0.1–1.0
Banana, fig, honeydew melon, mango, tomato	1.0–10.0
Apple, avocado, cantaloupe, feijoa, nectarine, papaya, peach, pear, plum	10.0–100.0
Cherimoya, passion fruit, sapote	>100

Source: Adapted from Kader, A.A., *Postharvest Technology of Horticultural Crops*, Publication 3311, Agriculture and Natural Resources, University of California, Davis, CA, 2002.

above their freezing point and below 5°C–15°C. This type of injury causes uneven ripening, decay, growth of surface molds, development of off-flavors, and both surface and internal discoloration. Similar physiological damage of a commodity is also observed when storage temperatures cause freezing leading to *freezing injury*, or excessively high temperatures leading to *heat injury*. Proper management of storage temperature is usually the key to success in minimizing quality damage of fruits and vegetables during storage.

Physical damage from impact bruising and surface injuries causes deterioration. Similarly, pathological breakdown due to bacteria and fungi results in product deterioration. Often, physical damage makes it easier for the bacteria and fungi to infect plant tissues. Therefore, postharvest systems should be designed to minimize physical damage to the commodity being handled.

Ripening Process

Many fruits are harvested from the plant when fully mature and ripe. Their quality is at its peak at the time of harvest and so they should be eaten within a short time after harvest. However, there are numerous other fruits (as shown in Table 17.5) that may be harvested when fully mature and then allowed to ripen during storage and transportation. To ripen these types of fruits, specially constructed ripening rooms are used where temperature, humidity, carbon dioxide, and ethylene gas concentration as well as airflow rates are controlled.

In a conventional ripening room, boxes of product are tightly stacked into a pallet (Figure 17.2). The pallets are raised above the ground, creating about 10–15 cm ground clearance. This spacing allows unimpeded air movement under the pallets. Tarps are used to cover air passages between pallets. In citrus degreening rooms, the pallet boxes are stacked up to six pallets high (Figure 17.3).

The storage temperature in a ripening room is maintained around 15°C to 21°C, with a relative humidity of 85%. Depending on the geographical location,

Table 17.5 Fruits Harvested at Different Stages of Ripening

Group A Fruits Harvested from the Plant When Dully Mature and Ripe	Group B Fruits Harvested from the Plant When Fully Mature and Allowed to Ripen during Storage
Strawberry, blackberry, raspberry, cherry, citrus (grapefruit, lemon, lime, orange, mandarin, tangerine), grape, lychee, pineapple, pomegranate, tamarillo	Apple, pear, quince, persimmon, apricot, nectarine, peach, plum, kiwifruit, avocado, banana, mango, papaya, cherimoya, sapote, guava, passion fruit

Source: Adapted from Kader, A.A., *Postharvest Technology of Horticultural Crops*, Publication 3311, Agriculture and Natural Resources, University of California, Davis, CA, 2002.

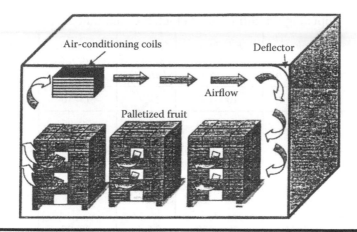

Figure 17.2   A typical ripening room with palletized fruit and air circulation.

the amount of insulation in the walls will vary. Table 17.6 shows the typical conditions used in ripening fruits.

The ripening process involves warming the room either using electric resistance heaters or indirect heat exchangers heated by hot water. The use of open flame for heating should be avoided because carbon dioxide emitted during the combustion process may interfere with the ripening process.

After the stored product in the ripening room is heated to the required temperature, the room should be maintained at that temperature. The stored product continues to respire and generate heat. Heat may be continually added to the storage room due to heat transfer from the outside and due to the operation of fans. Therefore, refrigeration is necessary to maintain the desired storage temperature. It should be noted that forced-air systems are better in removing heat from the product than conventional non-forced-air systems. The rate of heat removal is an important requirement for commodities that exhibit a dramatic increase in their heat content due to the respiration process. For example, with bananas, as the ripening process proceeds, the rate of heat removal increases five or six times that of the initial rate. The refrigeration system for such products must be properly designed to remove the extra heat generated within the room (Table 17.6).

To control the ripening process, the pulp temperature of the fruit is measured. A calibrated pulp temperature thermometer should be used to measure pulp temperature of fruits sampled from various locations within the ripening room.

Humidity in ripening rooms should be maintained at 85%–95% to keep moisture loss at a minimum. Humidifiers are often required to maintain the required humidity. On the other hand, the required high humidity in the ripening rooms has a negative effect on any moisture-sensitive packaging materials. Specifically, packages constructed of corrugated fiberboard may weaken due to high humidity. Therefore, caution should be exercised to avoid package collapse. A wet and dry bulb temperature psychrometer is useful to measure humidity in ripening rooms.

Table 17.6   Ripening Conditions for Selected Commodities

Commodity	Respiration Rate (mg CO_2/kg h)	Ethylene Concentration (ppm)	Ethylene Exposure Time (h)	Ripening Temperature (°C)	Storage Temperature (°C)
Banana	25–110	100–150	24	15–18	13–14
Honeydew melon	20–27	100–150	18–24	20–25	7–10
Mango	40–200	100–150	12–24	20–22	13–14
Orange degreening	22–34	100–150	24–72	20–22	5–9
Tomato	24–44	100–150	24–48	20–25	10–13

Source: Adapted from Kader, A.A., *Postharvest Technology of Horticultural Crops*, Publication 3311, Agriculture and Natural Resources, University of California, Davis, CA, 2002.

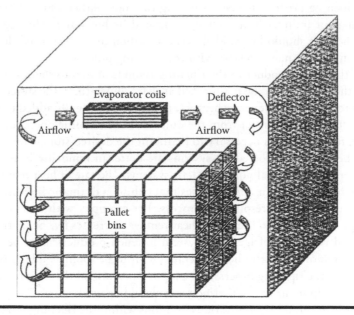

Figure 17.3   A typical degreening room used for citrus.

Airflow around pallet loads should not be impeded (Figure 17.3). Typically, a spacing of 10–15 cm between pallet loads ensures good air circulation. When evaporator fans are inadequate, additional fans may be necessary to distribute air in the ripening room in as uniform a manner as possible; in forced-air systems, the stacks are held tightly and the loads may be two or three pallets high.

While the requirements for airflow rates are low for distribution of ethylene or removal of carbon dioxide, higher airflow rates are necessary for removal of heat given off by the product due to the respiration process. With products such as bananas and avocados, an airflow rate of 0.3 m³/s kg is recommended. The pressure drop against which fans must operate to create this airflow rate is generally 0.6–1.9 cm of water column. This includes the resistance to airflow through evaporator coils and stacks. For low respiring products such as oranges, the recommended airflow rate is 0.1–0.05 m³/s kg.

A variety of methods are used to introduce ethylene gas into the ripening rooms. Ethylene gas may be obtained by using a generator that produces ethylene from ethyl alcohol, or pure ethylene gas may be used directly out of pressurized cylinders. Another method is to use ethylene gas diluted in nitrogen. Since ethylene gas at concentrations of 2.8%–28.6% is explosive in air, the electrical systems used in the ripening room must be explosion-proof. Ethylene can also be harmful to certain fruits and vegetables. Some typical symptoms of ethylene injury are shown in Table 17.7.

Carbon dioxide gas, at a concentration of 0.5%, can retard the ripening process. Regular venting of ripening rooms is necessary to maintain low carbon dioxide levels. The U.S. Occupational Safety and Health Administration requires that human exposure of carbon dioxide be limited to 0.5% and for no more than 15 min at 1.5%.

To illustrate the specific procedures used in ripening rooms, we will discuss four products—apples, bananas, mangoes, and tomatoes.

Apples

If apples are destined for long-term storage, then they should be harvested prior to their climacteric increase in respiration rate. As fruits mature, several internal changes take place, such as conversion of starch to sugars, an increase in titratable acidity, and a decrease in firmness. Thus, fruits harvested early may contain high levels of starch and acidity compared to fruits harvested at a later stage. Late-harvested apples are not good for long-term storage because they are more susceptible to physiological disorders.

If the starch content of apples is high, then a ripening protocol will improve their eating quality; for most varieties, the fruit should be warmed to 20°C–25°C and exposed to 10 ppm of ethylene for 24 h at 90%–95% relative humidity. After the 24 h exposure, the fruit is kept for 5 days at high temperature and high humidity to allow starch to convert into sugars. A decrease in firmness during the ripening process should be a sign of immediate removal of the fruit from the ripening room and it should be moved to a cold storage and then cooled to 0°C.

Bananas

Bananas harvested at the full maturity state exhibit best eating quality when they ripe. The fruit may be harvested at full maturity level if a controlled atmosphere is

Table 17.7 Symptoms of Ethylene Injury Observed in Vegetables

Commodity	Symptoms of Ethylene Injury
Asparagus	Increased lignification (toughness) of spears
Beans, snap	Loss of green color
Broccoli	Yellowing, abscission of florets
Cabbage	Yellowing, abscission of leaves
Carrots	Development of bitter flavor
Cauliflower	Abscission and yellowing of leaves
Cucumber	Yellowing and softening
Eggplant	Calyx abscission, browning of pulp and seeds, accelerated decay
Leafy vegetables	Loss of green color
Lettuce	Russet spotting
Parsnip	Development of bitter flavor
Potato	Sprouting
Sweet potato	Brown flesh discoloration and off-flavor detectable when cooked
Turnip	Increased lignification (toughness)
Watermelon	Reduced firmness, flesh tissue maceration resulting in thinner rind, poor flavor

Source: Adapted from Kader, A.A., *Postharvest Technology of Horticultural Crops*, Publication 3311, Agriculture and Natural Resources, University of California, Davis, CA, 2002.

used immediately after harvest, to delay the ripening process. However, bananas are very susceptible to chilling injury, and exposure to temperatures below 13°C for a few hours to a few days—depending upon the maturity and the type of the cultivar—can cause injury. Chilling injury is envisioned by surface discoloration, dull color of skin, failure to ripen, and browning of flesh. Therefore, it is generally recommended that the transport of bananas be carried out at 15°C ± 1°C in a controlled atmosphere (2%–5% oxygen and 2%–5% carbon dioxide).

Only when the fruit gets closer to the market is a ripening protocol used. For ripening, the fruit temperature is maintained at 14°C–18°C at a relative humidity of

90%–95%, with an ethylene concentration of 100–150 ppm for an exposure of 24–48 h. Sufficient air exchange is necessary to keep carbon dioxide concentrations below 1%. The rate of ripening is manipulated by varying the storage temperature.

As the fruit ripens, there is a three- to fourfold increase in carbon dioxide production, from 20 to 60–80 mL CO_2/kg h, whereas ethylene production increases from 0.1 to 2.4 µL/kg h.

Other changes during ripening of bananas include change of skin color from green to yellow, ease of peeling, increase in pulp-to-peel ratio, conversion of starch to sugar, and softening of the pulp.

Table 17.8 shows the changes for one lot of bananas as it goes through seven ripening stages. Different cultivars may show significantly different values from those shown in this table. However, the trends should be similar.

Mangoes

Mangoes exhibit several changes during ripening, such as a change in skin color from green to yellow. Chilling injury caused when the temperature of mature green mangoes drops below 12.8°C (in case of ripe mangoes below 10°C) has a major deteriorative effect on mangoes. Injury includes uneven ripening, poor color, and flavor development, higher susceptibility to decay, surface pitting, and browning of flesh.

The recommended ripening protocol for mangoes is temperatures of 20°C–22°C, relative humidity of 90%–95%, ethylene concentration of 100 ppm for 12–24 h, and carbon dioxide concentrations below 1%.

When treating a ripening room with ethylene, it is necessary to have periodic ventilation to avoid buildup of carbon dioxide.

Tomatoes

The main quality attributes of tomatoes are color, firmness, texture, and flavor. The flavor of tomatoes is influenced by the sugar and acid contents. A well-flavored tomato contains high sugar and relatively high acid contents. The maturity and ripeness of tomatoes at harvest may vary considerably, as shown in Table 17.9. As noted in the table, there are four stages of mature green. The breaker and more mature stages are also called vine-ripe tomatoes.

To prevent chilling injury, tomatoes should not be exposed to temperatures below 10°C. The effects of chilling injury include lack of uniform ripening, decreased flavor, softness, and mealiness when ripened, and increased decay.

Table 17.10 gives the number of days to reach full red color at a specified storage temperature. For example, if a breaker tomato is stored at 15°C, it will turn red in 2 weeks and would be good for at least one additional week of shelf life.

Table 17.8 Typical Strategies in Ripening in Bananas

Ripeness Stage	Skin color ("a" Value)	Pulp Firmness (lbf, 8 mm tip)	Soluble Solids (%)	Titratable Acidity (%)	pH	Respiration Rate (mL CO_2/kg h)	Ethylene Production Rate (µL/kg h)
1	−17 to −29	9–11	2.0–3.1	0.28–0.31	4.8–5.1	18.7–22.6	0.29–0.41
2	−14 to −28	8–11	2.2–3.3	0.28–0.32	4.8–5.1	14.9–25.7	0.29–0.44
3	−12 to −18	2–3	1.0–14.3	0.46–0.50	4.4–4.9	63.6–69.9	1.47–1.71
4	−4 to −9	1–2.5	18.9–20.5	0.36–0.40	4.5–4.7	56.0–62.2	1.28–1.82
5	−2 to −6	1.5–1.8	19.6–22.4	0.35–0.42	4.5–4.7	53.4–68.1	1.78–3.00
6	−0.5 to −3	1.4–2	19.4–20.2	0.35–0.41	4.3–4.5	70.1–80.6	0.68–68.9
7	−2 to 0.4	0.8–1.2	19.4–20.9	0.32–0.34	4.7–4.8	63.6–68.9	6.20–8.92

Source: Adapted from Kader, A.A., *Postharvest Technology of Horticultural Crops,* Publication 3311, Agriculture and Natural Resources, University of California, Davis, CA, 2002.

Table 17.9   The Ripening Process of Tomatoes

Class	Description
Mature green 1	Seeds cut by a sharp knife on slicing the fruit; no jellylike material in any of the locules; fruit is more than 10 days at 20°C from breaker stage
Mature green 2	Seeds fully developed and not cut on slicing fruit; jellylike material in at least one locule; fruit is 6–10 days at 20°C from breaker stage; minimum harvest maturity
Mature green 3	Jellylike material well developed in locules but still completely green; fruit is 2–5 days at 20°C from breaker stage
Mature green 4	Internal red coloration at the blossom end, but no external color change; fruit is 1–2 days at 20°C from breaker stage
Breaker	First external pink, red, or yellow color at the blossom end (USDA Color Stage 2)
Turning	More than 10% but not more than 30% of the surface, in the aggregate, shows a definite change in color from green to tannish yellow, pink, red, or a combination thereof (USDA Color Stage 3)
Pink	More than 30% but not more than 60% of the surface, in the aggregate, shows pink or red color (USDA Color Stage 4)
Light red	More than 60% of the surface, in the aggregate, shows pinkish red or red, but less than 90% of the surface shows red color (USDA Color Stage 5)
Red	More than 90% of the surface, in the aggregate, shows red color (USDA Color Stage 6)

Source: Adapted from Kader, A.A., *Postharvest Technology of Horticultural Crops*, Publication 3311, Agriculture and Natural Resources, University of California, Davis, CA, 2002.

Ethylene treatment is not effective in ripening partially ripe tomatoes, although it is useful in initiating ripening of mature green tomatoes. The protocol for ethylene treatment involves an ethylene concentration of 100 ppm, at a temperature of 18°C–20°C, humidity greater than 90%, and proper airflow. The response to ethylene treatment should occur within 3–3½ days to reach the breaker stage.

Table 17.10   Days to Full Red Color in Tomatoes after Harvest at Different Ripeness Stages

Days to Full Red Color at Indicated Temperatures						
Ripeness stage	12.5°C	15°C	17.5°C	20°C	22.5°C	25°C
Mature green	18	15	12	10	8	7
Breaker	16	13	10	8	6	5
Turning	13	10	8	6	4	3
Pink	10	8	6	4	3	2

Source: Adapted from Kader, A.A., *Postharvest Technology of Horticultural Crops*, Publication 3311, Agriculture and Natural Resources, University of California, Davis, CA, 2002.

Cooling of Commodities

The respiration process comprises a series of complex steps. As noted earlier, during respiration starch and sugars are converted into carbon dioxide and water. Intermediate products include organic acids. The process is influenced by the available oxygen. Oxygen concentration within a product is usually very similar to that of the surrounding atmosphere. If the surrounding atmosphere is deficient in oxygen, then anaerobic respiration occurs. The products of anaerobic respiration are ketones, aldehydes, and alcohols that cause death of a tissue.

Heat is evolved during respiration, causing the product temperature to increase. Furthermore, the respiration rate is influenced by the product temperature. Since higher respiration activity causes rapid physiological degradation, control of the respiration process by managing product temperature is a key to postharvest management.

Loss of water is irreversible in fruits and vegetables except in flowers. A loss of 3%–5% initial weight due to water loss shows wilting and shriveling of the product. Weight loss in fruits can be quite significant due to the loss of water as seen in Table 17.11. Water loss is due to gradient in vapor pressure. The loss of water is largely from the surface of a product, but it may also occur from the stem end and from any injury at the surface.

Typically, the inside of a fruit or vegetable is near saturation of water. The vapor pressure in the surrounding air depends on the specific humidity of the air. The vapor pressure gradient between the product surface and the surrounding air becomes the driving force for moisture loss from the product into the air. By quickly lowering the temperature of the product to its recommended

Table 17.11 Weight Loss in Fruits and Vegetables due to Moisture Loss

Commodity	Maximum Weight Loss (%)	Reason for Loss
Spinach	3	Wilting
Tomato	4	Shriveling
Leaf lettuce	3–5	Wilting decaying
Crape	5	Berry shriveling
Pear	6	Shriveling
Cabbage	6	Shriveling
Apple	7	Shriveling
Carrot	8	Wilting
Green pepper	8	Shriveling
Peach	11	Shriveling

Source: Adapted from Thompson, F.F. et al., *Commercial Cooling of Fruits, Vegetables and Flowers*, DANR Publication 21567, University of California, Davis, CA, 1998.

storage temperature, immediately after harvest, the vapor pressure of the product is lowered. Furthermore, the storage room air humidity is maintained at a high level, 90%–95%, minimizing the vapor pressure gradient between the product and the surrounding air. The rate of moisture loss for different commodities has been determined and expressed as transpiration coefficient as shown in Table 17.12.

The vapor pressure of air is calculated by the following formula:

$$P_v = \frac{H \times P_a}{0.622} \tag{17.1}$$

where
P_v is the vapor pressure (pascal)
H is the specific humidity (kg water/kg dry air)
P_a is the atmospheric pressure (pascal)
0.622 is the molecular weight of water divided by the molecular weight of air

Table 17.12   Transpiration Coefficient of Selected Commodities

Commodity	Transpiration Coefficient (mg/kg s MPa)	Range of Coefficients Reported in Literature
Apple	42	16–100
Potato	44	2–171
Onion	60	13–123
Orange	117	25–227
Grape	123	21–254
Tomato	140	71–365
Cabbage	223	40–667
Peach	572	142–3250
Carrot	1207	106–3250
Lettuce	7400	680–8750

Source: Adapted from Thompson, F.F. et al., *Commercial Cooling of Fruits, Vegetables and Flowers*, DANR Publication 21567, University of California, Davis, CA, 1998.

Let us calculate the weight loss in cabbage held at 0°C and exposed to air at 0°C with 90% relative humidity. The humidity ratio at 0°C at 100% relative humidity is 0.00380 kg water/kg dry air; therefore, the vapor pressure for cabbage is calculated using Equation 17.1:

$$P_{v,cabbage} = \frac{0.00380 \times 0.101}{0.622} = 0.000617 \text{ MPa}$$

and humidity ratio at 0°C and 90% relative humidity is 0.00340 kg water/kg dry air; therefore, the vapor pressure for air is

$$P_{v,air} = \frac{0.00340 \times 0.101}{0.622} = 0.000552 \text{ MPa}$$

Since the transpiration coefficient for cabbage from Table 17.12 is 223 mg/kg s MPa, we obtain the moisture loss rate as

Moisture loss rate = (0.000617 − 0.000552) MPa × 223 mg/kg s MPa

= 0.0145 mg/kg s

Therefore, the weight loss of cabbage, due to loss of water, is 0.0145 mg/kg s or 0.125% per day. Packing in a package such as a polymeric film with low water vapor permeability should assist in reducing the water loss from cabbage.

Table 17.13 Methods Used for Cooling Fruits and Vegetables

Cooling prior to storage or transport	Room cooling
	Forced-air cooling
	Hydrocooling
	Package icing
	Vacuum cooling
Cooling during transport	Top icing
	Channel icing
	Mechanical refrigeration

Source: Adapted from Kader, A.A., *Postharvest Technology of Horticultural Crops*, Publication 3311, Agriculture and Natural Resources, University of California, Davis, CA, 2002.

The activity of microorganisms is retarded at lower temperatures. A common cause of degradation, *Rhizopus* rot, does not grow below 5°C and most organisms grow at a very slow rate near 0°C.

The time between harvest and cooling is critical in some commodities. For example, strawberries deteriorate fast unless they are cooled within 1 h of harvest and sweet cherries should be cooled within about 4 h. Caution must be exercised if the cooled commodity needs warming, because this may cause surface moisture to form and result in cracking of surface layers.

The different methods used in cooling of fruits and vegetables are shown in Table 17.13. Various factors determine the cooling method for a given commodity. Some of these factors are as follows:

- Type of product, such as leafy, stem, root crop
- Product size
- Quantity of product to be cooled
- Package size and shape
- Type of packaging material
- Susceptibility to physical damage sustained by product
- Costs of operation, both capital and operating

Forced-Air Cooling

The use of cold air for cooling fruits and vegetables is one of the most common methods used in commercial operations. Here, air is first cooled by forcing it over evaporator coils of a refrigeration system; as the air is cooled it contacts the evaporator coils, because heat is transferred from air to the refrigerant flowing inside the coils.

The cooled air is then brought into contact with fruits and vegetables that are usually packed in containers with openings in the walls. As cold air is forced through the container with the product, heat transfer occurs between product and the air. While the product temperature decreases, the temperature of air increases. The warm air leaving the product is directed back toward the evaporator coils to be recooled.

The three most commonly used air cooling systems are tunnel cooling, serpentine cooling, and cold wall cooling.

Tunnel Cooling

In tunnel cooling system, rows of pallet boxes are set up in a cold storage room so that there is an empty channel, or tunnel, between them. As shown in Figure 17.4, typically, the two rows of loads are stacked on each side of the tunnel. The two ends and the top of the tunnel are covered with a tarp that guides the airflow through the boxes. A portable exhaust fan is operated to draw the air through the boxed product. The air is then directed toward an evaporator coil of a refrigeration system to be recooled and used again. The tunnel covers are operated as a batch system. The fan speed may be controlled manually reducing it toward the end of the cooling cycle in order to conserve energy.

Serpentine Cooling

In a serpentine cooling system, the product is typically packed in pallet-size bins with air vents in the bottom. When pallet bins are placed on each other, the tines, used by the fork lift to connect with a bin, help create an open channel for airflow, as shown in Figure 17.5. An even number of pallet bins are stacked along a wall.

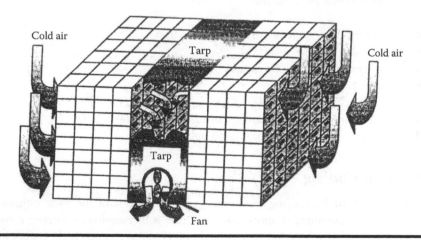

Figure 17.4   An air-blast cooler.

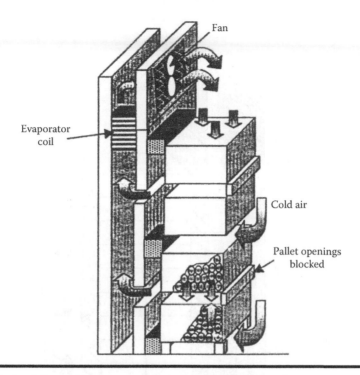

Figure 17.5   Schematic representation of a serpentine cooler.

Each alternate opening between the pallet bins is closed using tarp tape. The air is then guided through the product in a serpentine manner.

Typically, the bins are placed only 10–15 cm from the wall, so that there are sufficient air openings. Air speeds of 5–7.5 m/s are used through the bin openings. The airflow rate is typically 0.25–0.5 L/s/kg, with a cooling time of 10–12 h.

Cold Wall Cooler

In a cold wall cooler—commonly used for smaller lots of product—pallet loads are placed against a specially constructed wall, as shown in Figure 17.6. When a pallet load is placed against this wall, a lever is pushed to open a damper that allows air to be drawn into the plenum. This damper opening system may be arranged in such a manner that it opens only when it is actuated by the bin. Thus, pallets of different heights may be stacked along the wall and only the required number of dampers may be opened. Similarly, shelves may be used to place smaller boxes along the cold wall. After cooling, individual bins may be removed and replaced with new products. The cost of manual labor, shelves, and damper systems is higher than the other systems. However, there is a greater flexibility in the number of loads that may be cooled at a given time. Visual controls may be installed to alert the operator as to when a load should be removed.

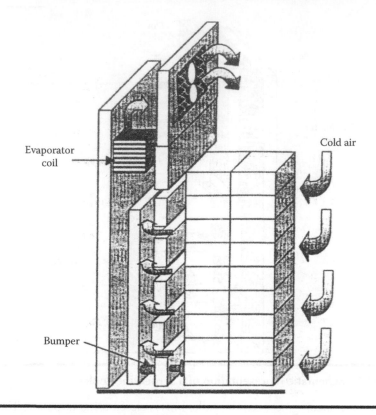

Cold air

Evaporator
coil

Bumper

Figure 17.6 Schematic representation of a cold wall cooler.

Cooling Rate in Forced-Air Cooling

As is evident from the illustrations shown in this section, in the forced-air cooling systems the width of a stack will influence the total distance the cold air must travel through the boxed product. If the stack is wider, then airflow must be appropriately increased or there will be a noticeable difference between the warmest and coldest product in the load at the end of the cooling period. Normally, these coolers are designed for a thickness of one pallet load (1.2 m).

The most common types of fans used in the forced-air systems are either axial or centrifugal. With the axial models, propeller fans are typically used, while with the centrifugal type, a squirrel cage is most common. Axial fans are well suited for low pressure drops of up to 4 cm of water column. If the pressure drops are greater, then centrifugal fans are preferred. The centrifugal fans are quieter to operate than the axial type.

In commercial practice, it is often necessary to control the rate of airflow. Often at the end of cooling, a reduced flow rate may be desired. Similarly, when the types of products are changed, different cooling rates may be necessary. A common way to obtain different airflow rates is to use a variable frequency motor to operate the fan. Alternatively, fans in series or parallels may be used.

The pressure drop for a forced-air cooling system depends on several factors such as the vent opening area in the side walls of a box, the alignment of vent openings between neighboring boxes, the number of boxes stacked through which air must pass through before entering the plenum, any packaging materials that may be used inside the boxes (e.g., individual fruit wrappings) that may obstruct airflow, pressure drop across the evaporator coils, pressure drop across the fan, and resistance to airflow in air ducts. The openings in the container side walls must be at least 14%. With corrugated fiberboard containers, up to 5% venting does not decrease their stacking strength. In practice, a cooling fan must be able to deliver the required airflow rate against at least 5 cm of water column.

Loss of moisture from a product being cooled has an important economic consequence. Some of the factors that influence moisture loss are initial product temperature, transpiration coefficient, wax coating or moisture-resistant packaging, cooling speed, and humidity of cooling air.

Moisture loss is greater in products that have higher initial product temperatures, whereas products that have high transpiration coefficients exhibit high moisture. For example, oranges have very low moisture loss, while carrots may lose 0.6%–1.8% of their initial weight. Moisture loss can be reduced by packaging and the use of high airflow rates that reduce cooling time.

If the cooling air humidity remains between 80% and 100%, then there is no major effect on moisture loss. Using air containing higher humidity will adversely influence the fiberboard container. The airflow should be stopped soon after the product has cooled to the desired temperature.

The dimensions of the air channels are important. If the air supply and return channels are designed to be too narrow, then air speed will be excessive and uneven. Air speeds should be kept below 8 m/s.

Hydrocooling

Cold water is widely used to cool fruits and vegetables. In commercial practice, hydrocooling is found to be most cost-effective in cooling fruits, root crops, and stem vegetables. Although leafy vegetables can be hydrocooled, they are more commonly vacuum-cooled. Hydrocooling is not recommended for grapes and berry fruits. To effectively transfer heat, water must contact as much of the product surface as possible, must flow with a certain velocity over the product surface, and must be cold and free of microorganisms that may promote decay.

There are three types of hydrocoolers as follows:

1. Batch shower system
2. Continuous shower system
3. Water immersion cooling system

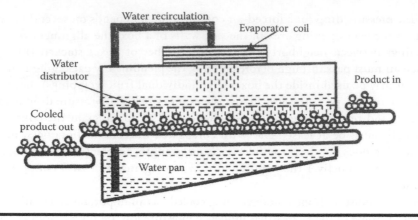

Figure 17.7   Schematic representation of a hydrocooler.

In a batch shower system, water cooled by contact with evaporator coils is conveyed to a distribution pan (Figure 17.7). The pan has small openings that create a uniform shower underneath. The product, in packages, is positioned under the shower. After contacting with the product, warm water collects in the water reservoir from which it is pumped to the top and recooled by contact with the evaporator coils. Water flow rate is maintained around at 480–490 L/min/m² for double-parallel bin depth.

In the continuous shower system (Figure 17.8), the product is moved on a conveyor belt. The shower system is created in a manner similar to the batch shower system. The speed of the conveyor is adjusted to cool the product to the desired temperature. In this system, damage from the impact of water on product should be minimized. The distance between the shower and product should be less than 15–20 cm.

In an immersion system, the conveyor belt contains cleats to aid product movement and move the product through a water bath (Figure 17.9). Products suitable for immersion cooling are those that have a density higher than water, such as cherries. Water must be pumped to create water speeds of 0.076 m/s past peaches. Water temperatures should be 0°C–0.5°C even for chilling-sensitive products, as long as the cooling is stopped when the required temperature is achieved.

The containers are vented at the top and bottom. Correct alignment of vent holes is important. The material used for packaging must be water-resistant. Wax-dipped corrugated fiberboard is commonly used, although use of waxed packaging materials is more costly and there are problems in recycling waxed fiberboard. Clean, potable well, or domestic supply water should be used at temperatures near 0°C with 100–150 ppm chlorine. The water in the cooler must be changed at least once daily.

Chlorination is achieved by adding either chlorine gas, sodium hypochlorite, or calcium hypochlorite. Chlorine is measured as hypochlorous acid (HOCl). The pH of the water must be between 6.5 and 7.5 for chlorine to be effective.

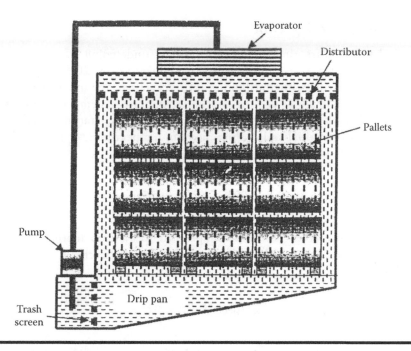

Figure 17.8 Schematic representation of a batch shower cooler.

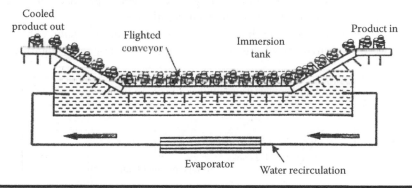

Figure 17.9 Schematic representation of an immersion water cooler.

Vacuum Cooling

Products with high surface-to-mass ratio, such as leafy vegetables, are most suited for vacuum cooling. The product is placed in a large cylindrical container and vacuum is drawn. As water evaporates from the product surface, it extracts heat from the product and cools it. If there is too much moisture loss, the use of a water spray prior to subjecting the product to vacuum is one method to counteract it.

Calculating Cooling Time

A typical temperature–time curve for a product undergoing cooling is shown in Figure 17.10. Here, at a uniform initial temperature of 20°C, a product is being cooled with a cooling medium at 0°C. As shown, the initial temperature drop is rapid, although as cooling continues, the rate of temperature decrease diminishes. It took 3 h for the product to decrease in temperature by half of the difference between initial and cooling medium temperatures, or reach 10°C, as shown in the following:

$$\frac{1}{2} \times (20°C - 0°C) = 10°C$$

Similarly, it took another 3 h for the temperature to undergo another half of the difference between new initial and final temperatures, or reach 5°C, as shown in the following:

$$\frac{1}{2} \times (10°C - 0°C) = 5°C$$

Thus, in this example, it takes 3 h for each half cooling time. Three half cooling times are equivalent to seven-eighth cooling. In commercial practice, products are often placed in coolers until they reach seven-eighths cooling.

Thus, seven-eighth cooking time, $t_{7/8}$, is the time required to reduce the difference between the initial and cooling medium temperatures by seven-eighths.

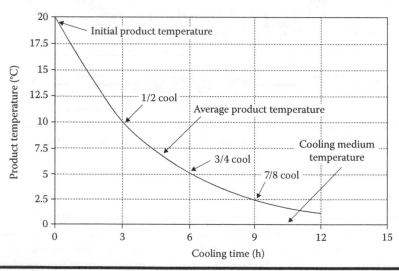

Figure 17.10 A cooling curve with product at an initial temperature of 20°C and cooling medium at 0°C.

Therefore, temperature after seven-eighth cooling is

$$\text{Temperature}_{7/8\,\text{cooling}} = T_{\text{m}} - \frac{7}{8} \times (T_0 - T_{\text{m}}) \qquad (17.2)$$

Table 17.14 includes some coefficients that may be used to estimate the expected seven-eighth cooling time. For example, if after 60 min in a cooler, the temperature of the product has dropped from 30°C to 16°C, and the cooling air temperature is held constant at 2°C, the product is half cooled, since

$$\frac{T_0 - T_{1/2}}{T_0 - T_{\text{m}}} = \frac{30°C - 16°C}{30°C - 2°C} = \frac{14}{28} = \frac{1}{2} \qquad (17.3)$$

where
T_0 is the initial product temperature
$T_{1/2}$ is the product temperature when half cooled
T_{m} is the temperature of the cooling medium

From Table 17.14, to determine the seven-eighth time, we multiply the half-cooled time by three. Therefore, it will take 180 min to reach seven-eighth cooling. After 180 min, the temperature of the product will be

$$\left(30°C - \frac{7}{8} \times (30°C - 2°C) \right) = 5.5°C$$

Table 17.14   Conversion Factors to Determine Seven-Eighth Cooling Time

$\dfrac{T_0 - T}{T_0 - T_{\text{m}}}$	$\dfrac{1}{2}$	$\dfrac{3}{4}$	$\dfrac{7}{8}$	$\dfrac{15}{16}$	$\dfrac{31}{32}$
Multiplication factor to determine seven-eighth cooling time	3	$\dfrac{3}{2}$	1	$\dfrac{3}{4}$	$\dfrac{3}{5}$

Source: Adapted from Kader, A.A., *Postharvest Technology of Horticultural Crops*, Publication 3311, Agriculture and Natural Resources, University of California, Davis, CA, 2002.

Notes: T_0 is the initial temperature, T_{m} is the cooling medium temperature, and T is the temperature at a given time.

The earlier calculations assume that the cooling medium temperature remains constant during the entire cooling period and the product temperature is uniform anywhere in the load. This assumption may not be valid in an air-blast cooler, where poor air distribution may cause nonuniformity in product temperature. However, in a hydrocooler, the product temperature will be more uniform as the water moves more evenly through the product load.

To determine the instantaneous cooling rate at any given time during the cooling process, the following formula may be used:

$$\text{Instantaneous cooling rate} = \frac{2.08 \times (T - T_m)}{t_{7/8}} \qquad (17.4)$$

where
$t_{7/8}$ is the seven-eighth cooling time
T is the instantaneous product temperature
T_m is the cooling medium temperature

Packaging of Fruits and Vegetables

Proper packaging is essential for maximizing the shelf life of fruits and vegetables. The different gas and water vapor permeabilities of polymeric films offer a wide range of opportunities for applications in packaging. Common examples are low-density polyethylene (LDPE), polystyrene (PS), polyvinyl chloride (PVC), polyvinylidene chloride (PVDC), and ethyl vinyl alcohol (EVOH).

Packaging materials are useful in various functions. For example, a package helps to contain a specific quantity of a food product. It provides a certain level of protection to the product from environmental factors such as light, oxygen, humidity, and temperature. The package also plays an important role in communicating information regarding the packaged product to the consumer such as quantity, recommended storage conditions, expected shelf life, instructions on how to prepare for consumption, and nutritional information.

Many requirements must be met in the design of packaging for foods. A food package must be nontoxic, provide a clean environment for the product, and be lightweight to minimize its effects on transportation costs. The package should make it easy for the product to be displayed on a grocery store shelf. It should also be easy for the consumer to open and, when necessary, reclose. These selection criteria should be met without the package becoming an environmental pollutant.

In the past, wood containers have been most commonly used to get fresh produce from harvest to the wholesale or retail market. Wooden crates and burlap bags are still in widespread use in many developing countries. Although wood provides good mechanical strength when used as a packaging material, it is not desirable for several reasons:

■ The heavy weight of wood means higher transportation costs.
■ Wood causes damage to packaged product due to splinters and sharp walls.
■ A package made of wood, jute, or other textiles harbors mold and other spoilage microorganisms that can infect the packaged product.
■ The large quantity of wood required for packaging contributes to deforestation.

In many countries with advanced agricultural systems, wooden containers have been replaced by containers made of plastic or fiberboard.

We will now briefly discuss the major polymeric materials used in packaging fruits and vegetables.

Low-Density Polyethylene

LDPE has the simplest structure of any polymer, with a repeating ethylene unit in its molecule. At high pressures of more than 138 MPa and temperatures above 150°C, ethylene promotes one of the carbon–carbon bonds to break and form a free radical. These free radicals join together to form polyethylene. This process is called free-radical polymerization. After polymerization, pressure is reduced in stages. The polymer is isolated as solid particles, which are then melted and extruded through a die, followed by cutting the extruded strands into pellets for shipping and later processing.

LDPE has good moisture-barrier properties and is therefore suitable for maintaining high relative humidity within a package. If the polymerization reaction is carried out at lower pressures, high-density polyethylene (HDPE) is obtained. HDPE is relatively less branched and more crystalline than LDPE. Although HDPE does not have good transparency, it is stronger than LDPE and so is used in manufacturing bags and liners.

Polypropylene

Polypropylene (PP) is obtained from polymerization of monomer propylene, from which either an oriented PP (OPP) or a cast PP is obtained. It is more transparent, stiffer, and tougher than LDPE.

Polystyrene

PS is obtained from the polymerization of styrene, in which the ethylene group attaches to the benzene ring of the monomer unit. An oriented PS is suitable for its gas-barrier properties. On adding hexane during the polymerization process, PS foam is obtained. Although PS foam has poor gas- and moisture-barrier properties, EPS trays are widely used when the product requires a rigid bottom support and is overwrapped.

Polyvinyl Chloride

PVC is obtained from the polymerization of vinyl chloride carried out at low pressure. In order to soften the film, a large amount of plasticizers is added. It has low gas and moisture permeabilities and is often used as an overwrap. Table 17.15 shows the gas permeabilities of PVC, LDPE, and HDPE.

Table 17.15   Gas Permeation of Selected Polymeric Materials

Material	Gas	Permeability ($\times 10^{-12}$ m^2/s)	Activation Energy (J/mol)
Polyvinyl chloride	Nitrogen	0.0098	69,100
	Oxygen	0.0376	55,600
	Carbon dioxide	0.1302	56,900
High-density polyethylene	Nitrogen	0.119	39,700
	Oxygen	0.334	35,100
	Carbon dioxide	1.402	37,600
Low-density polyethylene	Nitrogen	0.8	41,400
	Oxygen	2.39	40,100
	Carbon dioxide	10.45	38,400
Water	Nitrogen	32.8	15,800
	Oxygen	71.6	15,800
	Carbon dioxide	1647	15,800
Air	Nitrogen	17,200,000	3600
	Oxygen	20,300,000	3600
	Carbon dioxide	15,700,000	3600

Ethyl Vinyl Alcohol

EVOH is a hydrolyzed copolymer of vinyl alcohol and ethylene and provides excellent gas- and moisture-barrier properties. Because the EVOH film is hygroscopic, its moisture-barrier properties diminish as it absorbs moisture. Consequently, it is more suitable as a laminate.

In selecting a plastic film for packaging fruits and vegetables, the following characteristics should be considered:

- Moisture-barrier properties
- Gas-barrier properties
- Mechanical strength
- Antifog properties
- Sealability and machinability
- Cost

To prevent shriveling and wilting of fruits and vegetables, a film should provide a good moisture barrier as well as prevent anaerobic respiration. A certain amount of oxygen and carbon dioxide transmission is necessary for creating modified atmospheres inside the package to enhance the product's shelf life.

A package provides physical protection of the fruit and vegetable. Therefore, it should have appropriate mechanical strength.

Controlled-Atmosphere Storage

A large number of fruits and vegetables benefit from storage under controlled atmospheric conditions. Reduced level of oxygen and increased concentration of carbon dioxide in the immediate environment surrounding a fruit or vegetable retards its respiration rate. With reduced respiration rate, the storage life of the product is enhanced. Considerable research has been done to determine the most suitable concentrations for oxygen and carbon dioxide that extend the storage life of fruits and vegetables. Table 17.16 is a compilation of recommended conditions. The controlled atmosphere technology is well developed, and for certain products like apples it is used worldwide. In the following section, we will examine how the atmospheric conditions may be modified in the immediate vicinity of a product through packaging.

Modified-Atmosphere Packaging

We will use information on the recommended conditions for oxygen and carbon dioxide gas concentrations in designing modified-atmosphere packages. According to these conditions, the modified atmosphere contains lower oxygen and higher

Table 17.16 Recommended Storage Conditions for Controlled Atmosphere of Selected Fruits and Vegetables

Common Name	Scientific Name	Storage Temperature, °C	Relative Humidity, %	Highest Freezing Temperature	Ethylene Production	Ethylene Sensitivity	Approximate Shelf Life	Beneficial Controlled Atmosphere
Apple, non-chilling-sensitive varieties		−1.1	90–95	−1.5	Vh	H	3–6 months	CA varies by cultivar
Apple, chilling-sensitive	Yellow Newtown, Grimes Golden McIntosh	4	90–95	−1.5	Vh	H	1–2 months	CA varies by cultivar
Apricot	*Prunus armeniaca*	−0.5–0.0	90–95	−1.1	M	H	1–3 weeks	2%–3% O_2 + 2%–3% CO_2
Artichokes Globe	*Cynara scolymus*	0	95–100	−1.2	VL	L	2–3 weeks	2%–3% O_2 + 3%–5% CO_2
Asparagus, green, white	*Asparagus officinalis*	2.5	95–100	−0.6	VL	M	2–3 weeks	5%–12% CO_2 in air
Avocado Cv Fuerta, Haas	*Persea americana*	3–7	85–90	−1.6	H	H	2–4 weeks	2%–5% O_2 + 3%–10% CO_2
Banana	*Musa paradisiaca var. sapientum*	13–15	90–95	−0.8	M	H	1–4 weeks	2%–5% O_2 + 2%–5% CO_2

Common name	Scientific name							
Beans, snap; wax, green	Phaseolus vulgaris	4-7	95	-0.7	L	M	7-10 days	2%-3% O_2 + 4%-7% CO_2
Lima beans	Phaseolus lunatus	5-6	95	-0.6	L	M	5-7 days	
Strawberry	Fragaria spp.	0	90-95	-0.8	L	L	7-10 days	5%-10% O_2 + 15%-20% CO_2
Cabbage	Brassica	0	95-100	-0.9	VL	H	2-3 months	1%-2% $O2$ + 0%-5% CO_2
Chinese; Napa	Campestris var. perkinensis							
Carrots, topped	Daucus carota	0	98-100	-1.4	VL	H	6-8 months	No CA benefit
Carrots, bunched	Daucus carota	0	98-100	-1.4	VL	H	10-14 days	Ethylene causes bitterness
Cauliflower	Brassica oleracea var. botrytis	0	95-98	-0.8	VL	H	3-4 weeks	2%-5% O_2 + 2%-5% CO_2
Cherimoya; custard apple	Annona cherimola	13	90-95	-2.2	H	H	2-4 weeks	3%-5% O_2 + 5%-10% CO_2
Citrus, lemon	Citrus limon	10-13	85-90	-1.4			1-6 months	5%-10% O_2 + 0%-10% CO_2

(continued)

Table 17.16 (Continued)   Recommended Storage Conditions for Controlled Atmosphere of Selected Fruits and Vegetables

Common Name	Scientific Name	Storage Temperature, °C	Relative Humidity, %	Highest Freezing Temperature	Ethylene Production	Ethylene Sensitivity	Approximate Shelf Life	Beneficial Controlled Atmosphere
Citrus, orange	Citrus sinensis California; dry	3–9	85–90	−0.8	VL	M	3–8 weeks	5%–10% O_2 + 0%–5% CO_2
	Florida; Humid	0–2	85–90	−0.8	VL	M	8–12 weeks	
Cucumber	Cucumis sativus	10–12	85–90	−0.5	L	H	10–14 days	3%–5% O_2 + 3%–5% CO_2
Eggplant	Solanum melongena	10–12	90–95	−0.8	L	M	1–2 weeks	3%–5% O_2 + 0% CO_2
Garlic	Allium sativum	0	65–70	−0.8	VL	L	6–7 months	0.5% O_2 + 5%–10% CO_2
Ginger	Zingiber officinale	13	65		VL	L	6 months	No CA benefit
Grape	Vitis vinifera	−0.5 to 0	90–95	−2.7	VL	L	2–8 weeks	2%–5% O_2 + 1%–3% CO_2
Guava	Psidium guajava	5–10	90		L	M	2–3 weeks	
Lettuce	Lactuca sativa	0	98–100	−0.2	VL	H	2–3 weeks	2%–5% O_2 + 0% CO_2
Loquat	Eriobotrya japonica	0	90	−1.9			3 weeks	

Lychee, Litchi	*Litchi chinensis*	1–2	90–95		M	M	3–5 weeks	3%–5% O_2 + 3%–5% CO_2
Mango	*Mangifera indica*	13	85–90	–1.4	M	M	2–3 weeks	3%–5% O_2 + 5%–10% CO_2
Melon, honey dew, orange flesh	*Cucurbits melo*	5–10	85–90	–1.1	M	H	3–4 weeks	3%–5% O_2 + 5%–10% CO_2
Mushrooms	0		90	–0.9	VL	M	7–14 days	3%–21% O_2 + 5%–15% CO_2
Okra		7–10	90–95	–1.8	L	M	7–10 days	Air + 4%–10% CO_2
Papaya	*Carica papaya*	7–13	85–90		H	H	1–3 weeks	2%–5% O_2 + 5%–8% CO_2
Peach	*Prunus persica*	–0.5 to 0	90–95	–0.9	H	L	2–4 weeks	1%–2% O_2 + 3%–5% CO_2
Pepper, bell	*Capsicum annuum*	7–10	95–98	–0.7	L	L	2–3 weeks	2%–5% O_2 + 2%–5% CO_2
Persimmon Fuyu	*Diospyros kaki*	7–10	95–98	–0.7	L	L	2–3 weeks	2%–5% O_2 + 2%–5% CO_2
Persimmon Hachiya	*D. kaki*	10	90–95	–2.2	L	H	1–3 months	
Pineapple	*Ananas comosus*	5	90–95	–2.2	L	H	2–3 months	
Pomegranate	*Punica granatum*	5	90–95	–3.0			2–3 months	3%–5% O_2 + 5%–10% CO_2

(continued)

Table 17.16 (Continued) Recommended Storage Conditions for Controlled Atmosphere of Selected Fruits and Vegetables

Common Name	Scientific Name	Storage Temperature, °C	Relative Humidity, %	Highest Freezing Temperature	Ethylene Production	Ethylene Sensitivity	Approximate Shelf Life	Beneficial Controlled Atmosphere
Potato early crop	Solanum tuberosum	10–15	90–95	−0.8	VL	M	10–14 days	
Potato late crop	S. tuberosum	4–12	95–98	−0.8	VI	M	5–10 months	
Spinach	Spinacia oleracea	0	95–100	−0.3	VL	H	10–14 days	5%–10% O_2 + 5%–10% CO_2
Tomato mature green	Lycopersicon esculentum	10–13	90–95	−0.5	VL	H	1–3 weeks	3%–5% O_2 + 2%–3% CO_2
Tomato firm ripe	L. esculentum	10	85–90	−0.5	H	L	7–10 days	3%–5% O_2 + 3%–5% CO_2
Watermelon	Citrullus udgaris	10–15	90	−0.4	VL	M	2–3 weeks	No CA benefit

Source: Adapted from Thompson, F.F. et al., *Commercial Cooling of Fruits, Vegetables and Flowers*, DANR Publication 21567, University of California, Davis, CA, 1998.

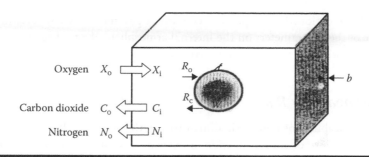

Figure 17.11 Schematic representation of a modified-atmosphere package.

carbon dioxide concentrations when compared with the ambient. Thus, concentration gradients are set up between the package atmosphere and the surrounding atmosphere. After a certain transient period during which the concentrations of these gases inside a package change, a final steady state can be reached. At steady state, the fluxes of carbon dioxide and oxygen, through the package wall, will be equal to the generation and consumption of these gases by the packaged commodity, respectively (Figure 17.11). This factor is kept in mind in the design of a package for modified-atmosphere applications.

Consider a commodity placed in a package constructed of a permeable film (Figure 17.11). The flux of two atmospheric gases, namely, oxygen and carbon dioxide, through the film may be written using the following expressions:

$$WR_x = P_x A \frac{(X_o - X_i)}{b} \tag{17.5}$$

$$WR_c = P_c A \frac{(C_i - C_o)}{b} \tag{17.6}$$

where
  W is the weight of the commodity, kg
  R_x is the rate of respiration of the commodity for oxygen, mL/kg h
  R_c is the rate of respiration rate of the commodity for carbon dioxide, mL/kg h
  P_x is the permeability of the package for oxygen, mL μm/m² h atm
  P_c is the permeability of the package for carbon dioxide, mL μm/m² h atm
  A is the area of the package, m²
  X_o is the oxygen concentration outside the package, %
  X_i is the oxygen concentration inside the package, %
  C_i is the carbon dioxide concentration inside the package, %
  C_o is the carbon dioxide concentration outside the package, %
  b is the thickness of the film, μm

The aforementioned two equations contain 11 parameters. We will examine the influence of these parameters on the internal atmosphere of a package by rearranging these equations in the following sections.

Gas Permeability Ratio

The gas permeability of polymeric films varies with each type. However, the ratio of the permeability of carbon dioxide to oxygen varies within a narrow range for most of the commonly used polymeric films. This ratio, denoted by β, is generally between 4 and 6. As a result of this narrow range, the suitability of many films in creating the required atmosphere becomes severely limited. We can see this limitation by dividing Equation 17.6 by Equation 17.5, yielding

$$C_i = C_o + \frac{1}{\beta}(X_o - X_i)\frac{R_c}{R_x} \tag{17.7}$$

The aforementioned is an equation of a straight line. We may draw this straight line on Figures 17.12 and 17.13 by using selected values for β. For example, in Figure 17.12, two straight lines are shown: one for a β value of 5 and another for a β value of 0.8. The first β value of 5 is a case for a LDPE film and a β value of 0.8 is for air. Note that any point on these lines represents the combined carbon dioxide

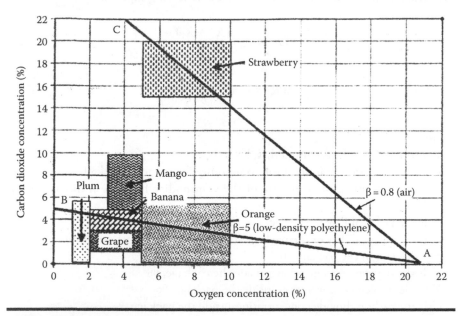

Figure 17.12 Recommended conditions of modified atmosphere for storage of vegetables.

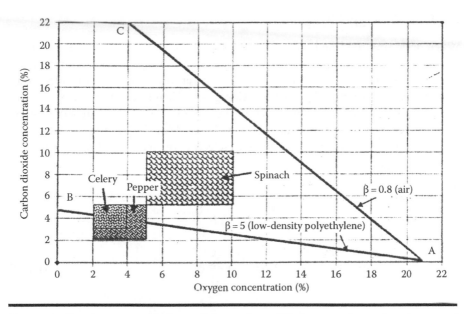

Figure 17.13 Recommended conditions of modified atmosphere for storage of fruits.

and oxygen concentrations that may be obtained inside a package. From Figures 17.12 and 17.13, it can be understood that LDPE is suitable for packaging products such as plum, orange, banana, pepper, and celery. All other products whose recommended atmospheric conditions lie outside the range of the straight line for LDPE will not benefit from using this polymeric film.

Respiratory Quotient

The ratio of carbon dioxide generation to oxygen consumption, shown as R_c/R_x in Equation 17.7, has an influence on the slope of the straight line A–B in Figures 17.12 and 17.13. This ratio is also called the *respiratory quotient* for the given commodity. The respiratory quotient is about 1 when the metabolic process uses carbohydrate as the primary substrate and unlimited excess of oxygen is available. On the other hand, if lipids are the primary source of oxidation, then the quotient is less than 1. This quotient is more than 1 if organic acids are the primary substrate of oxidation. Under anaerobic conditions, the respiratory quotient is less than 1, regardless of the substrate for oxidation. Using α for the respiratory quotient, we may rewrite the equation as

$$C_i = C_o + \frac{\alpha}{\beta}(X_o - X_i)$$

(17.8)

Package Design Variables

When designing modified-atmosphere packages, the commodity to be packaged defines the rate of oxygen generation, R_x, and the rate of carbon dioxide consumption, R_c. Furthermore, the commodity determines the recommended concentrations of oxygen and carbon dioxide gases. Therefore, the ratio of the gas concentrations, β, is fixed depending on the commodity. The calculated value of β helps select the type of polymeric film to be used for the modified-atmosphere package. We can illustrate the criterion for selecting the package film, by introducing a new factor, ϕ, such that

$$X_i = X_0 - \frac{R_x}{P_x} \phi \tag{17.9}$$

$$C_i = C_0 + \frac{R_c}{P_c} \phi \tag{17.10}$$

where

$$\phi = \frac{Wb}{A} \tag{17.11}$$

The value of factor ϕ increases or decreases depending upon the chosen values of W, b, or A. An increasing ϕ moves the design point toward B on line A–B. To increase ϕ either the weight of the commodity, W, or the thickness of the film, b, is increased or the area of the package, A, is decreased.

Variable Respiration Rates

The respiration rates of packaged commodities change as the atmosphere is modified. For example, as the carbon dioxide content in the package increases, the respiration rate of the commodity decreases. Similarly, an increase in the oxygen concentration in the package increases the rate of respiration. Thus, if the point on the A–B line moves toward B, the corresponding respiration rate increases. The consequence of this change in respiration rate is to moderate the effect of ϕ.

Effect of Temperature

Temperature influences the permeability of film as well as the respiration rate of the packaged commodity. The Arrhenius equation is commonly used to express these effects. We may write the Arrhenius equation to describe the temperature effect on permeability and respiration as follows:

$$P = P_0 \exp\left(-\frac{E_{aP}}{RT} \right) \tag{17.12}$$

$$R = R_0 \exp\left(-\frac{E_{aR}}{RT} \right) \tag{17.13}$$

To examine the effect of temperature changes on package atmosphere between two temperatures T_1 and T_{ref}, we may combine Equations 17.12 and 17.13 as follows:

$$\frac{R_x}{P_x} = \frac{R_{x,ref}}{P_{x,ref}} \exp\left[\frac{E_{aR} - E_{aP}}{R} \left(\frac{1}{T_1} - \frac{1}{T_{ref}} \right) \right] \tag{17.14}$$

The difference in activation energies of respiration and permeation determines the change in package atmosphere. The activation energies of permeability for some films are given in Table 17.15 and activation energies of permeability for respiration rates of selected fruits and vegetables are listed in Table 17.17. Temperature changes have minimal effect on the package atmosphere if the activation energies of respiration and permeation are similar in value.

Table 17.17   Respiration Rates of Broccoli, Cabbage, and Green Beans at Different Storage Temperatures

Produce	Atmospheric Composition	Respiration Rate (mL/kg h)				Activation Energy (J/mol)
		0°C	*5°C*	*10°C*	*20°C*	
Broccoli (Green Valiant)	Air	10	21	85	213	105,000
	1.5% oxygen, 10% carbon dioxide	7	11	15	33	50,950
Cabbage (Decema)	Air	1.5	—	4	10	63,150
	3% oxygen	1		3	6	59,800
Green beans (Blue Lake)	Air	—	17.5	28.3	59.6	54,900
	3% oxygen, 5% carbon dioxide	—	10.8	16.5	28	42,200

Source: Adapted from Kader, A.A., *Postharvest Technology of Horticultural Crops*, Publication 3311, Agriculture and Natural Resources, University of California, Davis, CA, 2002.

Water Condensation on Interior Package Walls

Water condensation on the inside of package walls is commonly observed in packages used for fresh fruits and vegetables. The effect of this layer of water on the permeability of carbon dioxide or oxygen through the package wall may be examined by reviewing the data of diffusion of these gases in water and polymers using the same units. As shown in Table 17.15, the values of permeability of gases in water are much higher than in polymeric films. This implies that thin layers of water, forming on the inside of the package film due to condensation, will have no appreciable effect on the internal package atmosphere.

Holes and Micropores in Package

Polymeric packages may be punctured during handling. Furthermore, special films with micropores are commercially used for packaging commodities. Gas flow through any hole in the polymeric film, created either accidentally or intentionally, is by a combination of convection and diffusion. The diffusive flow through a hole can be examined by noting the permeability of different gases in air. As seen in Table 17.15, gas permeability through air is several orders of magnitude higher than that in polymeric films. Therefore, presence of any holes in a film can seriously alter the permeability characteristics of the film. The line AG in Figure 17.12 represents a β value of 0.8, representing atmospheric air. Thus, a few holes created in an otherwise impermeable film can be made suitable for the recommended modified-atmosphere conditions for packaging strawberries.

Transient Atmospheric Conditions in Modified-Atmosphere Packages

Immediately after a product is placed inside a polymeric package and the package is sealed, gas concentrations within the package begin to change due to the respiration of the product and selective permeability of the film to different gases. Because the time required to reach a steady-state value of gas concentration is dependent on the free volume inside the package—the volume not occupied by the product—free volume must be determined. The larger the free volume, the longer it will take for the package atmosphere to come to an equilibrium. Furthermore, if the initial concentration of gases is similar to the desired modified-atmosphere conditions, the modified atmosphere will be created more rapidly.

The rate of change for concentrations of the main gases in the package, namely, oxygen, carbon dioxide, and nitrogen, is expressed using the following three differential equations:

$$V \frac{dX_i}{dt} = \frac{P_x A}{b}(X_0 - X_i) - WR_x \tag{17.15}$$

$$V\frac{dC_i}{dt} = \frac{P_c A}{b}(C_0 - C_i) - WR_c \tag{17.16}$$

$$V\frac{dN_i}{dt} = \frac{P_N A}{b}(N_0 - N_i) \tag{17.17}$$

The total pressure inside the package is governed by the ideal gas law; therefore,

$$P = (X_i + C_i + N_i)RT \tag{17.18}$$

In solving Equations 17.15 through 17.18, it is important to know how respiration rates are affected by gas concentrations. As the oxygen and carbon dioxide levels in the package change, so will the respiration rates. This requires the respiration rates to be expressed as functions of gas concentrations rather than constant values, as shown in Equations 17.15 and 17.16. This requirement makes the solution of the three differential equations more complicated. If constant rates of respiration are assumed, then these equations can be solved using a numerical technique, such as the predictor–corrector method or Runge–Kutta procedures.

As shown in Equation 17.18, the sum of the three gas concentrations, oxygen, carbon dioxide, and nitrogen, must remain the same, although individually, oxygen and carbon dioxide gas concentrations will change during the transient period. This implies that the pressure inside the package will change during this period, observed as a contraction of the package when the modified-atmosphere conditions are being set up.

Postharvest management of fruits and vegetables is a complex yet necessary task in minimizing losses of stored commodities. Fruits and vegetables remain biologically active after they are harvested from plants. Selection of proper storage factors such as temperature, humidity, gas atmosphere, packaging, and handling conditions requires careful attention. The physical and physiological breakdown of fruits and vegetables vary, depending upon the given species and cultivars. Scientific knowledge of the biochemical pathways leading to deteriorative products must be well understood, so that adequate practices are developed to minimize quality deterioration. Overall, the control of temperature and humidity remain by far the most important postharvest practices during storage and handling of fruits and vegetables. Many of the recommendations on postharvest handling and storage given in this chapter are an abridged version of a comprehensive treatment of this topic in Kader et al. (1985) and Thompson et al. (1998). The reader is referred to these two references for additional details.

Chapter 18

Food Preservation and Processing of Fruits and Vegetables

Foods are organic materials, mainly composed of water, carbohydrates, fats, and other organics such as proteins, vitamins, colors, flavors, etc., and some minerals as well in varied proportions. Humans and animals consume food either in a raw, a processed, or a formulated form for growth, health, and satisfaction.

All foods are subject to deterioration during storage. The rate of deterioration of untreated food may be very fast as in the case of fresh milk, fish, poultry, and meat that spoil within a day or two. It may be very slow in the case of cereals, pulses, and oilseeds, which can be stored safely for as long as 6–12 months or more depending on storage conditions. Fruits, vegetables, and root crops are semiperishables, as their storage life may vary from 1 to 3 weeks or more. Deterioration may occur in nutritional aspect, organoleptic, and other properties. Changes may also take place in color, flavor, texture, or in other quality attributes.

Major factors that influence the deterioration of foods are as follows:

- *Physical*: heat, cold, radiations, etc.
- *Chemical/biochemical*: oxygen, moisture, enzymes, and others
- *Biological*: micro- and macroorganisms

However, the most important causes of food deterioration are as follows: (a) microbial activities by bacteria, yeasts, and molds; (b) enzymatic; and (c) chemical and biochemical activities in food.

Moisture loss or gain, inappropriate storage temperature, storage time, oxygen, light, physical injuries, and others also influence the changes in quality of foods.

Most of the times, none of these parameters operate alone. As for example, the combined effects of moisture, heat, and air/oxygen are likely to accelerate the bacterial, enzymatic, and chemical activities simultaneously.

Therefore, the main purpose of processing of food is to protect food against all kinds of deterioration and preserve its quality as far as practicable.

Microorganisms and Other Organisms

Of the thousands of existing microorganisms, some cause spoilage and poisoning of food. Food-spoiling microorganisms widely spread in air, water, and soil, on poultry feathers, cattle skins, fruits, and vegetables—peels, grain hulls, nut shells, egg shells, processing machines, and almost everywhere unless they are sterilized. Hence, to reduce food spoilage first of all one has to ensure good sanitation conditions before and after processing.

Bacteria are unicellular organisms that can be broadly divided into three different shapes of individual cells, namely, spherical forms of cocci, the rod shaped bacilli, and spiral forms of spirilla and vibrios. Some of the bacteria, yeasts, and all molds produce spores that are very much resistant to heat and chemicals. Bacterial spores being most resistant, sterilization processes are to be designed to inactivate particularly these spores.

Generally, yeasts are spherical or ellipsoidal whereas molds are mostly complex in structure.

It is important to note that the bacteria and other microorganisms multiply at a tremendous rate under favorable environmental conditions.

Practically, bacteria, molds, and yeasts can attack all food materials and all of these favor warm and humid conditions. Each microorganism has (i) an optimum temperature for its best growth; (ii) a minimum temperature for its least or no growth; and (iii) a maximum temperature, above which all development is suppressed or stopped.

Mesophilic bacteria multiply fast at temperatures ranging from 16°C to 38°C; thermophilic bacteria can grow at a temperature as high as 80°C; and bacterial spores may survive for a long time at the boiling point of water. Only psychrophilic bacteria can grow at temperatures even below the freezing point of water.

Aerobic bacteria and all molds require oxygen for their growth whereas anaerobic bacteria can grow without any oxygen but facultative bacteria can grow under either aerobic or anaerobic condition.

Protozoa and phages are other food-spoiling microorganisms. Insects, mites, and rodents are also food-spoiling organisms.

Soil, water, air, and various organisms including animals and humans are the sources of food contaminations.

Heat and Cold

Leaving aside the effects of microorganisms, food may be deteriorated due to uncontrolled heat and cold. Heating above certain temperatures can destroy vitamins, denature proteins, and dehydrate the food materials. Generally, most of the foods are being handled within a temperature range of 10°C–38°C. Both chemical and biochemical reactions including enzymatic reactions are accelerated with the rise of temperature.

Foods are also damaged if exposed to low temperatures below certain levels. Some of the fruits and vegetables like tomatoes and bananas will undergo deterioration if stored at a refrigeration temperature of about 4°C as these are to be stored at about 10°C to retain their qualities. Deterioration of foods at lower temperatures is termed "chilling injury."

Moisture

Water in food is required for any microbial activity as well as chemical and biochemical reactions. These activities and reactions are always accelerated in the presence of high moisture leading to deterioration of food materials. Drying below certain level can affect texture as well as appearance of food. Presence of higher humidity over food can also cause spoilage of food owing to bacterial and mold growth.

Enzymes

Under natural conditions, the enzymes inherently present in most of the foods help in breaking down their higher molecules of proteins, lipids, carbohydrates, etc., into smaller molecules. In consequence, physical and chemical properties of food are altered. Some of the enzymes cause enzymatic browning. Some others are responsible for hydrolysis of fat into free fatty acids and for destruction of vitamins. Unless these enzymes are inactivated by addition of heat, chemicals, and exposure to radiations or by any other means, no food can be preserved in its natural state. Enzymes continue to catalyze chemical reactions within foods after harvest or slaughter. Some of these controlled catalytic reactions are desirable as in the case of continued ripening of tomatoes after harvest.

Chemicals/Oxygen

Various chemical and biochemical constituents of food materials may react with each other or with atmospheric oxygen resulting in destruction of vitamins A, C, and other nutrients with the change of color and flavor as well. Oxygen also favors mold growth on the surface of food, as molds are aerobic in nature.

Light

Vitamins A and C and riboflavin are destroyed along with the deterioration of color of food in presence of sunlight. Sensitive food materials are to be protected by using opaque packaging.

Storage Time

All microbial and enzymatic activities, chemical and biochemical reactions, and so also the effects of moisture, oxygen, heat, cold, and light can make progress with storage time. Proper processing, packaging, and storage conditions of food may prolong its life for a limited period only. That is why shelf-life dating of processed food has assumed so much importance. The shelf life of a food may be defined as the time taken by a product to come down to an unacceptable limit. The manufacturers are supposed to define a minimum acceptable quality (MAQ) for a food product before it is sold. The date of manufacture, packing/prepacking, and best before use by date, month, and year is to be given on the food product label.

Industrial Preservation Methods

Broadly, food preservation methods may come under inhibition, inactivation, and prevention from recontamination. Inhibition may include low temperature storage along with freezing, reduction of water activity, decrease of oxygen, increase of carbon dioxide, fermentation, and chemical addition with acidification, surface coating, and some others.

Inactivation follows the following technologies: sterilization/pasteurization, cooking, blanching, frying, pressure treatment radiations, etc.

Aseptic processing, packaging, hygienic storage, etc., may be called prevention technologies.

Though an ideal method of food preservation should prevent microbial, enzymatic, and chemical spoilage, till now no industrial method in use can take care of all the three main causes of food spoilage completely.

However, the common industrial methods of food preservation may be classified as follows:

- Preservation by addition of heat (canning, pasteurization)
- Preservation by removal of heat (cold storage, freezing)
- Preservation by removal of moisture (drying)
- Preservation by addition of chemicals (acid, salt, sugar addition)
- Preservation by fermentation
- Some special methods (application of high frequency current, microwave, dielectric, infrared radiations, and others)

Food Preservation by Thermal Treatment

Generally *higher temperatures* above the optimum are lethal to bacteria, yeasts, molds, and bacterial spores as well. Most of the bacteria are destroyed at a temperature between 80°C and 100°C. But many bacterial spores are not killed at 100°C, the boiling point of water for 30 min of heating. To ensure total destruction of microorganisms including spores, wet heat treatment at a temperature of 121°C is to be conducted for 15 min or longer. Hence, all microorganisms may be totally killed or their number can be diminished to any desired value by increasing the temperature of foods and holding these at the required high temperature for the desired length of time.

Most enzymes are also inactivated if these are subjected to about 85°C and above.

So heat treatment can effectively prevent both microbial and enzymatic spoilage. But at the same time, chemical reactions are accelerated at the elevated temperatures. To minimize chemical deteriorations, food materials should be heated just sufficiently in absence of oxygen so that microbial, enzymatic, and oxidative spoilage are prevented.

The well-established methods of food preservation by "canning" and "pasteurization" are based on heat treatment principle.

Preservation by Cooling

Low temperatures are not lethal to bacteria and other microorganisms. However at temperatures below 10°C, their growth is slow. Therefore, both their growth and multiplication are being retarded by lowering temperature and can be almost stopped if all the water in food were frozen. Psychrotrophs grow at temperatures 0°C and below. But all the water in some foods is not frozen even at –10°C because of the depression of freezing point owing to the presence of dissolved salts, sugars, and others.

Chemical and enzymatic reactions responsible for deterioration of color, flavor, vitamins, and nutrients are also retarded as the temperature is brought down.

Hence, retardation of microbial activity by removal of heat or cooling is the basic principle behind cold storage, refrigeration, and freezing preservation.

Preservation by Drying/Dehydration

Drying by removal of moisture is one of the oldest methods of food preservation.

Microorganisms cannot grow and multiply without sufficient water in their environment. During proper growth, microbes may contain more than 80% water. At any temperature, the higher humidity leads to higher population of microbes. Water requirement of microbes is actually related to water activity in the immediate environment whether in a state of solution or in a solid/particle state of food. Generally at room temperature, most bacteria need a water activity in the range of 0.9–1.00. But some yeasts and molds can grow slowly even at a low water activity of 0.65.

While drying of food water is also removed from microbial cells and in consequence their multiplication may be stopped. Partial drying may not be that effective compared to total drying. In general, bacteria and yeasts need more moisture than molds. Staling of bread and some semidried fruits are due to mold growth mainly.

In many enzymatic reactions being hydrolytic in nature, water is essentially required as a reactant. Moreover, all chemical reactions in food are retarded when the constituents tend to be in the solid state.

Preservation by Chemicals

Food preservation by chemicals is an ancient practice. Chemicals can preserve foods by controlling the growth of microorganisms. But chemical and enzymatic alterations may or may not be controlled by this method. Many chemicals may kill microbes or retard their growth but very few chemicals are permitted to add to some specific foods in prescribed low levels only. Some acids, sugars, salts, and preservative chemicals, namely, sorbic acid, sodium benzoate, sulfur dioxide, etc., are commonly permitted to add to food.

Microbes are sensitive to acids. Acids in certain concentrations may denature microbial proteins. Orange, tomato, apple, and citrus foods contain acids that act as natural preservative. Addition of phosphoric acid and citric acid to soft drinks is a common practice. But the degree of acidity allowed for any palatable foods is not sufficient enough to ensure sterility of food. Heat treatment in combination with acids is always more lethal to microbes.

It is a common practice to add sugar syrup to fruits and salt or brine solutions to vegetables and meat, poultry, and fish for their preservation. Living cells of microbes may contain more than 80% water. When microbes are present in some concentrated solutions of sugar or salt, water in microbial cells moves out of the cell through their membrane to the syrup or salt solution by osmosis. This dehydration known as plasmolysis of microbial cells retards the microbial growth.

Recently, sterilization of foods by treatment with microwaves, infrared radiation, dielectric radiation, high frequency radiation, and other special methods has been employed with a certain degree of success.

Food Preservation by Canning

The basic principle of food preservation by canning had been developed by a French scientist, Nicholas Appert, while France was at war with other European countries and there was an acute shortage of proper food supplies during 1795. The actual development of canning industry had to wait until the tin-coated steel container was found to meet the requirements of sanitary cans best.

Sanitary cans are identified by their overall dimensions, namely, overall diameter and height. Each dimension is expressed as a number of three digits as for example a U.S. No. 2 can designated 307 × 409 as 3 7/16 in. in diameter and 4 9/16 in. in height. Glass and other material containers are also being used for some foods for this industry.

Various Operations in Canning

In canning, the right choice of raw food materials is most important for the success of canning industry as all kinds of food cannot produce good canned products.

After harvesting and receiving the produce, the various operations generally performed in the canning line are as follows (Figure 18.1).

Grading/Sorting

In order to control the quality of canned foods, careful grading is a must. Grading before canning is done to have a uniform finished product and standardize the heat treatment process including other prior operations.

The method of grading depends on the kind of raw foods to be processed.

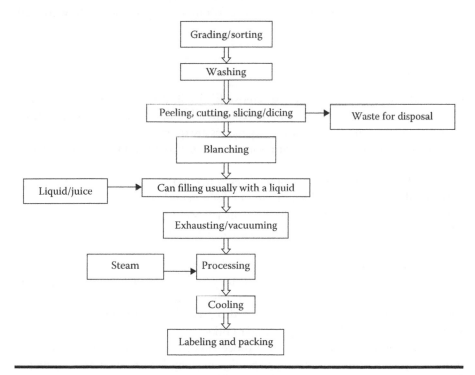

Figure 18.1 Flow diagram enumerating canning operations.

For fruits and vegetables, grading is usually performed for size, weight, maturity, and color. Grading for size, color, and defects is often done manually putting the raw foods on the moving belt, and workers sitting on both sides of the belt pick up the defective food materials while the sound ones are allowed to pass. The efficiency of the system depends on belt speed and workers' experience.

Vibrating copper screens with different sizes of circular openings are employed for size grading of most of the fruits and non-leafy vegetables as copper screens can withstand severe use and do not affect color of fruits as well.

Revolving double-roller graders can be used for spherical raw fruits and vegetables.

Oranges and apples are sometimes graded by weight.

Measuring the tenderness of the food materials can be used as a means of grading for maturity. This test can be easily done by the equipment known as "tenderometer" or "maturometer" where the force required to drive a plunger or knife through a fixed depth in the test food material is measured. From calibration curves showing force required against maturity of various foods, the maturity of the test material is determined.

Peas are often graded with their differential specific gravity in brine solution.

Sometime, *chemical methods* are used to determine the ratio between sugar and acid in case of oranges or the starch content in case of peas.

Electronic color sorters have been successfully employed for brightly colored tomatoes and apples.

Washing

Fruits and vegetables are washed to remove field soil, dirt, microbes, insecticides, and pesticides also. Laws specifying maximum allowable limits of these contaminating chemicals with the fruits and vegetables are quite rigid.

The importance of washing with water can be realized from the following facts:

■ Washing removes dirt and other foreign materials.
■ It reduces the bacterial load from raw food materials.
■ Washing increases the sterilization efficiency considerably.
■ It improves the appearance and quality in general.

Generally, different kinds of food are washed either by agitation or by sprays, etc.

In fruits and vegetables, soaking and spraying in combination can be applied. Common washers are revolving drums in which food products are tumbled and sometimes high-pressure water sprays are also used. Sensitive food like tomatoes is washed by dumping and floating in water tanks.

The selection of washing and other processing equipment for a particular kind and variety of a vegetable or fruit depends on many factors such as size, shape, fragility, etc.

Washing by using sprays is one of the effective and satisfactory methods. The spray efficiency is dependent upon

- Water pressure
- Water flow rate
- Spray nozzle distance from the food material to be washed

Leafy vegetables can be washed as well as conveyed by water flames.

In modern plants, laser, x-ray, infrared, and image analysis techniques can be used.

Peeling, Cutting, Slicing

Different kinds of trimming, cutting, peeling, slicing, dicing, coring, shredding, or pitting devices are required for various vegetables and fruits. Green beans may be cut into various desired shapes and sizes by a machine conveniently. Brussels sprouts are mainly trimmed manually by holding the base against a fast rotating knife. Mechanical peeling and coring are applied for products such as apples.

Peeling is necessary to separate nonedible parts from the edible parts of foods as far as practicable as edible parts should be canned only.

Various methods of peeling are in vogue, namely, *manual, mechanical,* and *lye peeling*. Earlier, *manual or hand peeling* was common in food processing industry. Different designs of knives have been developed for hand peeling. But hand peeling has always been both time-consuming and wasteful as 50% or above of the food materials may be lost during hand peeling of some fruits and vegetables.

Mechanical peelers of different designs have been in use for peeling, coring, and slicing of apples, pineapples, etc. Mechanical peelers can also peel carrots, potatoes, etc.

Root vegetables such as potatoes, beets, etc., can be easily peeled with their prior heat treatment with boiling water or steam.

Chemical (lye) peeling is one of the useful methods of peeling in food processing industry. Lye peelers are employed due to their efficiency and high capacity. Alkali peeling can be used for peeling of fruits such as peaches. Generally, 1%–2% sodium hydroxide solution maintained at a temperature 90°C below the boiling point is used as common lye in peeling of fruits whereas same lye of 10%–15% concentration may be employed in cases of root vegetables like beets, carrots, and potatoes. Beets and sweet potatoes with thick skins may be peeled with steam under pressure while they pass through cylindrical washing machines. This softens the skin and the underlying tissues. When the steam pressure is released, steam beneath the skin expands and causes skin cracking. Then water jets can be used to wash away the skins.

The lye-peeled foods must be thoroughly washed with hot water at 55°C–60°C to remove the alkali as soon as possible otherwise alkali will affect the edible parts. Water jets at 3–7 bar can be applied to spray water for washing the lye-peeled products in a rotary washer.

The main advantages of lye peeling are summarized as follows:

■ Less loss of edible materials
■ High capacity owing to fast handling
■ Less cost of peeling as lye solution may be recycled

Various designs of lye-peeling machines have been in commercial use as shown in Figure 18.2.

When the cost of chemical lye peeling is appreciable, hot water scalding followed by rupturing skins of some vegetables like tomatoes is being employed.

Flame peeling is used for peppers that are difficult to peel by other methods.

Blanching

Blanching is a short and mild hydrothermal treatment.

The vegetables that are not processed at a high temperature in canning must be heated to a minimal temperature to inactivate the inherent enzymes before processing, freezing, or storing. It also expels air and respiratory gases from the product, softens, and improves the food quality. This heat treatment is termed as blanching. If the two heat-resistant enzymes, namely, catalase and peroxidase, are inactivated, then other enzymes are supposed to be inactivated. Blanching conditions that inactivate 90% of the enzymatic activity is taken as optimum to avoid excessive softening.

Blanching is a must for vegetables that are to be frozen as freezing retards the enzymatic action only.

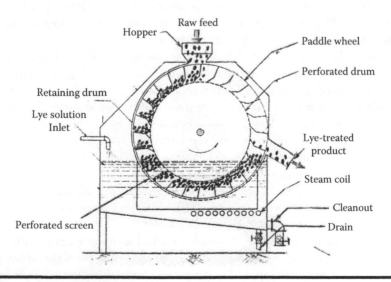

Figure 18.2 Fruit and vegetable lye peeler.

Many green vegetables and others are blanched or heat-treated at a temperature range of 45°C–75°C with hot water or steam.

Small vegetables may be sufficiently blanched in hot water within few minutes but large vegetables may require several minutes.

Conveyors filled up with the commodities are employed to carry through a heated long covered metallic tank that may be used for blanching of vegetables. Either hot water at 88°C–99°C or steam at atmospheric pressure is used in blanching machines.

Therefore, the major purpose of blanching is as follows:

- Softening the vegetables to fill in more weight in cans
- Stabilizing the green color of some vegetables like spinach
- Removing the undesirable odors, flavors, and other substances
- Facilitating preliminary operations like peeling, dicing, cutting effectively
- Additional washing of foods

However, the following points should also be considered: (a) change in texture, color, and flavor by heating; (b) loss of soluble solids in hot water blanching; (c) changes in nutrients and vitamins; and (d) use of more water and energy.

Fruits are not usually blanched as the heat treatment may finally cause sogginess and juice drainage after thawing. In place of heat treatment, some chemicals can be used either as an antioxidants or to inactivate oxidative enzymes.

Addition of Liquid

In general, solid foods are canned in either sugar or salt solution. Salt solution/brine is added to vegetables and oil or brine to fish, meat, and poultry whereas sugar syrups are added to fruits.

The benefits of this liquid addition can be summarized for the following reasons:

- Liquid prevents direct oxidation during heating.
- It helps in heat transfer during sterilization.
- It aids in checking disintegration of solid slices in the can.
- Liquid helps in retaining flavor as well as color of the processed food products.

Mostly sucrose (cane sugar) solution is used in canning. Sometimes, glucose or other sugars are also in use. The sugar concentration in syrup is mainly decided by the kind and quality of the processed fruit products.

For canning purpose, brine (sodium chloride) solution with salt content of about 1%–1.5% is generally used in vegetable canning and the purity of the salt should be in between 98% and 99%. Salt with iron compounds causes discoloration and calcium salts result in white precipitate and toughening of the food product. Special additives are being added to the product during filling such as firming agent calcium salts are applied for tomatoes or citric acid to decrease the pH less than 4.3 in some vegetables to arrest the growth of *Clostridium botulinum*.

Filling of solids may require both mechanical and manual operations.

Exhausting and Creating Vacuum

Exhausting is a short heat treatment of the product either before filling or both before and after filling. Heating causes expansion of the product, expulsion of all gases from the product, and reduction of air from the headspace.

The main purpose of exhausting is to create vacuum in the sealed can after processing and cooling. In canning, exhausting means heating the contents and the can by steam prior to sealing. The containers must maintain a reasonable vacuum of around 25 mmHg, which can be produced by either hot filling and steam injection or mechanical vacuum system. The degree of vacuum depends on steam pressure or temperature and steaming time. In practice, the food products, sterilized in glass jars, are exhausted under mechanically produced vacuum. Generally in mechanically produced vacuum, the food products are subjected to vacuum for a short period before can closure. Therefore, the air is withdrawn mainly from the headspace and partially from the product itself.

Various devices can serve the purpose of exhausting. The temperature of the contents with the can is raised to the desired level and then the can is to be sealed while it is hot. Heating the cans on conveyors is conducted through a long metallic tunnel heated with direct steam. Exhaust boxes are most suitable for fruits and vegetables in cans filled with brine or syrup solution. The major disadvantages of exhaust boxes are their bulkiness and their large steam requirement. The final vacuum in the can depends on the heating time and the temperature attained before sealing.

The major objectives of exhausting are as follows:

■ To expel air and dissolved gases from the food contents and headspace of the can
■ To ensure reasonable vacuum in the closed container after thermal processing
■ To arrest corrosion of tin plate of can by preventing oxidation of its food contents
■ To avoid undue strain on the double seam of the can during thermal processing and cooling
■ To produce concave can ends indicating safe and sound processed food product

Sealing/Double Seaming

When the can enters into the sealing machine, a jet of steam is injected on top of the contents driving out the air from the headspace and the can is closed immediately by a double seaming/sealing device. The term "double seam" implies that the seam is formed by two operations in the seamer. During the first operation, the can

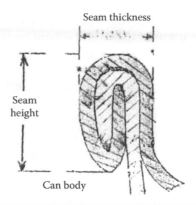

Figure 18.3   Cross section of a double seam.

end-seaming panel is rolled together and interlocked with the body flange. At the second stage, pressing it to the required tightness completes the seaming operation. A double seam joint may be seen in Figure 18.3. The double seamer machineries may have 4–12 seamer heads and at the speed of 100–1000 cans/min.

Processing

Sterilization by heat treatment known as processing is the most important operation in canning. Any food material while filling in cans contains microbes that cause deterioration of the food unless these are destructed.

Sterilization means complete destruction of microorganisms that often requires a wet heat treatment at a minimum temperature of 121°C for about 15 min.

Pasteurization refers to a comparatively mild heat treatment, generally at a temperature below the boiling point of water. Pasteurization process is employed to destroy the pathogenic organisms in milk, liquid eggs, and others to extend the shelf life of food products like fruit juices, beer, wine, etc. Milk is pasteurized at 62.8°C for 30 min to inactivate pathogens. Pasteurized products are not sterile as they may contain vegetative organisms and spores capable of further growing. Pasteurized products should be stored at a refrigeration condition; otherwise, they may be spoiled in a short time under ambient conditions. Pasteurized milk can be preserved under good refrigeration for more than 7 days but at room temperature it may be spoilt within a day.

Acidic fruit products with pH values less than 4.0 require *pasteurization* process only, and it is normally carried out in static batch vertical or horizontal retorts or in agitating cookers or in continuous pressure cookers by immersing the sealed can in boiling water or open steam at atmospheric pressure for relatively short times. For the product having a pH value 4.0–4.5, longer process time or acidification with the addition of citric acid may be required.

In canning, sterilization does not actually mean total destruction of all microorganisms; rather it aims at the achievement of sterility with respect to the most heat-resistant pathogenic and toxin-forming microbes present that may bring about spoilage under usual storage conditions of the canned foods. This is known as *commercial sterilization or processing* where absolute sterility is seldom achieved.

The temperature and time required for thermal processing of the products to be canned are determined with reference to the most heat-resistant food-spoiling microbes, namely, an aerobic *Clostridium*, anaerobic pathogen *C. botulinum*, putrefactive anaerobe 3679 (PA 3679), and few others found at the coldest point in the can. It is taken for granted that when the coldest point becomes sterile, any other point in the can that has been exposed to the higher level of temperature will also be sterile.

Factors Affecting Thermal Processing Conditions

- Number of microorganisms present in the contents of can
- Types of microorganisms and their spores
- Material of the container like tin-coated plate or glass
- Container size
- Effect of acidity (pH) of foods
- Influence of concentration of salt solution, sugar solution, and presence of colloids
- Consistency such as tightly packed or loosely packed can
- Influence of cooling rate after sterilization
- Effect of can rotation, if any
- Influence of initial temperature of the contents
- Influence of sterilizer temperature and time

Thermal Processing Time and Temperature

The two basic methods of calculation commonly used for in-container sterilization/thermal process time are (1) general method (Begelow method and its improved version) and (2) formula method. The commercial sterility method is also used in some cases.

Decimal Destruction Time (D Value)

The destruction rate of bacteria is nearly proportional to their number present in the system being heated. The thermal destruction of microorganisms and quality factors follow the first-order reaction. This refers to logarithmic order of death, which means that under constant thermal conditions the same percentage of bacterial population will be destroyed in a given time interval, regardless of the size of the surviving population. In other words, the logarithm of the surviving number of microorganisms during heat treatment or thermal processing at a particular temperature plotted against heating time will yield a straight line. These are known as survivor curves.

Suppose at a given temperature 90% of microbial population is destroyed in the first minute of heating, 90% of the remaining population would be killed in the second minute, 90% of what is left would be killed in the third minute, and so on.

In this context, the decimal reduction time or *D* value is defined as the heating time in minutes at a given temperature required to destroy 90% of the microbes in a population or concentration of quality factors.

Hence, *D* value reduces the surviving microbial population by one logarithmic cycle. The 90% reduction of microbes for example may mean the fall from 10^6 to 10^5 in one log cycle. Graphically, this represents the time range between which the survivor curve passes through one log cycle as shown in Figure 18.4.

$$\text{Mathematically,} \quad D = \frac{t_2 - t_1}{\log(a) - \log(b)} \qquad (18.1)$$

where *a* and *b* are the survival counts for heating times t_1 and t_2 min, respectively.

The *D* value is strongly influenced by the heating temperature as higher the temperature used smaller the *D* value required.

For example, the *D* values of microorganisms, namely, thermophilic bacteria and *C. botulinum* for pH > 4.5 at 121°C, are 2.5 and 0.1–1.0 min, respectively, whereas the *D* value of yeasts and molds for pH < 4.5 at 65°C is 0.5–1.0 min only.

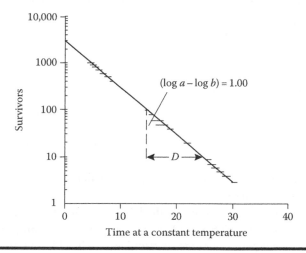

Figure 18.4 A typical survivor curve (*D* value).

Thermal Death Time

In food microbiology, another term thermal death time (TDT) is often used. It is defined as the heating time required causing microbial destruction at a given temperature. If TDT is measured with reference to a standard initial load, it represents a multiple of D value. Hence

$$TDT = nD$$

where n is the number of decimal reduction time.

Thus, subjecting the food through a thermal process that achieves 12 decimal reductions should destroy all the spores of *C. botulinum* in a gram. For food safety, the minimum thermal process should reduce the number of the spores by a factor of 10^{-12}. The time to reduce the number by 10^{-12} is known as TDT and therefore TDT is a measure of $12D$ values.

For canned foods, *C. botulinum* is the primary microorganism whose population is to be reduced. A dilemma persists in that exponential destruction never reaches zero, and there may be a surviving spore that may cause spoiling. In the worst possible case, the $12D$ thermal processing concept is taken to be its practical solution. At a reference temperature of 250°F (121.1°C), the D value for *C. botulinum* is 0.21 min. Therefore, if a food is subject to a process temperature of 121.1°C for 3 min, one should have safe and commercially sterile product.

Temperature Dependence (z Value)

The relationships between the D value and the temperature are expressed by

1. TDT method (D–z model)
2. Arrhenius kinetic method

TDT Method (D–z Model)

When thermal death rates of *C. botulinum* plotted as the decimal thermal reduction time at a given temperature vs. the temperature, T in °F on a semi-log graph paper, it yields a straight line on a temperature range used for food sterilization.

The significant influence of temperature on D values is usually represented by the thermal resistance curve.

This temperature sensitivity is clarified by a z value that represents a temperature range through which the D_T value curve passes through one log cycle (Figure 18.5). The temperature range that achieves a 10-fold reduction in D values

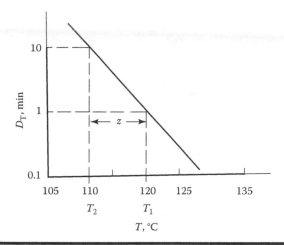

Figure 18.5 **Thermal death time curve: decimal reduction time vs. temperature.**

is called Z value. In other words, the term Z represents the temperature range in °F for a 10:1 change in D value.

$$\text{Mathematically,} \quad (\log D_{T2} - \log D_{T1}) = \frac{T_1 - T_2}{Z} \qquad (18.2a)$$

$$\text{or,} \quad Z = \frac{T_1 - T_2}{\log D_{T2} - \log D_{T1}} \qquad (18.2b)$$

With this knowledge, one may represent the destruction at any temperature in terms of equivalent destruction at the reference temperature.

The Arrhenius equation may be represented by

$$D_T = D_0 \times 10^{(T_0 - T)/Z} \qquad (18.3)$$

where
D_{T1} and D_{T2} values correspond to temperatures T_1 and T_2
D_0 is the D value at a reference temperature of T_0 normally at 250°F (121.1°C) for thermal sterilization

With this knowledge, one may represent the destruction at any range of temperature in terms of equivalent destruction at the reference temperature. As the standard reference temperature, $T_0 = 250$°F (121.1°C), on which thermal processes are generally compared, the Arrhenius equation becomes

$$\frac{D_T}{D_{250}} = 10^{(250-T)} \qquad (18.4a)$$

$$\text{or,} \quad \frac{D_T}{D_{121}} = 10^{(121-T)} \qquad (18.4b)$$

Arrhenius Kinetic K–E Model

The effect of temperature on specific reaction rate, k, 1/s may be predicted by the K–E model also. Following the rate of destruction of microorganisms per unit time is proportional to the number of organisms

$$\left(\frac{dN}{dt}\right) = -kN \qquad (18.5)$$

Rearranging and integrating this equation,

$$\ln\left(\frac{N}{N_0}\right) = -kt \qquad (18.6)$$

$$\text{or,} \quad \left(\frac{N}{N_0}\right) = e^{-kt} \qquad (18.6a)$$

As D is the time during which the original number of viable microbes is reduced by 1/10 as expressed by

$$\left(\frac{N}{N_0}\right) = \left(\frac{1}{10}\right) = e^{-kD} \qquad (18.7)$$

Taking $\log_{10}$ on both sides and solving for D

$$D = \frac{2.303}{k} \qquad (18.8)$$

Combining the aforementioned equations,

$$t = D \log\left(\frac{N_0}{N}\right) \qquad (18.9)$$

If log (N_0/N) is plotted against t, a straight line should be produced and it is experimentally verified for vegetative cells and to some extent for spores.

Lethality/Sterilization Value—F Value

The F value is measured by the total time required to accomplish a stated reduction in a population of vegetative cells or spores. A thermal process is considered safe if the slowest heating point of a canned food reached an acceptable F value. When the standard temperature, $t = 250°F$ is used in Equation 18.9 and F_0 is substituted for t

$$F_0 = D_{250} \log\left(\frac{N}{N_0}\right) \tag{18.10}$$

$F_0 = t$, min, a process time at 250°F that produces the same degree of sterilization as the given process at its temperature T.

Combining these equations, F_0 of the same process at T may be expressed by

$$F_0 = t10^{(T-250)/z} \quad \text{at } T \text{ °F} \tag{18.11}$$

$$\text{or,} \quad F_0 = t\,10^{(T-121.1)/Z} \quad \text{at } T \text{ °C} \tag{18.11a}$$

where F_o is a value in minute for the thermal process at a constant temperature $T°$F or $T°$C for a given time t minute. The F_0 and z values may vary with the type of food for sterilization against *C. botulinum*.

With reference to the equations $D_T/D_{250} = 10^{(250-T)}$ and $D_T/D_{121} = 10^{(121-T)}$, the ratio D_T/D_{250} or D_T/D_{121} at any given temperature can also be referred to as lethality and can be envisaged as the equivalent destruction in minutes at 250°F (121.1°C) achieved by 1 min at any other temperature. Then all the lethalities can be added up to have an equivalent time at 250°F (121.1°C), and this sum of the lethalities is known as F_0 value; as it is necessary to achieve 12 decimal reductions at this temperature, the F_0 value of 3 should provide a safe food against *C. botulinum*.

C. botulinum, which affects low acid foods, is a common threat to public health, and hence the destruction of their spores is used as the minimal criterion for thermal processing. It is arbitrarily accepted that the minimum thermal process should be at least as severe as to reduce the population of *C. botulinum* through 12 decimal reduction time (D). It is known that D value is 0.21 min at 121.1°C for *C. botulinum*. Therefore, F_0 value $= 12 \times 0.21 = 2.52$ min. The minimal thermal process lethality, F_0, required is thus 2.52 min.

The effect of temperature on specific reaction rate, k, may be predicted by the K–E model as follows:

$$k = \frac{A\exp}{(-E/RT)} \tag{18.12}$$

where
  A is the frequency factor, 1/time
  E is the activation energy, kJ/mol
  R is the gas constant, 8135 kJ/(mol K)
  T_0 is the reference temperature, K
  T is the temperature at any time, K
  N is the number of viable organisms or sterility level at a given time, t
  k is a constant (1/time)
  N_0 is the original number of microbes before sterilization at $t = 0$

Thermal Processing Methods and Equipment

Thermal processing is normally conducted at a temperature of 100°C or below in case of acid (pH < 4.5) foods such as fruits, and tomatoes. But processing is generally done at a temperature higher than 100°C for low acidic or nonacidic vegetables, fish, and meat.

Thermal process time for canned foods also depends on can size of the food product. Whole tomatoes in U.S. No. 2 can (307 × 409) may require thermal processing at 100°C for 45 min for static retorts. Sterilization time can be significantly reduced in processing by using agitated cans in a retort. The mixing inside the can increases the heat transfer efficiency so that desired temperature at the coldest point can be achieved in less time and thereby food quality is also improved.

In aseptic canning, the sterilization time can be drastically reduced to seconds or less than a second. In this method, food is continuously pumped through a plate-type heat exchanger or other heat exchanger where it is heated up so rapidly at a high temperature of about 150°C in 1–2 s for sterilization. This rapid sterilization refers to ultrahigh temperature–short time sterilization (UHST). The sterile foods are cooled, kept in aseptic containers, and sealed under sterile conditions so quickly that these foods retain their nutrients, tastes, and other qualities well.

However in general, cooker is employed for sterilization temperatures below 100°C and retort or autoclave (pressure vessel) is used for temperatures above 100°C–121°C.

Cookers

Different types of cookers are used in the canning industry. Batch cookers may consist of open tanks filled with boiling water into which the processing cans in metal crates can be lowered by cranes. In some cases, a long metal tank may be filled about 75% of its volume with water heated by steam tubes. The cans in metal crates may be lowered and carried through the heated water from one end of the tank to the other by overhead cranes. Live steam can also be used. The main advantage of the hot water process is that it allows processing at any desired temperature like pasteurization of canned fruit juices at about 71°C.

Retorts

Different designs of retorts are available for various purposes including *static, rotatory, steam,* and *water-heated systems with or without air overpressure device.* The saturated steam or hot water with over pressure is employed as heating media depending upon the container used. *Steam/air retort* system where a mixture of steam and air is in use for glass and plastic containers and a constant overpressure of air is needed to ensure the integrity of the package on heating. The steam and air must be mixed properly so that pockets of cold air/steam mixture are not formed and a fair uniform temperature distribution is achieved in the retort. *High-temperature radiations by flames* have been introduced to shorten the processing time and improve the food quality.

Batch/static—vertical and *horizontal retorts*: Retorts are usually of vertical and horizontal types. The retorts made of thick-walled cylindrical steel shell are practically pressure vessels or autoclaves that are used to sterilize canned food products. They are fitted with inlets for steam, water, and air and outlets to vent air during retort come-up and drain the retort at the end of the cycle. A pocket is provided on the side of the vessel to incorporate instruments such as thermometer, pressure gauge, and temperature-recording probe. The instrument pocket is fitted with a continuous steam bleed. The safety devices such as pressure relief valve and safety valve are also equipped with the pressure vessel. The lid of the vertical retort is hinged at the top and placed on tightly by a number of bolts. The swinging door of the horizontal retort is usually located on its end for opening (Figures 18.6 and 18.7). Both these systems are static in operation.

In the retorts, food cans are heated at temperatures above 100°C using steam under high pressure. The cans in crates are placed into the vertical and horizontal retorts and heated with steam under operating pressure at the processing temperature for the required length of time. After processing, the steam is turned off and a mixture of cooling water and air may be allowed to enter into the retort to cool the cans to a temperature of about 40°C and then the cans are removed from the retort.

The major advantage of the horizontal retort is that it may be fitted with two swinging doors at both the ends and cans in crates on steel tracks fitted with the retort can be quickly loaded and removed as well. But it requires more floor area and leaves much unutilized space inside in comparison with vertical retorts.

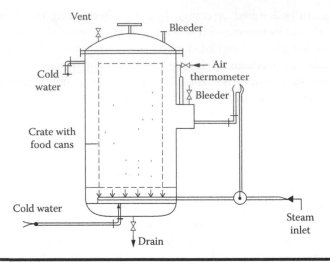

Figure 18.6   Vertical still retort.

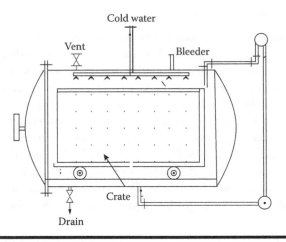

Figure 18.7   Horizontal still retort.

Both these systems being static in operation, it takes relatively long process-ing time due to low rate of heat penetration to the center of the canned food. Agitating the cans, the heat transfer rate can be enhanced for both. In this system, cans are placed in a cage inside the retort and the cage is revolved during steriliza-tion keeping the cans in constant motion for thorough agitation.

Any one of the systems like steam alone or steam along with air or water with steam can be employed for heating retorts. Steam alone is mostly used for metal-canned products whereas air in combination with steam or hot water may be employed for glass containers. Water-heating retorts are mainly used for processing of food in glass jars, which may require a heat exchanger outside the retort for heating inlet water. The containers/glass jars are so arranged that ensure uniform distribution of hot water and good contact between the hot water-heating media and the product.

In agitating retorts, cans are placed in a cage inside them and the cage is revolved during sterilization keeping the cans in constant motion for thorough agitation.

Cooling

After thermal processing, heated cans should be cooled down as early as possible to avoid any significant changes in texture, color, flavor, and nutritional values (known as stack burning) and prevent growth of hemophilic bacteria as well. The processed hot cans in the center of a pile may take a few weeks to cool down to room temperature. A can temperature of about 40°C may be good to keep the cans dry and free from rust without spoiling the contents. The sprays of potable cold water (chlorinated to the level of 1–2 ppm) may be employed for cooling the processed cans.

Labeling

The cans after being cooled and dried are labeled, coded, and packed to make them ready for the market. Generally, label means any tag, brand, pictorial, or other descriptive matter printed, marked, attached to the container, cover, lid, or crown of any food package. The purpose of a food label is to inform the consumers about the products, special features over others, and fulfillments of legal requirement.

Spoilage of Canned Foods

Any processed canned food is subject to deterioration under usual storage conditions.

Two types of deterioration generally occur: microbiological spoilage and changes brought about by chemical and physical agencies. More important is the microbial spoilage of canned foods. There are characteristic differences in odor, taste, and appearance of spoiled canned foods. Besides others, the terms generally used in abnormal cans are swell and leaker.

In swell, the can ends are tightly bulged owing to the formation of gas by microorganisms unfit for consumption as they may be poisonous due to the presence of pathogens.

In leaker, a can may leak mainly because of faulty seaming, pin holing by corrosion of the can, and bursting of the can by pressure of the gas developed inside the can by microbes or by formation of hydrogen gas through corrosion.

Another common spoilage of canned food is because of its discoloration that may be due to browning, other chemical reactions among the constituents, and reaction between dissolved iron from can and other compounds of food. Using proper enamel on the inner surface of the can, it can prevent the last type of discoloration.

Food Preservation by Cooling

Microbial, enzymatic, chemical, and biochemical changes can be retarded significantly by removal of heat or cooling for food preservation of food. Except psychrophilic bacteria that grow even at 0°C, most bacteria, yeasts, and molds grow best at temperatures between 16°C and 38°C. Most of the spoilage microorganisms grow above 10°C. Cooling even by severe freezing only reduces the microbial activities and population but cannot destroy the bacteria completely.

The two common types of cold preservation are *refrigeration* and *freezing*. The temperatures maintained in small and commercial refrigerators are at 4.4°C–7.2°C. Commercial refrigeration called cold or chill storage may use slightly lower temperature depending on the type of food to be refrigerated. Perishable foods can be preserved by cold storage for days or weeks. The advantages of refrigeration or cold storage over freezing are less energy consumption to cool a food, less insulation requirement, retention of food texture as well as flavor, and no need of thawing before use.

Air is to be circulated to dissipate heat of respiration accumulated during cold storage. Most of the foods are preserved well at a refrigeration temperature when the relative humidity of air ranges from 80% to 95%.

When some fruits are harvested fully mature and ripen at their peak quality, these fruits are to be consumed in a short period after harvest. On the other hand, a large number of fruits like banana, apple, mango, etc., are harvested at certain maturity level and they may be allowed to ripen during storage and transportation. These fruits are ripened in a controlled ripening room maintained at a temperature of about 14°C–22°C for a required length of time with a relative humidity of about 85%–95% at a proper flow rate of air and a desired concentration of CO_2 and C_2H_4 gas depending on the type of fruit.

Usually for a short period of safe storage temperature above 0°C is used, whereas for long-time preservation a storage temperature below 0°C is applied. It is discussed earlier that many tropical fruits are best preserved between 10°C and 15°C as they suffer from *chilling injury* at low temperatures. Bananas can be preserved well at around 15°C while fruits like apples from temperate regions are preserved well at about 0°C–1.5°C.

Cold Storage

As regards cold storage using warehouses, there are two types of construction generally available, namely, curtain wall and insulated types.

A curtain wall type may be taken as one inner building into the outer one or as an inner and an outer shell. Either granular or sheet materials may be used as insulation in between the two shells.

An insulated-type warehouse is a common building, the walls and floors of which are insulated. Glass wool, wood, cork, corkboard, sawdust, straw and their boards, paper, etc., are the common insulating material.

In place of refrigerants like freon or ammonia, a brine-circulating system may be employed to maintain cold storage rooms at temperatures above freezing point to avoid a chance of contamination due to pipe leakage.

Freezing of Foods

Freezing refers to cold preservation of food maintained in a frozen condition. A temperature of –18°C or below is required for good freezing. Frozen foods can be preserved for several months or years by deep freezing without any significant change in texture, flavor, color, and shape. Freezing is one of the best food preservation methods to maintain the quality of foods during long storage. That is why frozen foods have been widely accepted. As rate of freezing affects the quality of frozen foods significantly, quick freezing is considered to be superior to slow freezing. Quick freezing may be defined to signify the rate of freezing at which the foods are cooled through the zone of maximum crystal formation (+32°F to 25°F) in 30 min or less.

The main advantages of quick freezing method are as follows:

- The smaller ice crystals formed during quick freezing cause much less damage to the food cells.
- The freezing time is much less.
- The food products are so rapidly cooled below the level of temperature that the microorganisms may not have a chance to grow.

The quick freezing methods may be classified into *direct immersion methods, air-blast methods, plate contact methods,* and *vacuum freezing methods.*

Out of these methods, *air-blast freezers* are usually used for their easy construction despite being not so efficient and economical. In this method, the food product is packed into its final packaging, which may be retail packs or 10 kg boxes. Tunnel freezing system is most commonly used where rapid cold air blast at a temperature of –30°C to –40°C and a velocity of 400–600 ft/min may be applied. It may consist of a slow moving mesh belt or wire mesh trays loaded on trucks passing through a long tunnel circulated with very cold air.

Direct immersion method of freezing ensures rapid cooling of foods in sufficiently cooled refrigerant brine at –3°F for vegetables or 50% inverted sugar syrup solution at about 0°F for fruits. Adhering brine or sugar solution can be removed from the frozen product by a specially designed centrifugal device.

In *liquid nitrogen freezing*, the food material is placed on a mesh belt that passes through a tunnel, where liquid nitrogen is sprayed onto it. This is an expensive method and it is suitable for a small load of sensitive fruit where a premium price can be obtained for its good quality.

When a highly moist food material is subjected to high-vacuum, rapid vaporization of water from food takes place with the loss of latent heat of vaporization from the food itself. In consequence, food material is cooled down and finally frozen. *Vacuum freezing method* can produce a very good quality food product but its commercial production cost may be high for cheap products. To select from the various types of vacuum freezing equipment available, the following main factors may be considered for their suitability: versatility for freezing of loose or packed material, speed of freezing, defrosting requirement, initial cost, material handling cost for batch or continuous system, packaging cost, operating cost of power, maintenance, etc., and total space requirement.

Apart from these, rooms freezing and fluidized bed freezing methods are also used in freezing.

Food Dehydration

Food dehydration industry has been initiated during the World War I when it has facilitated the supply of dehydrated food of small volume and mass in bulk conveniently to the army. On average, dehydrated fruits, vegetables, fish, or meat have around 1/10th the volume and 1/16th the mass of the raw foods. The dehydrated fruits, vegetables, coffee, cocoa beans, milk, and other milk products have been in commercial production for a long time. The standard of dry food product requirements with respect to organoleptic, nutritional, and functional properties for food dryers should be met with low cost. It is a challenge to retain original color, flavor, texture, taste, and other quality attributes in many dried food products. Removal of moisture from wet foods by thermal drying is an energy-intensive operation mainly because of the fact that latent heat of water vaporization is 2.26 MJ/kg at 100°C. In practice, the total energy consumption may go up to 3–6 MJ/kg water evaporated depending on the type of dryer. Initially, 20%–30% free water can be easily removed by mechanical pressing/expression, centrifugation, or osmotic dehydration before thermal drying. For economy, milk and soluble coffee can also be concentrated by using efficient multiple-effect evaporators before spray drying. Apart from the basic process engineering requirements, food dryers have to meet certain standards for food quality, hygiene, and safety.

Fruits and Vegetables: Preservation and Processing

Fruits and vegetables are clubbed together because of their many similarities in composition, storage, and processing practices. Botanically some of the vegetables are considered as fruits. Fruits are the mature ovaries of plants that house their seeds. Tomatoes, cucumbers, and other vegetables may be classified as fruits. However according to food processors, those plant items taken as a main course of meals are usually considered as vegetables and those are frequently eaten alone are considered fruits. Fruits are mostly sweetish and acidic in nature. Particularly, citrus fruits such as orange, lemon, and grape contain high percentage of citric acid. Bananas, mangoes, pineapples, papayas, and others are known as tropical and subtropical fruits due to their origin from these regions. Fresh fruits and vegetables generally contain as high as 70%–85% water, but their protein and fat contents are usually within 3.5% and 0.5%, respectively. The legumes, peas, and some beans are characteristically high in protein content, whereas oilseeds are rich in oil or fat. As a whole, vegetables and fruits can be considered as sources of digestible carbohydrates like starches, sugars, and indigestible fibrous celluloses, hemicelluloses, as well as pectic substances. Fruits and vegetables are also the sources of some minerals and vitamins such as vitamins A and C. Citrus fruits are characterized by their high vitamin C content. Green leafy vegetables and tomatoes are the good sources of vitamin C. Potatoes also contain a low level of vitamin C. They have characteristic colors and flavors as well. Vegetables have nonvolatile acids, namely, citric, oxalic, malic, and succinic acids to contribute certain flavors. Onion and garlic have very strong flavor characteristics due to certain sulfur-containing volatiles. The radishes and red cabbage contain flavonoids. Green vegetables have a pigment like chlorophyll and yellow/orange and red colors of carrot and tomato are due to carotenoids and lycopene.

Physiological Changes

When fruits and vegetables are in the stage of maturation in the field, every day some physical and biochemical changes take place. At some point of time, they will attain the peak quality with respect to texture, color, and flavor. As this peak quality exists for a short period only, there is an optimum harvesting time for any vegetable or fruit. For example, harvesting and processing of peas, tomatoes, corn, etc., are to be carefully scheduled rigidly to maintain this peak quality. They continue to respire even after harvest and overall starch and sugar present in the tissue are converted to carbon dioxide and water in presence of oxygen in the surrounding atmosphere. After harvest during storage, sweet corn at a room temperature may lose about 20%–25% of its total sugar within a day and peas can lose as high as 40%–50% total sugar in a day under the same conditions. At the same time, almost all fruits and vegetables release heat of respiration. The heat of respiration at 15°C for fruits like apples and strawberries are 41–92 and 210–273 W/Mg, respectively, whereas vegetables such as cabbage and green peas in pod release respiratory heat of 66–169 and 529–599 W/Mg, respectively. For further details, refer to Chapter 17. That is why harvested fruits and vegetables should be cooled as early as possible to retard the rate of deterioration.

Classification

Some of the common vegetables and fruits can be classified as follows:

Cole crops: broccoli, cabbage, cauliflower; *Cucurbits*: cucumber, pumpkin squash
Bulbs: garlic, onion, leek; *Fruit vegetables*: eggplant, okra, and tomato
Leafy vegetables: amaranth, lettuce, spinach; *Leguminous vegetables*: beans, peas
Perennial vegetable: asparagus; *Root vegetables*: beetroot, carrot, and turnip
Tuber crops: potato, sweet potato, tapioca

Other vegetables: mushroom, lettuce, spinach, cabbage are also used in salads

On the basis of the distribution of fruits in nature, they can be classified into *tropical*, *subtropical*, and *temperate fruits*.

The following are examples of some tropical, subtropical, and temperate fruits:

Major tropical fruits: Banana, guava, jackfruit, mango, papaya, pineapple
Minor tropical fruits: Cashew, apple, litchi, passion
Subtropical fruits: Citrus—lemon, lime, orange, mandarin; non-citrus—avocado, fig, litchi, olive; berries—grapes, strawberry, raspberry
Temperate fruits: Pomes—apple, pear; stone—apricot, cherry, peach, plum

Depending upon their respiratory system and ethylene synthesis during ripening, harvested fruits are also classified into *climacteric* and *non-climacteric types*.
Some climacteric and non-climacteric fruits are as follows:

Climacteric	Non-Climacteric
Apple, apricot, avocado, banana	Blackberry, cashew apple, cherry
Fig, guava, jackfruit, melon	Cucumber, grape, grape fruits, lemon
Papaya, peach, pear, plum, tomato	Lime, lychee, orange, pineapple, strawberry

Respiration

Respiration is a process that always takes place in all living cells and releases exothermic energy, CO_2, and H_2O through the breakdown of stored carbon compounds. Normally, carbohydrate substrates like starch or lipids are present in their tissues.

These are further broken down to smaller molecules of simple sugars. Finally, sugars are oxidized in presence of O_2 in accordance with the following generalized equation:

$$C_6O_{12}O_6 + 6O_2 = 6CO_2 + 6H_2O + 686 \text{ kcal}$$

Two systems of respiration, aerobic and anaerobic, may occur in harvested produce. The amount of respiratory heat varies with different commodities. The oxygen concentration in the plant tissues may be almost the same as that of atmospheric oxygen. Therefore, respiration results in a depletion of mass, reserve energy, shelf life, and quality of the produce. The rate of respiration is usually expressed as the amount of carbon dioxide evolved per unit mass.

Two systems of respiration, aerobic and anaerobic, may occur in harvested produce. If O_2 is absent or insufficient, living cells begin a process known as anaerobic respiration producing alcohol, aldehydes, and ketones with the development of off-flavor and sometimes toxicity in vegetables.

Overall respiration influences storage, packaging, and cooling conditions greatly. The heat of respiration, if not dissipated, would cause rise in temperature that in turn accelerates both the respiration rate and microbial spoilage.

Transpiration/Water Loss

Besides respiration, another important physiological change, namely, transpiration or water loss, also occurs in fruits and vegetables. Transpiration is a process of water loss from a fresh produce leading to a loss of mass, texture, appearance (causing shriveling and wilting), and other qualities. Most fruits and vegetables with a loss of 5%–10% moisture are visibly shriveled. The transpiration rate depends on many parameters. Among them, structural and surface conditions, the surface to volume ratio, atmospheric pressure, storage temperature and relative humidity, and rate of cooling after harvest are important. The continuous loss of water vapor from fresh fruits and vegetables to the surrounding environment occurs by the diffusion mechanism from plants, which can be mathematically expressed as follows:

$$R_w = A P_w \Delta P_w \qquad (18.13)$$

where
R_w is the rate of water loss from produce, mol/s
A is the surface area of produce, m^2
P_w is the effective permeance of the produce surface for movement of water vapor under prevailing conditions, mol/(s Pa)
$A P_w$ is difference in partial pressure of water vapor between inside of produce and environment, Pa

Transpiration not only causes moisture loss but also leads to shrinkage. Shrinkage is responsible for surface cracks through which microbes can attack easily. The moisture loss from some food materials causes an appreciable change in textural quality and in consequence these commodities lose their crispness and juiciness. The continuous loss of moisture by some harvested vegetables owing to transpiration, respiration, and subsequent drying of the cut surfaces causes wilting of leafy vegetables, and loss of both plumpness and mass of fleshy vegetables.

Ripening

Ripening is a series of irreversible qualitative and quantitative changes occurring toward the end of the growth period of fruits. The changes during ripening may be physical/structural, chemical, and biochemical including enzymatic. The various structural changes that occur during fruit ripening are (a) in cell wall thickness; (b) in the amount of intercellular spaces leading to softening; (c) in color and texture; (d) transformation of chloroplasts into chromoplasts; (e) decrease or disappearance of epidermal hairs; (f) thickening of cuticles; and (g) lignifications of endocarp among others.

A very important physiological change taking place during respiration is in the production of ethylene gas. Different fruits produce ethylene gas at different rates. Ethylene gas influences the maturation of fruits and vegetables with the chemical change in green color of chlorophyll and the formation of yellow or red color owing to the development of anthocyanin and carotenoids, respectively. This ethylene production can be controlled by storage temperature, environmental oxygen, and carbon dioxide composition.

Fruits may be fleshy as well as dry. They are also classified into *climacteric* and *non-climacteric* fruits.

Respiration climacteric is a stage of growth phase of the fruit marked by a sudden increase in metabolic activity. The term "climateric rise" is used to describe the sudden upsurge in the evolution of CO_2 that occurs during ripening of certain fruits like apples. But different fruits show the phenomenon with considerable variation in pattern. Banana may respire CO_2 at 10–50 mg/(h kg fruit) at 15°C–25°C in the pre-climacteric stage, whereas it may respire at high rate of about 50–250 mg/(mg/h kg) during climacteric phase. The ripening stage can be advanced or retarded by controlling the atmosphere surrounding the fruits during storage. The respiration rate also hastens as temperature increases.

The *climacteric fruits* can be harvested unripe as they undergo normal ripening when separated from plant. On the other hand, non-climacteric fruits are not generally ripened further once separated and may enter into senescence after being harvested. Moreover, climacteric fruits generally produce higher amount ethylene and carbon dioxide during ripening stage and they can be stimulated into ripening by exposure to an external source of ethylene.

Some fruits are particularly sensitive to ethylene just before the climacteric rise in respiration and ethylene would have little effect on fruits once ripening is initiated. Bananas are too sensitive to the effect of endogenous ethylene, and the harvesting of green banana fruit hastens ripening. Particularly during long transportation and storage, the temperature is lowered to a level where ripening is not initiated by exposure to ethylene. As the optimum storage temperature is about 13°C for bananas, it should be protected from the effects of ethylene following harvest about 2–3 days only until cooled to the optimum transportation temperature. Apart from temperature, ripening of green bananas can be controlled in an atmosphere of an optimum combination of low O_2 and high CO_2.

In cases of *non-climacteric fruits*, usual production of ethylene and carbon dioxide remains substantially low and at the ripening stage also no increased rate of evolution of these gases occurs. Many other biochemical processes occur during the ripening process that influence the four main quality attributes, namely, texture (softening), flavor, color, and appearance. The variation in ethylene production rate from fruit to fruit may be controlled by storage temperature, atmospheric oxygen, and carbon dioxide concentration.

Fruit ripening is a highly coordinated, genetically programmed, and an irreversible phenomenon with subsequent physiological, biochemical, and organoleptic changes that finally lead to the development of soft edible ripe fruits with desirable quality attributes. Carbohydrates play an important role in the ripening process. The major classes of polysaccharides that undergo modifications during ripening are starch, pectin, cellulose, and hemicellulose. Pectin is the main component of primary cell wall and middle lamella, contributing the texture and quality of fruits. Pectin-degrading enzymes, pectin methyl esterase and lyase, are generally responsible for fruit tissue softening. Fruit pectin extracted from guava, apple, citrus, and others are used for commercial purposes. Pectin content varies with different fruits. As for example, the percent pectin contents (fresh weight basis) of apple, guava, mango, and pineapple pulps contain 0.5%–1.6%, 0.25%–1.2%, 0.65%–1.5%, and 0.04%–0.45% pectin, respectively.

In some cases, the conversion of certain compounds during ripening may be undesirable.

The water loss from a fresh produce may cause some major deteriorative changes such as alteration of texture, loss of crispiness, and juiciness besides its weight loss.

Some of the detrimental effects of ethylene-related qualities are summarized later.

Ethylene Effect	Symptom	Fruit
Physiological disorders	Chilling injury internal browning	Persimmon, avocado pear, peach
Toughness	Lignification	Asparagus
Off-flavors	Volatiles	Banana
Softening	Firmness	Apple, mango, melon

Senescence refers to the undesirable degradative changes that occur in all plants and harvested plant parts, which eventually die. Mostly senescence leads to detrimental changes in product quality, as its most evident symptom is the degradation of chlorophyll of harvested vegetables.

Dormancy is a phase of suspended growth of vegetables when their respiration rate and other physiological changes are appreciably slow making them suitable for a longer storage.

These physiological processes are influenced by the environmental factors, namely, temperature, relative humidity, airflow rate, and atmospheric composition.

The physiological damage named *chilling injury* occurs generally in tropical and subtropical food produce when they are stored at their freezing point or below 5°C–15°C. The chemical compositions of some fruits and vegetables are presented later for convenience. It causes discoloration, off-flavor, and uneven ripening of the chill-injured commodities. Almost similar physiological damage called heat injury may be found if the commodities are stored at temperatures above the specified or recommended storage temperatures.

Physical injury made by mechanical impact and other physical operations causes deterioration as microorganisms easily infect physically damaged commodities.

Moreover after harvest fruits and vegetables undergo many important chemical changes in carbohydrates, acids, pectin, and others that alter their food quality attributes.

There is an optimum cold storage temperature for each type of vegetable or fruit that generally varies from about 0°C to 15°C. Generally, tropical fruits and vegetables should be stored at temperatures above 45°F whereas temperate zone produce should be stored at about 33°F. In actual practice, a fairly specific optimum storage temperature is being used for each type and variety of fruits and vegetables.

It should be made clear that there is difference between freezing preservation and cold storage of whole fresh produce. In freezing, the lower the temperature the better would be the product while in cold storage of fruits and vegetables for fresh produce market, the fresh produce should not be frozen.

Precooling of Fresh Produce

After harvesting, as the produce is generally warm, they should be cooled as early as possible to remove the heat generated by immersing in cool water. However, several improved methods of cooling available are *hydrocooling, forced air cooling, tunnel cooling, serpentine cooling*, and *vacuum cooling. Hydrocooling* appears to be quite effective for cooling some fruits and vegetables. Some hydrocoolers consist of a continuous belt conveyor passing under sprays of cold water at about 33°F–34°F. Fruits and vegetables in wooden boxes, placed on the conveyors, are flooded with cold water by the sprays to cool down the produce as required.

After cooling as before the produce in crates may be placed in cold storage or in refrigerated wagons and trucks for transport to the market. During cold storage or refrigerated transport, the contents should attain the optimum storage temperature and maintain them at that temperature.

For a long distance, shipment of fruits and vegetables refrigerator cars should be used. But refrigerated cars may not be necessary for transport of the produce over a small distance.

Processing of Fresh Produce

Canning of Vegetables

As stated earlier, the quality of processed vegetable and fruit products depends upon proper selection and grading of the fresh produce. After preliminary sorting, fruits and vegetables are graded for uniform quality with respect to size and color mainly. In India, grading is generally done manually except in a few cases where special grading machines are used.

Some vegetables such as asparagus, beans, cauliflower, carrots, mushrooms, peas, potatoes, tomatoes, and others are usually canned. The canning of peas is a common practice.

Canning of Peas

Only good varieties of peas are selected for canning to satisfy the following:

- The variety should be productive or high yielding.
- Most of the kernels in pods with the vines should be matured uniformly.
- The green color of peas must be retained after canning.
- Peas should be of very good texture, flavor, and taste.

The kernels can be removed from the pods by the shelling machines in the processing of plant and pea kernels are canned as follows.

Grading

The peas are allowed to pass through a cleaning machine to remove vines and pods. Then these are graded into different sizes by a grader.

Peas are further graded for "fancy" and "standard" using specific gravity separators. The peas are carried through two tanks filled with two respective brine solutions of specific gravities 1.04 and 1.07. The peas that float on brine of 1.04 and 1.07 specific gravities are known as "fancy" and "standard," respectively.

The maturity of peas can be found by measuring their alcohol-insoluble starch contents. Maturity is also measured by using an instrument "tenderometer" as discussed earlier.

Sorting

Peas are then sorted on belt conveyors to separate pods or any other foreign matter.

Blanching

The main purposes of blanching are to soften the material, remove the veneer juice from the skin, fix the green color, and improve the flavor. Blanching of peas is carried out in hot water at about 88°C–93°C for varying periods of time depending on their maturity and quality. The length of blanching may vary from about 2 min for tender peas to 10 min or more for starchy peas. Hot water blanching can be accomplished continuously in a machine. The blanched peas are then thoroughly washed.

Final Sorting

The washed and blanched peas are to be sorted on a broad belt to remove splits, defective peas, and undesirable materials.

Filling and Sealing Cans

Prior to filling in cans, peas are usually heated. Sulfur-resistant lacquered cans should be used. The measured quantity of hot peas is filled and a mixture of boiling salt, sugar, and water solution is added into the can. The mixture may contain about 2% salt and 2%–5% sugar.

Usually exhausting of filled cans is not necessary.

Then filled cans are closed by double seaming machines.

Thermal Processing

Generally, processing of peas in 307 × 409 cans at a temperature of 240°F for about 35 min may be conducted for sterilization.

Large starchy peas may require longer processing time.

Cooling

Hot cans should be thoroughly cooled as rapidly as possible so that processed peas are not too softened to form a cloudy solution because of starch gelatinization.

Labeling, Coding, and Packing

The processed cans are then labeled, coded, and packed to make them ready for the market.

Canning of Tomatoes

A simplified process flow diagram for canning of peeled tomatoes is presented in Figure 18.8, which is self-explanatory.

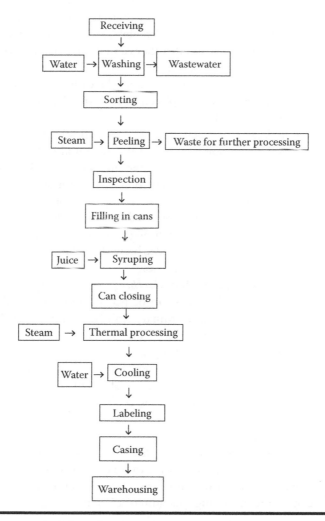

Figure 18.8   A typical flow diagram of tomato canning.

Canning, Bottling, and Freezing of Fruits

Canning, bottling, and *freezing* are the common long-term fruit preservation methods. All these methods have certain advantages and disadvantages. The kinds of fruits suitable for canning may not necessarily be used for freezing.

For *canning,* fruits like berries, cherries, and plums are graded whole whereas apricots, pears, guavas, mangoes, and pineapple are usually graded after cutting into halves and slices. Some widely grown fruits like mango, orange, papaya, pear, pineapple, and certain amounts of apple, apricot, berries, cherries, grape fruits, litchi, peach, plums, etc., are canned to supply the defense forces mainly and the general market as well.

High standards of *cannery hygiene* are needed to avoid any spoilage due to contamination, and to achieve that the following protective measures must be applied: (i) cleaning is to be monitored visually and microbiologically; (ii) entire production and storage areas must be free from pests so that they cannot cause damage to raw materials; and (iii) cooling water has to be chlorinated to avoid all post-process spoilage.

After delivery to the factory, the fruits should be canned as early as possible. For short-time storage, they should be placed in cool and dry environment. Cold storage at 5°C–8°C may be used.

Peeling of fruits is done manually with knife, mechanically by peelers, or chemically using about 2%–10% hot solutions of NaOH for around 1–2 min. Jets of water can remove the loosened skins and traces of adhering caustic solutions. Mechanical peelers may peel apples and pears while their cores may be removed simultaneously.

Enzymatic or nonenzymatic browning or both may occur on the surfaces of many cut fruits. Placing the peeled pears and apples in dilute brine solutions of ascorbic acid can prevent their brown color development during exposure to air. Some soft fruits like strawberries and raspberries may be discolored to brown hues during thermal processing.

Artificial colors can be used in these cases.

Some fruits may require proper hot *blanching* before filling as it reduces the processing time and softens solids that can be packed and readily filled into the cans. Regarding *choice of cans* usually ordinary tin-plated cans are used for apples, while fruits like rhubarb, strawberries, and plums should be filled in lacquered cans to prevent their acids to react with the tin plate.

Filling in cans may be done mechanically or manually. Controlled weights of fruits with hot cane or beet sugar syrup or glucose syrups of about 40°–45° Brix at temperatures 80°C or above or fruit juice at a temperature must be filled in before closing the cans.

Exhausting accomplished by any method is to remove air and other entrapped gases from the headspace before can closing.

Can closing may be preceded by steam-flow closure to sweep air or entrapped gases and create a partial vacuum over the headspace. Cans are closed by placing the lid on any can and sealing it with the formation of a double seam. Acidic fruits being corrosive in nature need a high vacuum of 25 mmHg or above.

Processing of fruits is the most important operation and it is usually carried out by immersing the sealed can in boiling water or steam at atmospheric pressure using batch or continuous retorts or agitating cookers. The processing temperature and time depend on the type and pH of the fruit product. Acidic fruits with pH values less than 4.0 are usually pasteurized. The pH values of some canned packs of fruits such as apples, rhubarb, and strawberries may vary from 2.8 to 3.3, 3.2 to 3.6, and 3.2 to 3.8, respectively.

Strawberries filled in fully lacquered cans should be pasteurized at 100°C for about 5–10 min to reach 85°C at the center of the can.

But prunes require longer processing time to destroy the bacteria and acquire their right texture.

Quick *cooling* of finished products in cans to right temperatures is important so that they are properly dried. The filled cans should be stored in cool and dry environment.

Canning of Apples

Apples should be *graded* to uniform size, shape, and good quality.

The washed and graded apples move to a machine that will peel and core apples in a single operation.

After *peeling*, the apples are quickly treated with dilute 1.5% brine solution in a tank to prevent oxidative enzymatic browning.

Then they are cut into properly sized sections and these sections are again kept in dilute brine for some hours to keep the color, texture, and other quality of the canned apples. The apples can be mildly *blanched* with open steam or hot water for just sufficient time in a stainless steel blancher. The apples segments are subject to water spray onto a conveyor belt to remove the brine and discolored pieces.

Hot *filling* of slices at about 70°C or above and the cans are topped up with boiling water to keep a headspace. After filling, the cans are usually *exhausted* for sufficient time to remove all air/gases. Double seaming operation is done to seal the hot cans tightly.

Then the sealed cans are thermally *processed* in a retort to raise the temperature of the can center above 70°C.

Canned products are cooled as quickly as possible and stored under proper environment.

Canning of Rhubarb

Rhubarb is extensively cultivated in Europe and the United States for its long, fleshy red stalks. In these countries, it is used as fruit, stewed with sugar, in pies and preserves. The leaves of rhubarb are rich in oxalic acid. It is also used in medicine. A typical process flowchart for canning of rhubarb is presented here for general understanding (Figure 18.9).

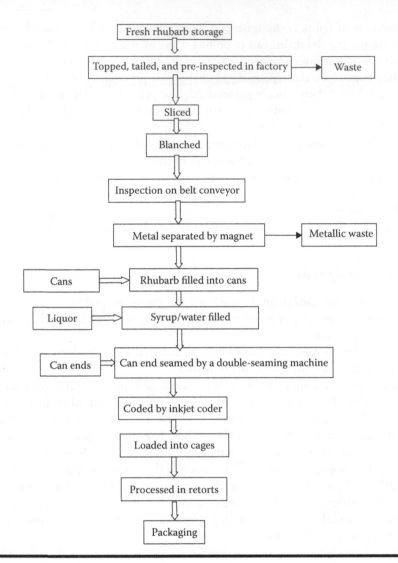

Figure 18.9   A flowchart for rhubarb canning.

Bottling

Bottling is a relatively small-scale industry and can rather be carried out at home due to its simplicity. Good quality colored fruits in transparent containers is quite appealing to the consumers.

Same methods are being followed in the *preparation* of fruits as in canning. The jars are thoroughly water-washed and jars are immediately used directly for filling.

After *filling* proper quantity of fruits, the jars are generally filled with water, fruit juice, or syrup leaving a certain headspace and then steam is injected before the cap is placed to *evacuate* any remaining air.

The *processing* may be carried out with hot water in a steam-heated tank or steam/air mixture so that the temperature is raised slowly to the processing temperature to avoid shattering of glass jars. *Water cooling* is not generally applied. In case it is used, hot water must be slowly replaced by cold water and the water level should not cover the caps. Rapid cooling and excessive movement of filled jars after processing should be avoided to prevent breakdown. Bright sunlight should not be allowed to enter into the storeroom so that it may not bleach or turn brown the color of the processed fruit during *storage*.

Freezing

Fresh good quality fruits free from any microorganism and cross contamination should be used for freezing as quickly as possible.

Fruits may be continuously washed rapidly to (i) reduce the bacterial load and foreign matter; (ii) improve the appearance; and (iii) minimize leaching, color, and flavor. A close *inspection* and *sorting* help in removing any adhered foreign material as well as blemished pieces.

Fruits are not generally blanched prior to freezing. They can be packed in sugar or syrup before freezing. A quick freezing process is essential to keep the quality of the fruits and they are brought below the freezing point as quickly as possible to form as small ice crystals as possible to retain their cellular structure and quality. Soft fruits such as berries are best preserved by this method.

Dehydration of Fruits and Vegetables

The influence of various parameters on food preservation has been discussed in Chapter 20. As stated earlier, the most important parameter in food dehydration is water activity as dried food can be stored or preserved well for a longer time under reduced water activity. Each bacteria, yeast, as well as mold has an optimal or a minimal water activity for its growth.

The water activity, a_w, of fresh fruits and vegetables varies mostly from 0.95 to 0.99. The dried fruits containing 18%–24% moisture have a_w of 0.7–0.8, whereas the dried vegetables of 14%–24% moisture content have a_w of 0.7–0.78.

As it is beyond the scope of this book to present all aspects of drying of a large number of fruits and vegetables, the discussion is restricted to dehydration of only few common fruits and vegetables. The usual operations/processes followed in fruit and vegetable dehydration technology are shown in the flow diagram (Figure 18.10), which is self-explanatory.

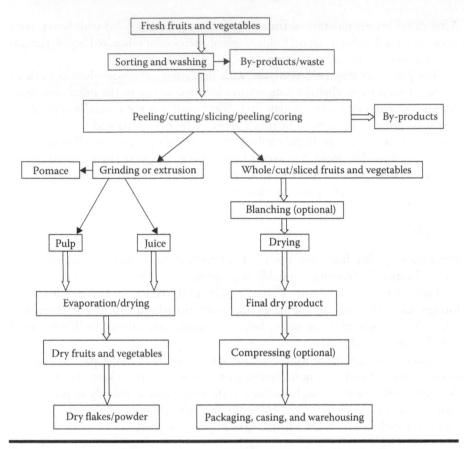

Figure 18.10 General operations involved in fruit and vegetable dehydration.

Dehydration of Vegetables

The vegetables provide desirable texture, color, taste, and flavor to our daily diet along with their nutritive values. Vegetables form important items particularly for vegetarians. The appearance, taste, and flavor of the meals can be varied with the proper choice of vegetables served.

The dehydrated vegetables like onion flakes, onion, garlic, and chili powder; mashed potato powder; and others have been in use. Some other dehydrated vegetable powders have also increasing market. As a whole, dehydrated vegetable would have a good market provided their desired qualities are preserved.

Generally, dehydration technology involves the following steps.

The main equipment generally employed for dehydration are graders, washers, peelers, blanchers, dryers, and packaging machines.

Mechanical Dehydration

The major operations involved in dehydration of vegetables may be summarized later.

Preparation

Preparation for dehydration of vegetables may include washing, peeling, destoning, slicing, dicing, shredding, etc.

Blanching

Almost all vegetables to be dehydrated are blanched to arrest the undesirable enzymatic activities and rehydrate/refresh the dehydrated products more readily.

Blanching may be accomplished with either hot water or steam. Steam blanching may lead to lower leaching loss and cleaner product. Blanching should be carried for a just sufficient time to inactivate peroxidase enzyme completely. In case of cabbage, catalase is to be inactivated in place of peroxidase.

Sulfiting

The exposure of vegetables with sulfite or bisulfate to retard the oxidative changes during storage has now become a common practice. Sometimes, a suitable mixture of sulfite and bisulfate is also used. The sulfiting prolongs the shelf life and protects their carotene and ascorbic acid contents. Moreover, higher drying temperature can be used for the sulfited vegetables to increase the drying efficiency as the treatment protects them from scorching damage. The dried potatoes, carrots, and cabbage may contain 500–800, 1000–1500, 1000–2000 ppm SO_2, respectively. After cooling, the sulfited vegetables should not have a perceptible taste/flavor of sulfur dioxide.

Drying

While drying vegetables in tray dryers equipped with galvanized screen trays or trays made of plane, wooden strips can be used. For cabbage, potatoes, and carrots, the tray load may vary from 0.75 to 1.5 lb/ft². Heavy loading may result in nonuniform dried products.

The final moisture content of most of the dried food materials should be brought down to a level of about 5% to be compatible with good quality.

The drying ratios for potatoes in terms of the ratio of the unpeeled raw vegetables received to the dried finished product or the ratio of the prepared raw product to the finished dried product may be about 8–10:1 for the former and around 5–6:1 for the latter cases, respectively.

Generally, dehydrated vegetables do not attain their original size and weight on refreshing in water and cooling. These ratios vary with type, variety, maturity, etc., of vegetables. Tomato may attain less than 50% of its original fresh weight whereas corn may come back on final cooling to more than its original fresh weight. In some cases, dehydrated vegetables are compressed to incorporate more weight of product per unit volume of the container.

Dehydration of Tomatoes

The tomato is a true fruit but it is usually consumed as a vegetable. Tomato is low in fat, free from cholesterol, and rich in vitamins A, C, K, and is a source of fiber and protein. It is also a natural source of bright color pigments, red lycopene, and reddish yellow carotenoid. Lycopene is recognized as a preventive for various cancers and it has a good antioxidative property as well. It has two forms: *cis* and *trans* isomers. The *cis* isomers may be better absorbed than their all-native *trans* form mostly existing in raw tomato. The thermal processing of tomato converts *trans* isomers of lycopene into the better form of *cis* isomer. Dehydrated tomato products are of three major groups: tomato paste and its other products, tomato powder, and dehydrated tomato. Tomato paste is usually used as an ingredient for products like sauce and tomato powder. Dehydrated tomatoes are generally rehydrated before their use; otherwise, they may be added to recipes. Raw tomatoes may be either partially dried to about 35% or further dried to a level of about 10% (w.b.).

Preparation

In the selection of raw materials, the varieties chosen for dehydrated tomato, tomato paste, and tomato powder should be uniform in size and maturity. Other important quality parameters that are taken into consideration are

- Firm thick-walled flesh of soluble solids with 4°–5° Brix
- Bright red color, good flavor
- Free from any kind of foreign matter, mold, and bacteria rot

Roller conveyor is used for full view of the raw materials so that it can easily remove the defective tomatoes manually.

The sorted and graded tomatoes should be washed with clean, potable, and chlorinated water with hypochlorite in the range of 50–300 ppm to avoid microbial contamination. High-pressure water jets and agitation in the water tank are also employed for effective washing and cleaning.

The clean tomatoes are fed to the chopping machine before drying.

Dehydrated Tomato

In the conventional heated air-drying system, the preheated air at a temperature of 55°C–105°C and an air velocity of about 30–90 m/min with varying relative humidity (depending on weather) is allowed to pass over the product. The drying time may vary from 10 to 20 h to dry the product to about 30%–40% (w.b.) depending upon the load.

The long drying operation at high temperature in presence of air may cause oxidative heat damage of the product. In this drying system, case hardening and discoloration may cause an inferior product quality and low drying efficiency as well.

Tomato Paste

Extraction, Juice Refining, Concentration, Sterilization, and Packaging

In hot break treatment, the washed and sorted tomatoes are crushed to form the pulp at room temperature and the pulp is immediately heated to 65°C–85°C for inactivation of pectin esterase and cellulase enzymes to prevent breakdown of pectin. The lower break temperatures of about 60°C–65°C can be used to retain more color and flavor. Alternatively, the washed and sorted tomatoes may be quickly preheated before chopping and crushing to produce the pulp.

In extracting and juice refining, the preheated tomato pulp is screened through a series of sieves having diameters between 0.4 and 0.7 mm to remove 3% skin and seed from the pulp so that 75%–85% juice rich in lycopene content is extracted.

The refined juice is now ready for the concentration and it may be concentrated in a series of batch-type single effect vacuum evaporators or continuous multiple-effect vacuum evaporators to produce tomato paste of about 24% total sold (TS). The paste is stored in closed tank.

Tomato Paste to Dried Powder

The tomato paste of 35%–40% TS (usually prepared by hot break method) is generally dried to about 97% TS for the production of tomato powder.

Roller/drum drying, foam mat, or spray drying may produce tomato powder.

Spray drying seems to be the most suitable method to produce high-quality powder economically. But conventional spray dryers mostly suffer from the problems of deposit losses, quality degradation, and others. Recently, high-capacity spray dryers have been so designed to handle the hygroscopic dried product and finally pack it without any contact with the surrounding air. The tomato powder is cooled and conditioned in the presence of dehumidified cool air before packing the finished product.

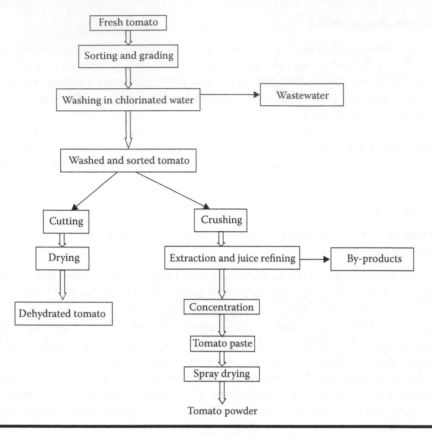

Figure 18.11 A flowchart for preparation of dehydrated tomato and tomato powder.

The tomato paste is generally hot filled into cans at about 90°C. The loaded cans are double seamed without any further heat treatment and allowed the filled cans to pass through water spray to cool them early before packing. A simplified process flow diagram for preparation of dehydrated tomato and tomato powder is presented in Figure 18.11.

Dehydration of Potatoes

Potatoes are dehydrated by different methods to produce dices, granules, and flour. The various popular snack products like chips, crisps, and others are also prepared from potatoes. In vegetable dehydration industry, potatoes contribute various common products such as dices, flakes, granules, and flour including snack products like chips, crisps, and others. Prior to drying, all processes may follow sorting, washing, peeling, inspecting, dicing/slicing, blanching, sulfiting, precooking, etc., operations according to the final products.

Potato dices: The clean good quality potatoes are cut and diced to some specific shapes and sizes. The diced potatoes can be blanched in steam or hot brine solution before drying. Generally, tray/cabinet, tunnel, or conveyor dryers are used for hot air drying of wet material to a level of 8%–9% moisture (w.b.). Conveyor dryers are generally used to dry the fresh-diced potatoes with a load of 75–90 kg/m² in two or more stages. The air temperatures of 60°C–75°C and 50°C–60°C are commonly used in the first and second pass of conveyor drying.

A general flow diagram for dehydration of diced potato is presented in Figure 18.12.

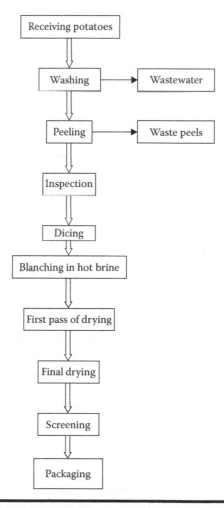

Figure 18.12 A typical flowchart for diced potato.

Potato flakes: The dried mashed potatoes are prepared from potato flakes of 6%–7% moisture content. The potato flakes can be ground, milled, and screened to produce potato flours.

After washing, peeling, precooking, cooling, and mashing of the product, the mashed material contains potato pulp with 77%–82% water (w.b.) along with other additives.

The drum dryers are useful and economical for the drying operation mainly due to their low consumption of steam (1.0–1.5 kg steam/kg water evaporated) and high thermal efficiency (75%–85%). That is why single- or double-drum dryers are commonly used to produce potato flakes where the surface of the cylinder is heated to about 150°C by steam.

Dehydration of Fruits

Fruits are commercially important and nutritionally necessary food commodity being a part of nutritionally balanced diet. An important factor that affects their economic value is the relatively short ripening period and reduced post-harvest life.

Fruits are widely distributed in nature. Either sun or mechanical drying depending upon the climatic conditions and other economic considerations is used for them. Sun drying of fruits is still common in most of the developing countries. As sun drying is totally weather-dependent and its production capacity and its efficiency are comparatively low, mechanical or artificial drying has gradually been replacing sun drying.

Sun Drying

Large fruits are generally cut into two or four pieces. Small fruits, such as grapes and berries are dried whole. The whole fruits may be treated with dilute lye solution of sodium hydroxide or sodium carbonate to remove any bloom and waxy substances from the surface and to make some hair cracks for rapid dehydration. Fruits are usually exposed to the burning sulfur fumes prior to sun drying in the yard because this treatment helps in producing lighter-colored products. Sulfuring device mainly consists of an enclosed chamber where sulfur is burnt on the floor. Fruits should be exposed to prescribed amount of sulfur fumes; otherwise, excess dose may cause softening and deshaping the fruits. Then fruits on trays can be dried in the sun.

Mechanical Drying

Washing, dipping, cutting, peeling, sulfuring, and sometimes blanching may generally be used for pretreatment of fruits. Steam blanching of many cut fruits may result in faster drying and lighter-colored products. Blanching time depends on the kind, variety, maturity, and size of the fruit. Most of the cut fruits on trays are subjected to burning sulfur fumes in a closed and sealed chamber. Sorption and retention of sulfur dioxide in the product are dependent upon many parameters like fruit maturity, sulfur dioxide concentration, and sulfuring temperature. Sulfurated blanched fruits should be cooled down to lukewarm condition to avoid poor sulfur dioxide absorption and breakdown of tissues of soft fruit-like apricots.

The sulfurated cut fruits can then be dried on a large scale in a tunnel dryer at an air temperature of about 140°F–150°F with airflow of 600–1000 ft/min. After cooling, the dried fruits should be packed in an impermeable packaging material.

Dehydration of Grapes for Raisins

Generally, some suitable varieties of grapes are dehydrated for raisins.

Still drying of grapes in the sun is usually a traditional practice being followed around the world where sun is abundantly available. The grapes may be spread on metallic, wooden, paper trays or on plastic sheets in the vineyards. The sun drying takes 2–3 weeks or more depending on the weather.

Before drying, the grapes may be dipped into an alkaline solution of 0.2%– 0.3% caustic soda (NaOH) at about 90°C for a few seconds (followed by a water rinse) to remove waxy layer from their outer skin and speedup the drying rate and then they may be exposed to burning sulfur to have light-colored raisins. A combined chemical treatment of about 0.3% NaOH and 0.5% K_2CO_3 emulsified with 0.4% olive oil at an elevated temperature is also practiced. These treatments may remove the waxy bloom from the surface; they modify their waxy coat structure to increase moisture permeability and allow rapid drying as well.

Various other chemical treatments including treatment with ethyl esters of fatty acids and free oleic acid at 1.5% (w/w) are also applied depending upon the grape variety.

In mechanical drying of grapes, the process begins with the treatment of grapes with 0.25% hot chemical lye solution, washing with water to remove remaining lye, spreading on trays, and exposing to the burning sulfur at about 1.5–2.5 kg of sulfur/ton of fresh grapes in the sulfur house for about 4 h. Tunnel dryers are usually used to dry grapes at a moderate temperature for about 24–48 h to a level of 9%–14% (w.b.) moisture. The SO_2 content of the final dry raisins should not exceed 2000 ppm.

The processes after grape harvesting do not vary much. In some cases, the grapes are dried naturally without any further treatment; otherwise, they are pretreated prior to drying to accelerate the dehydration process.

Natural raisins in California are sometimes produced without any treatment. The grapes are spread on paper or other sheets in the sun for around 2–3 weeks to dry them from about 70% to 15%, at which stage the paper sheets may be rolled

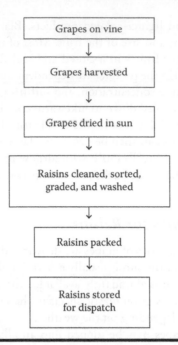

Figure 18.13 A flow diagram for natural raisin production.

with the dried grapes to produce bundles. These are kept for a few more days till they are dried to a safe level for storage in large containers. The raisins so produced are cleaned to remove leaves, stems, etc. Then, the remaining lighter stems may be removed by using vacuum separator. After washing and grading, other food producers pack the raisins to several sizes for use.

A typical flow diagram for the production of natural raisins is presented in Figure 18.13.

Quality Parameters of Dehydrated Foods

The dried foods may be safe from microbial as well as toxic chemicals points of view but their acceptance depends on variety and maturity, color, flavor, texture, taste, nutritional value, and other aspects. Selection of pea variety and maturity is so important that it may control the quality of dehydrated pea. Some of the *functional properties* such as protein denaturation play a key role to influence the corn-wet milling quality and wheat-flour milling quality as well.

Shrinkage or capillary collapse is a phenomenon that induces the change in volume during dehydration, which may occur due to various reasons such as moisture loss in drying, ice formation while freezing, and pore formation by puffing. Shrinkage occurs, as biopolymers cannot support their weight in absence of sufficient water. Shrinkage may reduce the drying rate and affect the quality of the dried foods significantly. Its impact on the dried food product is due to its reduction in wettability, rehydration capacity, and change in texture. The density, texture, rehydration, and other properties of dried food products mainly depend on the drying process. Heated air-drying usually produces comparatively a dense food product with an impermeable surface whereas quick freeze-drying usually leads to a porous product with much better rehydration capacity and superior quality. Generally, a food material dried at 80°C produces a dense product having poor rehydration capacity whereas the same food dried at 5°C results in a porous product with good rehydration capacity. Some of the mechanisms that play an important role to control the shrinkage are surface tension, plasticization, and electrostatic and gravitational forces.

Caking and *stickiness* of powders are desirable for a product like tablet and undesirable for free-flowing materials. These tendencies during spray drying depend on some factors such as viscosity of concentrated solution, particle size, surface tension, and exposure time.

Fall in diffusivity and development of shrinkage combined together causing the surface/skin impervious to interior moisture movement during rapid drying of materials is known as *case hardening*. The surface shrinkage may also result in crust formation, *cracking/checking*, and *warping*. Slow drying rate induces least concentration as well as temperature gradient during drying that can minimize case hardening, crust formation, checking/cracking, and other effects as well. Food or grain cracking may lead to easy insect infestation or mold attack during storage, loss of volatiles, consumer satisfaction, and organoleptic and other product quality attributes. High thermal and moisture stresses developed by grain drying with high initial moisture and high temperature cause fissures in grain. Development of rice grain fissures during drying results in kernel breakage, poor milling quality and fetches low price as well. Usually, tempering in between drying passes or after drying improves grain-milling quality. Stress cracking that may be further enhanced by microwave drying may further enhance stress cracking.

Rehydration: Dehydration toughens skins and makes it difficult to permeate water into dried products. The moisture/water loss from plant cells changes some of the important physical properties that affect the rehydration of dried food products with the following effects: (i) permeability change; (ii) osmotic pressure loss; (iii) pH change; (iv) crystallization of polysaccharide gel in cell wall; and (v) coagulation of protoplasmic proteins. Both intrinsic and extrinsic parameters of rehydration are thus affected.

It is an industrial practice to prick peas before drying so that rapid dehydration and better rehydration of fresh green preen peas may occur. Dehydrated carrots are rehydrated well at pH values between 2 and 12 and they rehydrate to the maximum extent at pH 12. Carrots are better rehydrated in distilled water than that in brine solution whereas color in cherries rehydrated in sugar solution is superior to that of cherries rehydrated in water.

Retention and degradation of natural color: Drying at high temperature for a long period results in degradation of original color of many food products. Food colors can be retained by (i) minimizing thermal level and exposure time; (ii) using short drying period with pH adjustment; (iii) applying HST; and (iv) freeze-drying. Enzymatic browning causes color degradation that can be minimized by inhibiting the phenol oxidase enzyme. Photooxidation of pigments caused by light in combination with oxygen may also result in discoloration.

Retention and development of volatile flavor: Release and chemical change of volatiles may affect the natural aroma of the food products after drying. The levels of odorous compounds/volatiles such as hexanol, 2-hexanal, 3-hexanal, 2-heptanone, and others are decreased after air-drying of bell pepper. Generally, off-flavors may be developed when moist immature or mature peanuts are dried at temperatures above 35°C–40°C due to change in concentration of ethyl alcohol, ethyl acetate, acetaldehyde, etc. Various parameters, which affect retention or loss of volatiles during spray drying, are atomizer pressure, rotation speed, spray angle, drying air temperature, air temperature profile, feed concentration, feed composition, foaming of feed, and others. The important parameters that need to be considered are drying air temperature and feed solution concentration along with spray-dryer type and atomization method. A detail on color and flavor chemistry is available subsequently in Chapter 19.

Chapter 19

Food Chemistry
for Technologists

Water (H_2O) and water content play a number of important and crucial roles in food chemistry and processing. It has great influence on food quality. Fruits, vegetables, fish, meat poultry, beverages, and many processed foods contain a high percentage of water. Apart from quality attributes, the quantity of water has a great economic impact as foods are mostly sold on weight basis.

Water can exist in solid (ice), liquid (water), and gaseous (water vapor) states. Its chemical reactivity varies widely from one state to another. Chemical reactions between water and some high molecular constituents of food such as fats, starch, sugars, and proteins that break down into smaller molecules are known as hydrolysis. Thus, hydrolysis brings about substantial changes in chemical as well as functional properties of food. On hydrolysis, fats may be converted to free fatty acids (FAs) and glycerol: starch to dextrin and glucose; sucrose to glucose and fructose; and proteins to peptides and amino acids. Water is a polar molecule and it acts as a solvent for other polar compounds of food. H_2O molecules being dipolar in nature when they are exposed to microwaves, water molecules oscillate at a very high frequency generating frictional heat.

Water Activity (a_w)

It is the water activity, a_w, of a food and not its moisture content that determines the lower limit of available water for microbial growth. Very few intrinsic properties are as important as a_w in predicting the survival of microorganisms in a food product. It also reveals the energy status of water in a food system. Control of water activity, a_w, is important to maintain the stability of foods. That is why water activity is considered to be a very powerful tool in understanding the reactivity of water in foods and it is taken as the single most important property of water in food that determines its stability in food systems. Water activity, a_w, is a measure of the degree to which the water is bound and therefore of its availability to act as a solvent and participate in chemical or biochemical reactions and help the growth of microorganisms. The effect of water activity on microbial growth is influenced by various parameters like pH, food composition including nutrients, preservatives, and oxygen supply. It plays a key role in the safety, quality, processing, storage life, texture, and sensory properties of foods. Hence, a_w can be used to predict the stability and safety of food in respect of microbial activities, rates of deteriorative chemical reactions, and some physical properties as well.

Thermodynamically, water activity, a_w, is defined as the ratio of fugacity (or escaping tendency) of the species in the test condition (in a food sample), f, to the fugacity of the species in its pure state, f_o:

$$a_w = \left(\frac{f}{f_o} \right) \tag{19.1}$$

Fugacity is not measurable but for water in foods, the fugacity is closely approximated to the vapor pressure and therefore for all practical purposes a_w is taken to be equal to equilibrium relative humidity (ERH) under isothermal condition.

Mathematically a_w in food can be expressed by

$$a_w = \left(\frac{p}{p_o} \right) \tag{19.2}$$

or

a_w = (ERH in a food)/100

or

a_w = (moles of water)/(moles of water + moles of solutes in food)

Hence,

$$a_w = \left(\frac{f}{f_o} \right) = \left(\frac{p}{p_o} \right) = \frac{ERH}{100} \tag{19.3}$$

where p and p_o are water vapor pressures in food and in pure state, respectively, under the same conditions.

Theoretically, both a_w and ERH can vary from 0 to 1. In practice, a food approaches to equilibrium with the relative humidity of the surroundings.

Water activity in food is related to the energy with which water is bound or to the work required to remove an infinitesimal quantity of water from a sample. Water activity, a_w, is not determined by the total quantity of water in a sample, but only by that which is least tightly bound. This binding energy, or work, is known as water potential or chemical potential of water. Water activity or energy of the water can be computed by the following equation that reveals that a_w is also influenced by temperature:

$$\mu = \left(\frac{RT}{M_W}\right)\ln(a_w) \qquad (19.4)$$

where
μ is the energy, J/kg
T is the sample temperature, K
R is the gas constant, 8.314 J/(mol K)
M_W is the molecular weight of water

The relation between a_w and moisture content at a particular temperature is known as sorption isotherm. With the increase in a_w, the moisture content also increases mostly. A large number of isotherm models have been developed. Of them, the following Guggenheim–Anderson–de Boer (GAB) model is a widely accepted model for foods over a wide range of a_w from 0.1 to 0.9 a_w:

$$M_e = \frac{C_1 k m_o a_w}{(1 - ka_w)(1 - ka_w + C_1 ka_w)} \qquad (19.5)$$

where
C_1 and k are constants
m_o is the monolayer moisture content

At moisture content and a low water activity, a_w, water molecules are considered to be distributed in a monolayer. The water in monolayer with a low vapor pressure does not form ice at a freezing temperature and it is firmly bound with food constituents that are not easily removed by drying. At high relative humidity and a_w, water is added to food in multilayers that is more reactive, which is called free water that behaves like pure water. Activities of different microorganisms (bacteria, mold, yeast, as well as enzymes) as influenced by a_w and moisture content in food are summarized in Figure 19.1.

This figure clearly reveals the stability of foods as a function of a_w. Very few intrinsic properties are as important as a_w in predicting the survival of microorganisms in a food product.

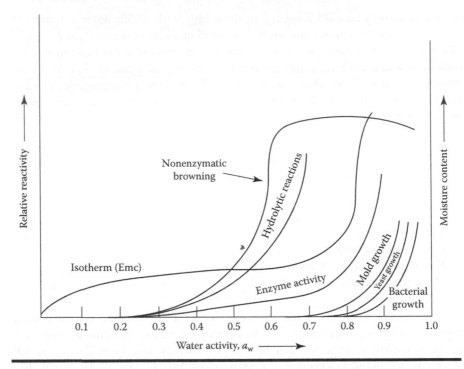

Figure 19.1 **Relation between water activity (a_w) and relative food chemicals reactivity sorption isotherm.**

The temperature dependence of a_w is predicted by the Clausius–Clapeyron relationship:

$$\ln\left(\frac{a_{w1}}{a_{w2}}\right) = -\left(\frac{\Delta H}{R}\right)\left(\frac{1}{T_1} - \frac{1}{T_2}\right) \tag{19.6}$$

where
  a_{w1} and a_{w2} are the water activities at T_1 and T_2 K, respectively
  ΔH is the heat of sorption
  R is the gas constant

The effects of a_w on microbial growths are also influenced by various other parameters like pH, food components including nutrients, preservatives, and oxygen supply.

Ultimately, water activity plays a crucial role in the safety, quality, processing, storage life, texture, and sensory properties of food materials.

Hence, impermeable packaging is essential to protect food material from its environmental humidity; otherwise, unwanted chemical and microbial activities would spoil its desirable quality attributes.

Glass Transition

Glass transition (t_g) is an alternative system of understanding the influence of water on the texture of food. Glass is a property related to the structure of food whereas water activity is a property of solvent. Both these properties are important to understand food–water relations in regard to its quality and storage stability under various conditions. The changes that may take place in foods during processing include glass formation, crystallization, stickiness, gelatinization, and collapse.

Two types of phase transition can occur in foods during heating or thermal processing. The first one is an isothermal change of physical state of the food material from solid to liquid or liquid to gas or vice versa with the release or absorption of latent heat such as melting, evaporation, condensation, and crystallization. Another phase transition occurs from amorphous to glass state without release or absorption of latent heat. The glass transition is dependent upon temperature, moisture, and time as well. Generally, food materials remain in an amorphous state and are brittle below the glass transition temperature. But glasses are not crystalline with a regular structure. Kinetically, glass temperature can be defined as the temperature at which the viscosity of a material reaches 10^{13}–10^{14} Pa s. This transition greatly influences the food stability as food should remain stable at the glassy state. Therefore, it may not be able to take part in chemical reactions (Rahaman, 1999a,b).

The glass forms at a characteristic glass transition temperature that is less than the eutectic temperature. At eutectic point, the water is not freezable.

The shift from glassy to rubbery state causes variation in texture of starch and protein owing to the change of temperature and/or moisture level.

Starch, proteins, and other polymers have a different glass transition temperature, t_g. Below the level of t_g the polymers remain in a glassy state, and above the t_g they go to the rubbery state resulting in a large change of their textures.

The glass transition can be achieved by the change of product composition such as water content, pH, and proper selection of any emulsifier, surface-active agent, etc. The t_g in food can be elevated with the addition of high-molecular-weight solutes.

Some of the important properties governed by t_g of foods and their ingredients can be stated as follows: microbial activity, enzymatic activity, nonenzymatic browning, oxidation, aroma retention, collapse, caking, and crystallization.

Dispersions

Dispersion can be classified according to sizes or physical states of dispersed particle.

In a *pure solution*, the molecules or ions have dimensions below 1 nm. When the size of the particles exceeds 0.5 μm in a system of two phases, it is known as *suspension*. Foods can remain in various dispersion states.

In a *colloidal dispersion/state,* the size of the particles ranges from 1 nm to 0.5 μm and the particles remain dispersed for a long time without precipitation. In this state, either a solid or a liquid or a gas is dispersed in another phase. Often foods remain in a colloidal dispersion where a liquid, solid, or gas phase is dispersed in another phase. Milk is homogenized to decrease the size of the fat droplets to retard the gravity separation. Viscosity affects the stability of dispersion. Colloids may be hydrophilic (solvent loving) or hydrophobic (solvent repelling) in nature. Skim milk and egg yolk are common hydrophilic colloids.

When a solid is dispersed in a liquid it is called *sol,* as in the case of skim milk proteins are dispersed in water in a state of sol. Sols exhibit characteristic optical properties. These are either opaque or clear in appearance and have low osmotic pressure. Addition of salts to sols causes precipitation.

An *emulsion* is a colloidal dispersion of a liquid in another immiscible liquid such as a salad dressing, milk, and cream. Emulsions may be oil in water or water in oil. Butter and margarine are the examples of the latter type. Emulsions can be stabilized by the use of stabilizers such as pectin, plant gums, and some cellulose derivatives.

Foams are dispersion of gas or air bubbles in a liquid that are in a continuous phase. Whipped egg, whipped cream, and ice cream are typical foams. Glucosides, surface-active lipids, and some proteins are used as foaming agents to help in the formation of foams.

Sols with a fairly high concentration of dispersed solids change spontaneously into *gels.* During the change of a sol to gel, dispersed solid polymers in a sol form a network of interconnected strands that trap water and turn into a semisolid form of a gel. Gel is a semisolid structure where polymers such as starch, gelatin, or pectin have trapped a large amount of water. The strength, elasticity, and brittleness of a gel depend on concentration of a gelling agent, acidity (pH), salt content, and temperature. Jelly, puddings, and yogurt are the common examples of gels.

During processing and storage of food, it is necessary to maintain these dispersions so that the phases are not separated resulting in an unacceptable product.

Carbohydrates

Carbohydrates are one of the main classes of food components together with lipids and proteins. Carbohydrates represented by the empirical formula $C_n(H_2O)_n$ are vastly distributed in nature as sugars, starches, celluloses, and many other complex compounds. Primarily, they are the products of photosynthesis and are readily available as a source of food in cereals, roots, tubers, sugarcane, sugar beet, and they provide about 4 kcal/g in the diets. Carbohydrates can be broadly divided into sugars and polysaccharides on the basis of their wide difference in chemical and functional properties. Most of the foods contain carbohydrates consisting of simple monosaccharide to complex polysaccharides. The monosaccharides include soluble reducing sugars, namely, glucose, fructose, and galactose; the disaccharides include reducing lactose sugar (having glucose and galactose units) and reducing maltose sugar (consisting of two glucose units).

The monosaccharide sorbitol, the disaccharide sucrose made of glucose and fructose units and trisaccharide raffinose consisting of glucose, fructose, and galactose units are nonreducing in nature.

Sugars

Sugars not only act as sweeteners, rapid energy-producing agents, but also function as texturizing agents, plasticizers, and flavor-producing agents. Sugars can be broadly classified into monosaccharides (glucose, fructose, and galactose) and oligosaccharides (sucrose, maltose, lactose, raffinose, and stachyose). The general formula of oligosaccharides, $C_n(H_2O)_{n-1}$, is different from that of monosaccharides, $C_n(H_2O)_n$. As for example, molecular formulae of glucose and sucrose are $C_6(H_2O)_6$ and $C_{12}(H_2O)_{11}$, respectively. Sugars add sweetness to food though they are not uniformly sweet. The relative sweetness of different sugars, namely, sucrose, fructose, glucose, and maltose, are assigned as 1, 1.1–1.5, 0.7, and 0.5, respectively.

Glucose and fructose are present in ripe fruits, honey, and fruit juices. Fructose is present in honey and fruits in high concentration but mostly it is found in foods in small amount. Maltose composed of two glucose units is found in starch-degruded products. Galactose, a component of milk sugar, is not found in free state in natural foods. Lactose, made up of galactose and glucose, is present in milk and dairy products only. In the presence of lactobacilli, lactose is fermented to lactic acid and CO_2.

Sucrose is widely used in various food products. It is commercially manufactured from abundantly available sugarcane and sugar beet. Sucrose is formed when glucose and fructose molecules combine together through the elimination of a molecule of water. Except sucrose, most of the sugars being reducing in nature are hygroscopic. Naturally, foods containing reducing sugars are difficult to bring to dryness. Sugars may be in either amorphous or crystalline form. The texture of sugar cookies depends on the crystallinity of sucrose. Hard or soft candies are formed according to the development of crystalline or amorphous sugars in confectionary products.

Presence of the carbonyl ketonic or aldehydic group in sugars makes them *reducing sugars*, which are chemically more reactive. Sucrose can be hydrolyzed to the component monosaccharides glucose and fructose by heat and acid treatment with the consequent increase of reactivity of these reducing sugars. On heating sugars at a high temperature in presence of a proper catalyst, *caramelization* reactions can be brought about to produce brown coloring and flavoring caramel compounds for food use. For the production of caramel flavor, concentrated sucrose syrup is caramelized whereas for coloring purposes in beverages glucose syrups treated with dilute sulfuric acid partially neutralized with ammonia are used.

Nonenzymatic Maillard browning: The *Maillard reaction* takes place when an aldehydic or ketonic group of a reducing sugar combines with an amino group of proteins, peptides, or amino acids, producing sugar amines.

Actually, a complex set of reactions among sugar and amines undergo resulting in *nonenzymatic browning*, which is called Maillard browning reaction.

$$
\begin{array}{ccccccc}
HC{=}O & & HC{=}NR & & H & & H_2CNHR \\
| & & | & & | & & | \\
HCOH & & HCOH & RHN{-}C & \rule{1cm}{0.4pt} & & C{=}O \\
| & & | & & | & & | \\
HOCH & + \; RNH_2 & \leftrightarrow \quad HOCH & \leftrightarrow & HCOH & | & \leftrightarrow \quad HOCH \\
| & & | & & | & O & | \\
HCOH & & HCOH & & HOCH & | & HCOH \\
| & & | & & | & & | \\
HCOH & & HCOH & & HCOH & | & HCOH \\
| & & | & & | & & | \\
H_2COH & & H_2COH & & HC & \rule{1cm}{0.4pt} & H_2COH \\
& & & & | & & \\
& & & & H_2COH & & \\
\end{array}
$$

D-Glucose   Schiff base   Glucosylamine   Amodori product

It is a simplified Maillard browning reaction with D-glucose and an amine through the intermediate products of Schiff base and glucosylamine into brown amadori product (Oliver et al., 2006). Generally, this reaction causes the formation of thermally induced brown color and flavor as well.

Usually these reactions occur during baking, roasting, or toasting at a lower level of moisture of the products in the course of thermal processing. The development of color and flavor in bread crust, roasted coffee, cocoa, and potato chips are a few examples. Normally, the Maillard reaction takes place at a low moisture level as in the case of toasting and baking. This reaction is accelerated with the increase in reducing sugar.

The Maillard reaction products may sometimes be undesirable also. Fruit juices usually undergo Maillard browning reactions depending upon their composition, concentration, and storage conditions.

In general, nonenzymatic browning reactions can be inhibited or controlled by (a) lowering the processing temperature as low as possible; (b) controlling the moisture of the finished product; (c) excluding oxygen-reducing oxidative reactions; and (d) using chemical inhibitors like sulfites, thiols, and calcium salts in fruit processing.

Ascorbic acid in fruit juices may be destructed at normal processing temperature leading to brown discoloration under either aerobic or anaerobic condition. Ascorbic acid degradation in juice concentrates like kiwifruit can occur readily even at room temperature during storage.

Tin and phenol are also responsible for pink discoloration in canned fruits.

Caramelization of sugars can take place at comparatively high temperatures under acid or alkaline conditions in absence of amino acids, resulting in nonenzymatic browning, and unpleasant burnt as well as bitter flavors.

Starch

Starch is a polysaccharide, naturally present in foods. Plant starch generally contains 20%–30% amylose but it may range from 0% to 80%. Starch in plant is available as granules in crystalline and amorphous forms. It is also a glucose polymer linked by α1–4 and α1–6 linkages. Starch primarily consists of two forms of polymer such as amylose and amylopectin. Amyloses are linearly formed mostly by α1–4 glucose units with molecular weights ranging from 2×10^5 to 2×10^6.

Much larger amylopectins are highly branched polymers with both α1–4 and α1–6 glucose units. Amylose is not truly soluble in water; rather it forms hydrated micelles. The structure of amylose helps in gelling of cooked and cooled starches. On the other hand, amylopectin contributes to the thickening characteristics of starch preparation but does not contribute to gel formation.

Gelatinization

The extent of gelatinization of starch may determine the structural and textural properties of breakfast cereals, pasta, rice noodles, and other food products. Generally, cereal–starch granules contain 12%–14% (w.b.) moisture under ambient conditions and they are insoluble in cold water. After soaking in water at temperatures below 50°C, cereals attain equilibrium moisture of about 30%–35% (w.b.) and swell reversibly. When starch is heated in water, the viscosity increases and eventually the formation of a gel or paste takes place. This is the final stage in the process known as gelatinization, although other changes, such as loss of crystallinity, swelling of granules, and leaking of amylose, occur in the starch granule before a change in rheological properties is detected. The term *gelatinization*, thus, does not have a precise meaning. Gelatinization includes collapse of molecular order, native crystallite melting, irreversible swelling, loss of *birefringence*, and starch solubilization in an attempt to define gelatinization. During heating of a suspension of starch in water, diffusion of water into the starch granules causes an irreversible swelling with the increase in granule volume and disrupts hydrogen bonding and consequently water molecules get attached to the hydroxyl groups in starch. The amylose diffuses out of the swollen granules during the course of gelatinization and makes up a continuous gel phase outside the granules. The temperature at which this phase transformation occurs with a rapid swelling of starch granules irreversibly and a loss of *birefringence* is called the *gelatinization temperature* and the phenomenon is known as *gelatinization*. If the water content in granules is abundant, gelatinization occurs normally at a temperature between 60°C and 70°C. Thus, gelatinization is a process caused by heat in presence of water. The parameters, namely, heating temperature, heating period, amount of moisture, pH of the starch–water mixture, variety of starch like corn or rice or wheat starch, granules size, and ionic or nonionic solutes, mainly influence gelatinization temperature. Gelatinization temperature decreases as granule size decreases.

The pathways of gelatinization can be summarized as follows: (i) starch granules hydrate and irreversibly swell to many times their original volume; (ii) granules lose their birefringence; (iii) clarity of mixtures improve; (iv) consistency increases rapidly and reaches its maximum; and (v) linear amylose molecules dissolve and diffuse from ruptured granules and matrix disperses uniformly to form a gel.

A starch gel or paste is a thermodynamically unstable system, and with time a gelatinized aqueous starch system increases in crystallinity and in firmness. To describe the changes in a starch gel after heating and during cooling and subsequent storage, the term *retrogradation* is used. This term means recrystallization of both amylopectin and amylase. It may so greatly influence the texture of the product during storage that the consumer might discard it. A familiar example to the point is the loss of quality in staling of bread during storage.

To measure the relative viscosity of starchy flour water suspension as it is heated at a constant rate, the *amylograph* is used. In this test, the change in viscosity is measured on heating of starch granules during gelatinization and irreversible swelling due to breakdown of crystalline structure of starch granules. A suspension of flour and water is prepared and is heated from 30°C to 90°C in a rotating bowl at a heating rate of 1°C–2°C/min. A paddle inside the bowl is attached to a force-measuring device, which records relative viscosity as *Brabender amylograph* units (AU). A recent alternative to the amylograph is the *rapid viscoanalyzer* (RVA), which measures the viscosity of a small amount of a flour suspension using a disposable plastic paddle rotating in the suspension, heated at a constant rate. Force–time data are directly recorded on a PC.

The main carbohydrates present in vegetables and fruits are in the form of starch and sugars and these starch granules may swell and gelatinize in the presence of high moisture and heat.

The other carbohydrates present in young cell walls are *cellulose, hemicellulose,* and *lignin*. With the aging of plant, cell walls become thick with higher hemicellulose and lignin contents. These contribute fibers. The main carbohydrates present in vegetables and fruits are in the form of starch and sugars.

Cellulose is considered to be an insoluble dietary fiber having zero calorific value and water-binding capacity in food. Celluloses are polymers of 10 or more number of glucose units united together by β1–4 linkages.

Mild heat, moderate acid, alkali, or enzymes do not affect celluloses during food processing and in human digestive system.

Pectin, the cement-like material mainly found in the middle lamella, helps hold plant cells to one another. These water-insoluble pectic substances on hydrolysis produce water-soluble pectin, which can gel with sugar and acid.

In order to increase structural rigidity after processing of some softened fruits and vegetables, namely, tomatoes, apples, etc., usually calcium salts are added before canning or freezing.

Proteins

Proteins are the prime components of the living cells and play various important roles in their structures and functions. Proteins are naturally constituted by 20 amino acids, which act as basic components of the polymeric structure. Once proteins are ingested, amino acids are released by enzymatic digestion and absorbed into the body. Hence, protein quality largely depends on its amino acid content and digestibility.

Red meats such as beef, lamb, and pork (without fat) contain about 18%–20% protein (w.b.).

The protein content of cow milk is about 3.5%. The two parts of milk proteins, namely, casein and whey, contain about 80% and 20%, respectively.

Egg and edible part of fish account for 13%–14% and 10%–21% protein. Most of the fresh vegetables may contain less than 1% protein whereas protein content of fresh peas may be about 6%–7%. Though potatoes contain around 2% protein, it can be considered to be of superior quality due to the presence of relatively high levels of lysine and tryptophan.

Cereals may contain 6%–19% protein. The protein contents of rice and wheat correspond to 7%–8% and 11%–14% whereas seeds may contain 14%–16% protein. Legumes are always rich in protein and it can be as high as 20% protein.

Chemical Structure

Proteins are complex macromolecules made up of successive amino acids that are covalently bonded together in a head-to-tail arrangement through substituted amide linkages called peptide bond. Hence, proteins contain both basic amino (NH_2/NH_3^+) and acidic carboxylic groups ($COOH/COO^-$) attached to the same α-carbon atom. When the α-amino group of one amino acid reacts with the α-carboxyl group of another, the peptide bond is formed. Proteins are polypeptides and made up of the same building blocks of 20 amino acids and their molecules have the property of acquiring a distinctive three-dimensional configuration. The following are the schematic versions of the general formulae of amino acids, peptides, and proteins.

$$\text{Amino group} \qquad \overset{\displaystyle H}{\underset{\displaystyle R}{H_3N^+ - C - COO^-}} \qquad \text{α-Carboxyl group}$$

Side chain group $\qquad$ R $\qquad$ α-Carbon

Structure of amino acids

$$\overset{+NH_3}{\underset{H}{R_1 - C - COO^-}} + \overset{+NH_3}{\underset{H}{R_2 - C - COO^-}} = \overset{+NH_3}{\underset{O}{R_1 - C - C - N - C - COO^-}} + H_2O$$

Formation of a general structure of a peptide bond

$$\overset{H}{\underset{H_3N^+}{R_1 - C - C}} \overset{H\ \ H\ \ O}{\underset{O\ \ \ \ R}{- [N - C - C]_n}} - \overset{R_2}{\underset{H\ \ H}{N - C - COO^-}}$$

A general structure of a peptide bond (protein)

Peptides are present naturally in foods, arising mainly from partial degradation of protein peptide chains. Enzymatic hydrolysis of food yields peptides. The peptides have multifunctions such as antioxidants, antimicrobial agents, and they also help in the development of flavors like sweetness and bitterness.

Protein molecules are great in size, complexity, and diversity as well and they are characterized by the nitrogen content. Plants can synthesize proteins from inorganic nitrogen sources of NH_3, NO_3, and NO_2 but animals cannot do so. Proteins found in vegetable and animal sources vary widely in concentration. Though protein content of foods is considered to be important from a nutritional point of view, their functional roles such as emulsification, texture, and gel formation are also significant.

Their unique characteristics mainly depend on the chain length and the mix of amino acids that make up the sequence. Despite variation in amino acid composition, the amount of nitrogen in food proteins remains almost constant at about 16% and that is why for determination of protein content the value of nitrogen is multiplied by a factor of 6.25. Actually, the factor varies from 5.7 for wheat to 6.38 for dairy.

The amino acid components differ mainly in reactivity and functional groups. Native proteins create a specific three-dimensional structure that decides the functionality of a protein in food. Some proteins like egg whites can develop foam by agitation with water and air. When gliadin and glutenin proteins in wheat flour come in contact with the added water in a mixture for dough preparation, they are transformed into gluten, responsible for extensible, cohesive, and elastic structure that plays a key role to the texture of dough products.

Classification

Proteins can be grouped into catalytic proteins and structural proteins on the basis of their function. They may also be classified into simple and conjugated proteins based on their composition. The simple proteins can be further subdivided according to their solubility in water, salt solution, acid/alkali, and alcohol solutions. For example, globulins are soluble in water whereas albumins are soluble in both water and salt solutions. Plant seed proteins belong to the acid-soluble glutelins. But most hormone and enzyme proteins fall under the group of either albumins or globulins.

Properties

Proteins exhibit the identical properties of electrolytes. They are affected with the change of pH. When milk proteins are subject to low pH of around 4.5, the casein proteins are precipitated but at the same pH whey proteins still remain in solution. Both denaturing heat treatment and a low pH of about 4 are necessary to precipitate the whey protein.

Proteins are subject to hydrolysis. Hydrolysis of proteins can be brought about in presence of enzymes, acids, or alkalis and almost a complete hydrolysis can take place by a thermal treatment of protein with 6 M HCl at about 120°C for 8–10 h.

The charge on proteins may be affected by salt with the change in their solubility and salt can allow better protein–protein interaction and gluten formation.

Enzymes are nothing but proteins and they can act as biocatalysts in chemical and biochemical reactions. Most of the enzymes are denatured by a proper heat treatment and their catalytic actions are inactivated. Either yeast or mold or bacteria is being added in the food industries as a source of enzyme for various food products.

Denaturation

The delicate ordered three-dimensional structure of protein can be altered effectively by either heat or some chemical treatments without rupture of peptide bonds. The alteration of native conformation brings about the changes in some specific identifying properties of protein and this phenomenon is known as denaturation. The effect of temperature is so pronounced that for each 10°C rise the increase in denaturation rate is about 600-fold, and the effect of denaturation can be reduced by lowering the working temperature.

The denaturation assumes great importance in food processing, as most of the thermally processed foods may be heat-coagulated or denatured. Denaturation is an irreversible change in the special arrangement of a native protein molecule that alters chemical and physical properties of protein. It may cause many changes in protein like decrease in solubility, losses in catalytic and hormonal properties.

Gel Formation

Some dry proteins have the capability of absorbing water and they can form gels. When gelatin and casein are coagulated by the catalytic action of rennin, the formation of protein gel takes place.

Essential Amino Acids

Out of the 20 amino acids in body tissue, the nine, namely, histidine (essential for infants), leucine, isoleucine, lysine, methionine, phenylalanine, threonine, tryptophan, and valine, cannot be synthesized by humans and must be supplied in the human diet. These are called essential amino acids. Histidine is also essential for infants. The following are the chemical structures of a few essential amino acids.

$$
\begin{array}{ccc}
\underset{\text{Valine}}{
\begin{array}{l}
H_3C \\
\quad \diagdown \\
\quad HC-CH-COOH \\
\quad \diagup \quad | \\
H_3C \quad NH_2
\end{array}}
&
\underset{\text{Leucine}}{
\begin{array}{l}
H_3C-CH_2 \\
\qquad \diagdown \\
\qquad CH-CH-COOH \\
\qquad \diagup \quad | \\
H_3C \qquad NH_2
\end{array}}
&
\underset{\text{Lysine}}{
\begin{array}{l}
\qquad\qquad\qquad NH_2 \\
\qquad\qquad\qquad | \\
NH_2-CH_2-CH_2-CH_2-CH_2-CH-COOH
\end{array}}
\end{array}
$$

Lipids

Lipids consist of a diversified class of compounds that are insoluble or sparingly soluble in water but are soluble in nonpolar solvents, such as benzene, chloroform, ether, and carbon disulfide. Lipids are widely distributed in foods. They can be broadly divided into simple lipids and compound lipids. Triglycerides and waxes belong to simple lipids whereas other lipids come under compound lipids. Of them, triglycerides and phospholipids are quite common in food system.

Chemical Structure

The chemical reaction between glycerol and FAs to form triglyceride and water is shown as follows where R1, R2, and R3 represent different FAs.

$$
\begin{array}{lll}
\underset{\text{Glycerol}}{
\begin{array}{l}
CH_2OH \,+\, HOOCR_1 \\
| \\
CHOH \,+\, HOOCR_2 \\
| \\
CH_2OH \,+\, HOOCR_3 \\
\text{Glycerol} \quad \text{Fatty acids}
\end{array}}
& = &
\underset{\text{Triglyceride}}{
\begin{array}{l}
H_2C-COOR_1 \\
| \\
HC-COOR_2 \,+\, 3H_2O \\
| \\
H_2C-COOR_3 \\
\text{Triglyceride}
\end{array}}
\end{array}
$$

The triglycerides from animal sources with a higher percentage of saturated FAs (saturation) normally found in solid state at room temperature are called fats and the plant triglycerides rich in unsaturated FAs (instauration), usually available in liquid state at room temperature, are known as oils. Mostly liquid oils have shorter chain of FAs and more number of double bonds while solid fats are supposed to have longer chain of FAs with more saturation. In practice, lipids, fats, and oils mean same thing in food system. Most of the vegetable oils (except coconut and palm oils) contain more installation whereas solid animal fats (except fish oil) contain less instauration.

The group of aliphatic long straight chain carboxylic acids, $CH_3(CH_2)_nCOOH$, is called saturated FAs. Mostly, they are linear with some exceptions. Generally, FAs containing C_{12} to C_{24} are distributed in most of the vegetables and animal fats. Of them, lauric, myristic, palmitic, and stearic acids are common.

Unsaturated FA may contain one to six double bonds. The FAs in triglycerides vary in chain length or carbon number as well as in number of unsaturated double bonds. The chemical reactivity and functionality of a fat mainly depend upon its carbon number and double bond number in FA. Oleic, linoleic, and linolenic acids are present in many fats and oils. The structure of oleic acid can be presented by $CH_3(CH_2)_7CH=CH(CH_2)_7COOH$. Unsaturated oleic, linoleic, and linolenic acids are found in olive oil, safflower oil, and linseed oil to the extent of 75%, 60%–80%, and 50%–60%, respectively. Names and structures of few saturated and unsaturated FA in foods are enlisted as follows.

Saturated FAs	Structure	Unsaturated FAs	Structure
Myristic	$CH_3(CH_2)_{12}COOH$	Oleic	$CH_3(CH_2)_7CH=CH(CH_2)_7COOH$
Palmitic	$CH_3(CH_2)_{14}COOH$	Linoleic	$CH_3(CH_2)_4CH=CHCH_2CH=CH(CH_2)_7COOH$
Stearic	$CH_3(CH_2)_{16}COOH$	Linolenic	$CH_3(CH_2CH=CH)_3(CH_2)_7COOH$

Linolenic acid and some other unsaturated acids derived from oleic acid that must be supplied in the diet from plant sources are called *essential FAs* as these cannot be synthesized by animals.

Food lipids are characterized by their high energy content of about 9 cal/g. Vitamins A, D, E, and K, cholesterol, fat-soluble pigments like chlorophyll, carotenoids, etc., also belong to the class of lipids.

Oils and fats have various functions to play. They are used as a good medium of heat transfer in frying and there is practically no substitute for frying oil. Fats and oils can act as lubricants.

Fats can also provide flavor. Meat flavor is developed by degradation of its fat during cooking. Oils are effective tenderizing agents and fats can induce flakiness in pastry products. Fats can provide opacity as well as glossiness to the food products.

Rancidity

As regards lipid oxidation, rancidity, and stability, the storage stability of fat or fatty food is required to keep their odor and fresh taste till consumption. In lipid oxidation, the reaction of double bonds in lipid with oxygen causes the formation of off flavors usually during storage. The oxidation of FA causes undesirable flavors that may lead to unacceptable quality of food. The storage life of breakfast cereals, potato chips, and edible oils is limited due to their oxidative changes in presence of atmospheric oxygen leading to the development of off flavors, discoloration, the loss of essential FAs, and fat-soluble vitamins. This is called *oxidative rancidity*. The presence of iron or copper ions in cans can act as catalyst for lipid oxidation. Maintaining low storage temperature, removal of air or oxygen from food, usage of opaque or color, or impermeable packaging or vacuum packaging can be considered as preventive measures against lipid oxidation. The change in unacceptable quality of food may also occur owing to the hydrolysis of its oils and this phenomenon is known as *hydrolytic rancidity*. The hydrolytic rancidity owing to the reaction of triglycerides with water to form free FAs (FFA) can be accelerated by catalysts, high temperatures, acids, and bases. Raw milk undergoes hydrolysis with the formation of off flavor. The oil in raw rice bran is also hydrolyzed into FFA rapidly due to the catalyst lipase inherently present in it. Higher temperature can further accelerate hydrolytic rancidity.

Heating of fats and oils during frying causes degradation and polymerization of glycerides. It produces smoke, dark color, off flavors, and higher viscosity.

Vitamins

It is realized that besides carbohydrates, fats, proteins, minerals, and water, small amounts of vitamins are necessary in the diets of higher animals for growth, maintenance, and reproduction. But all animals do not require all vitamins.

The functions of vitamins in various animals are different. They can regulate metabolism, assist in growth, and help convert carbohydrates and fats into energy.

The vitamins can be divided into fat-soluble and water-soluble vitamins. The vitamins A, D, E, and K being fat-soluble are found in foods along with lipids or fatty foods such as butter, vegetable oils, fats of meat, and fish. These vitamins are thermally more stable than water-soluble B and C vitamins and are usually less lost during cooking and thermal processing of foods. Citrus fruits such as lemons, oranges, etc., are rich in vitamin C, which is lost significantly during processing. It is interesting to note that apart from vitamin A, the fruit guava contains almost five times more vitamin C than an orange; it is rich in fiber, folic acid, and minerals such as K, Cu, and Mn. Amla contains about 20 times more vitamin C than an orange. Vitamin C is an antioxidant that fights free radicals responsible for many grave diseases. Vitamin A being an alcohol ($C_{20}H_{29}OH$) and having its function in the retina of the eye is termed retinol. Therefore, prolonged deficiency of vitamin A may lead to night blindness.

Vitamin D termed calciferol is best available in fish liver oils. The deficiency of vitamin D in children during their growth of bones may cause rickets.

Vitamin E named tocopherol is widely distributed in milk fat, egg yolk, meat, nuts, wheat germ, soybean, and corn oils. Its function as an antioxidant prevents the formation of peroxides and it may have some role related to aging. The activity of vitamin K may be fairly reduced during processing, packaging, and storage of foods.

The K vitamins are napthoquinone derivatives. The green leafy vegetables, particularly spinach, cabbage, and lettuce, are their best sources. The deficiency of K vitamin may lead to prolonged clotting of blood.

Vitamins C and B complex are water-soluble. The various B vitamins are termed vitamins B-1, B-2, B-6, B-12, or by their chemical names.

Few Chemical Structures

The vitamins under B complex group are as follows: thiamin (B-1), riboflavin (B-2), cobalamin (B-12), folic acid, biotin, pantothenic acid, and few other vitamins are necessary for lower animals.

Pyridoxine (B-6)   Niacin

Most of the water-soluble vitamins are components of essential enzymes. Generally, B vitamins acting as coenzymes play some roles in metabolism of the living cells. So, a regular supply of these vitamins is needed for the normal function of the body. These vitamins are mostly lost during cooking and thermal processing.

Cereals, yeasts, and liver are the good sources of these vitamins.

The loss of vitamins and minerals occurs during various stages of processing right from washing, peeling, to thermal processing. In grain milling, the removal of bran and germ causes the loss of fats, proteins, vitamins, and nutritional value. Particularly, the water-soluble vitamins and minerals are easily lost during washing and cooking. The most heat-sensitive vitamins are thiamin and ascorbic acid or vitamin C and their losses during thermal processing like baking and canning can be significant.

Vitamins A and E being fat-soluble may be lost due to oxidation of lipids. Vitamin A and riboflavin are light-sensitive and they may undergo chemical changes on exposure to sunlight. When milk packaged in transparent pouches is exposed to light, a significant loss of riboflavin occurs during storage. Generally, minerals being insensitive to heat are not lost during thermal processing.

Color

Color is one of the key parameters for the acceptability of foods. Sometimes desirable food colors are developed during processing as in the case of baked products, while in some other cases food colors are degraded in the course of thermal processing in canning of green beans or green peas. Hence, retention of original colors of processed fruits and vegetables is highly desirable for the attraction and acceptability of consumers. The pigments, chloroplasts, and other chromoplasts are found in fruits and vegetables within their cell protoplast and vacuole. These pigments belong to the following main groups: *chlorophylls*, *carotenoids*, and *flavonoids*.

The fat-soluble *chlorophylls* in the chloroplasts of green leaves are mainly responsible for synthesis of carbohydrates from carbon dioxide and water. Natural green color of chlorophyll in plants is due to the presence of the fat-soluble green pigment, which is chemically unstable. The stability of chlorophyll is affected by Mg^{2+} ions, acid, and heat. On heating in acidic media, green chlorophyll present in bright green vegetables is converted to olive green–colored pheophytin. Practically, an irreversible loss of this color of green vegetables takes place after thermal processing.

Carotenoids are constituted by eight isoprene units and most of them have 40 carbons in their chemical structure as shown as follows.

Isoprene group

C_{40} carotenoids (eight isoprene groups)
Lycopene

Carotenoids can be classified either by their chemical structure or by their functionality.

Carotenoid can have only elements C and H like β-carotene or lycopene; otherwise, they may also have O such as lutein. By their functionality in plants, they are considered as a primary carotenoid as it is necessary in the photosynthetic process like P-carotene or lutein, while a secondary carotenoid is not directly associated with the plant survival like a-carotene or lycopene. *Carotenoids* have a great impact in the food industry as natural colorants. They have the color ranging from a pale yellow through a strong orange to a dark red. Tomatoes, carrots, oranges, pineapples, and other plant products are rich in fat-soluble orange red carotenoid pigments. The carrot, being rich in vitamin A, has also K, Ca, P, S, and Na. Its P-carotene helps purifying blood and fighting infection. Carotenoids can inhibit some diseases like cancer. The common carotenoids in foods are β-carotene (orange), γ-carotene (orange), δ-carotene (orange), α-carotene, lutein, lycopene (red), and canthaxanthin. Few forms of fruit and vegetable carotenoids are shown as follows.

α-Carotene (orange)

β-Carotene (orange)

They remain chemically stable to a certain extent during thermal processing but owing to the presence of double bonds they are subject to the oxidative changes on exposure to air (Bast et al., 1998; Bertram and Vine, 2005; Britton, 1988). Air drying at high temperature causes degradation of carotenoids in dried foods. Sometimes, β-carotene with vitamin A can be added to food products to improve both color and nutritional value.

β Carotene

COH + COH

Retinal

Chemical structure of β-carotene and its splitting into two molecules of vitamin A: The β-carotene splits into two molecules of vitamin A. There are two pathways for carotenoid conversion to vitamin A. However, central cleavage, in the 15:15′ double bond of the carotenoid backbone, leads to two molecules of retinal (vitamin A). Carotenoid as vitamin A regulates the expression of several genes and functions against some forms of cancer at the stages of initiation as well as progression. It is involved in human growth and vision and eye health.

The colors of fat-soluble carotenoids may range from yellow, orange, to red. Orange carotene of carrot, peach, apricot, and citrus fruits, red lycopene of tomatoes and watermelon belong to the carotenoids group only. Some carotenoids act as precursors to vitamin A. Carotenoids are heat-resistant but they are sensitive to oxidation resulting in loss of color and vitamin A during food processing. The effect of lycopene on tumor cell reveals that lycopene resists different cancer cells such as prostrate cancer cells and others (Ajlouni et al., 2001).

Lycopene

β-Carotene

Lutein

Zea xanthin

The colors of apples, grapes, berries, and cherries ranging from red to blue are due to the presence of water-soluble pigments from the *anthocyanin* family. These colors can be changed with the change in pH. Under alkaline media, the usually red anthocyanin turns into pale blue or gray. They are also subject to discoloration in presence of iron or tin ions. Beet colors originated from *betalains* are also having similar properties. The color concentrates extracted from beets, grapes, or other sources of anthocyanins can be used for other foods. Their colors are also subject to change on heating, oxidation, and with the variation of pH.

Another related family of white pigments, *anthoxanthins*, found in potatoes and wheat flour turn yellow in alkaline media. They turn brown or black in contact with iron.

Some colorants available from natural sources that can be used as food ingredients are termed *uncertified colors*. Caramels produced from sugar browning or Maillard reaction are used in cola beverages.

The artificial colors termed *certified colors* are water-soluble in the form of dyes. The six safe dyes with a common name yellow No. 5 are approved for food uses.

Generally, artificial colors are more stable and consistent during processing and storage.

Flavor

Color or appearance of a food certainly attracts the consumer but finally flavor determines its acceptability. Flavor is a phenomenon that causes an overall sensation of taste, odor, or aroma, hot or cold feeling, appearance, and texture. Flavor may or may not have any nutritional value but it greatly influences the acceptance of food. Hence, flavor in food is an essential quality component.

In climacteric fruits, minute amounts of carbohydrates, lipids, proteins, and amino acids are converted to volatile flavors and the flavor formation increases after the respiration climacteric and it continues till harvesting of fruits.

The flavor components responsible for characteristics in fruits may be present as low in concentrations as ppm or pbm (parts per billion). The aroma and flavor chemical profiles of different fruits may be grouped under the following heads.

Acids: Among the carboxylic acids, the nonvolatile citric as well as malic acids are the most plentiful and widely dispersed in fruits and they are followed by tartaric, fumeric, ascorbic (vitamin C), and traces of others. The radish has vitamin C along with Fe, Na, and Ca, but cooking destroys vitamin C mostly but not minerals. The red cabbage contains flavonoids.

Esters: Numerous esters are the important natural components of fruit flavor and their variations characterize the fruit type.

Carbonyls: The aroma and flavor of many fruits mostly depend on the presence of certain carbonyls such as benzaldehyde in stone fruits, acetaldehyde in oranges, furfural in strawberries, and 5-hydroxy-2-methyl furfural in pineapples.

Lactones: Lactones having functional carbonyl group impart characteristic aroma in some fruits like peach and apricot. Banana volatiles are mostly composed of ethanol, *n*-butanol, isobutanol, 2-pentanol, hexanol, ethyl acetate, butyl acetate, isobutyl acetate, and others. Flavor developed during processing may be desirable or undesirable. Mostly, flavor is degraded with the progress of time and thus often it shortens the shelf life. Either natural or artificial flavor can be added to food as an ingredient. Most artificial flavorings formulated are to be at a minimum dose rate of 0.1%. The typical doses of majority of fruit aroma concentrates are supposed to be used at about 0.5%. Natural flavoring agents generally being used in foods are spices, essential oils, etc. A large number of synthetic flavors are also in use as food additives.

Along with appearance, taste, and smell, the textural aspects such as hardness, elasticity, brittleness, chewiness, and gumminess also contribute to food acceptability and flavor. Despite the individual contribution of color, odor, and texture of food, the composite flavor sensation depends on the interaction of all these sensations in the mouth called mouth feel.

A number of parameters including temperature influence the sense of taste. The sense of taste is divided into four basic tastes, namely, sweet, salty, sour, and bitter.

The only inorganic compound sodium chloride, NaCl, provides pure salty taste.

The sweet taste comes from the carbohydrates like sugars and few other organic compounds such as saccharin and high intensity sweeteners. The sweetness varies with the sugars or sweeteners.

The proton donors cause sourness that is a function of pH of any acidic food. But at the same pH_2, CH_3OOH, acetic acid is sourer than HCl. Longer-chain FAs are more sour than shorter-chain FAs.

Bitter taste is produced by some alkaloids such as quinine and nicotine. Caffeine, a component of coffee and tea, is bitter in taste. Tannin being a polyphenol also imparts bitterness and astringency in black tea. Coffee is a complex mixture of chemicals that provides significant amounts of caffeine and chlorogenic acid. Chemically, caffeine is 1,3,7-trimethylxanthine. Recent studies show that coffee drink may help prevent some chronic diseases including type 2 diabetes, Parkinson's, but at the same time it may be associated with significant increase in blood pressure. Two to three cups of filtered coffee having 250–350 mg caffeine per day may be allowed for adults. Unfiltered coffee containing cafestol may significantly increase the cholesterol level.

Caffeine

Other taste sensation like heat of pepper comes from capsaicin or pipeline and cooling sensation from menthol. A compound called flavor enhancer that brings about better flavor is monosodium glutamate.

Allicin is one of the major organosulfur compounds in garlic (as shown later), considered to be the biologically active, which has antimicrobial and anticancer activities.

$$CH_2=CHCH_2S-SCH_2CHCH_2$$
$$\overset{\|}{O}$$

Allicin

Along with the appearance, taste, and smell, the textural aspects such as hardness, elasticity, brittleness, chewiness, and gumminess also contribute to food acceptability and flavor. Despite individual contribution of color, odor, and texture of food, the composite flavor sensation arising from the interaction of all these sensations in the mouth called mouth feel is also important.

Chemical structures: A few structures of flavoring compounds of fruits and others are shown as follows.

$$
\begin{array}{cccccc}
H_3C\text{-}\underset{\underset{H}{|}}{C}=O & H_3C\text{-}(CH)_4\underset{\underset{H}{|}}{C}=O & H_3C\text{-}\overset{\overset{O}{\|}}{C}_{OC_5H_{11}} & CH_3C\overset{\overset{O}{\|}}{}_{OC_8H_{17}} & C_4H_9C\overset{\overset{O}{\|}}{}_{OC_5H_{11}} & \\
\text{Acetaldehyde (butter)} & \text{Hexanal (apple)} & \text{Banana} & \text{Orange} & \text{Apple} & \text{Lemonene}
\end{array}
$$

Enzymes

Enzymes are proteins that act as biocatalysts to activate the rate of chemical and biochemical reactions. Enzymes are naturally distributed in cells and tissues of plants and animals. Yeasts, molds, and bacteria are all the sources of enzymes in food. They may bring about both desirable and undesirable chemical changes in foods. In food processing industries, the enzymes may play key roles for ripening cheese, fermenting starch to sugar and finally alcohol, or developing certain flavor and texture in fruit and vegetable products. Malted barley is used in brewing, and addition of rennin to milk for cheese making is a common technique.

At the same time, enzymes are also responsible for discoloration of cut apples, potatoes, mushiness of the over ripe fruits, and rancidity of milk fat.

The catalytic properties of enzymes are due to their capability of specific activity. Certain enzymes are dependent upon the cooperation of some nonproteins—organic and inorganic compounds, generally called cofactors.

Generally, the majority of enzymes in any living system have a temperature of maximum activity between 30°C and 40°C in the pH range from 4.5 to 7.5, while undesirable enzymes are often inactivated or denatured to lose their catalytic activity by heating at 55°C–85°C, changing pH to lower levels or adding some salts. The oxidative browning/blackening caused by the catalytic activities of polyphenol oxidase (PPO)/polyphenolase may be prevented by adding a reducing agent like ascorbic acid or creating an anaerobic environment.

The important factors that influence the catalytic reaction rate of enzymes are temperature, pH, enzyme concentration, and substrate concentration among others.

Enzymatic browning: Both nonenzymatic and enzymatic browning may take place in foods.

Nonenzymatic browning has already been covered under sugar earlier.

As regards *enzymatic browning/blackening*, this phenomenon is commonly observed on the cut surfaces of guavas, bananas, apples, potatoes, as well as in other fruits and vegetables, mainly due to catalytic action of the enzyme phenolase. When the phenolic substrates present in these foods are exposed to air/oxygen in presence of phenolase enzyme, rapid browning of the cut surfaces occurs by the oxidation of phenols to orthoquinones, which rapidly polymerize to a brown pigment melanin. Other phenolic substances like tyrosine, chlorogenic acid, and caffeic acid also act as substrates. Thus, the pigment melanin is formed during processing of fruits and vegetables.

As stated, PPO is widely distributed in fruits and phenolase can catalyze many oxidative reactions with the change in phenolic compounds naturally present in many fruits. This enzymatic browning can be prevented by elimination of oxygen, which can be accomplished by processing fruit products under vacuum or in the presence of an inert nitrogen gas. The reducing agents usually used in the fruit processing industry are sodium bisulfite and ascorbic acid to inactivate the enzyme as well.

The enzymes may be irreversibly inactivated at a pH of about 3 or less by the addition of citric, malic, or phosphoric acids. Browning can also be prevented by exhausting or by application of vacuum.

In raw and processed fruits and vegetables, the chemical and biochemical activities cause brown, yellow, red, and black discoloration. The browning/blackening reactions that occur in raw fruits and vegetables as a result of abiotic stresses are generally considered to be due to enzymatic oxidation of phenolic compounds. PPO is inducible upon biotic or abiotic wounding. Blackening of the skin of raw peaches and nectarines may be related to abrasion damage during fruits handling. The formation of ferric complexes with anthocyanins may contribute to the black discoloration.

The development of pink color in canned lychee juice may be due to both enzymatic and nonenzymatic reactions of phenolic and other compounds during pre-processing, thermal processing, and storage.

The addition of tartaric acid or 30% sugar syrup containing 0.1%–0.15% citric acid attaining a pH level of 4.5 controls this undesirable pink discoloration.

Chapter 20

Food Engineering Operations

Fluid Mechanics

Newtonian Fluid Flow

Introduction

The fluids flow under gravity without retaining their shape and they may exist in gaseous, liquid, or solid states at different temperatures.

The behavior of an ideal or Newtonian fluid is that the rate at which the fluid material deforms, γ is proportional to the applied force, τ. For an ideal fluid, the shear stress, τ, Pa(N/m^2) increases linearly with the increase of shear rate, γ, 1/s. When shear stress is plotted against shear rate, a straight line is produced passing through the origin where its constant slope gives the value of Newtonian viscosity or dynamic viscosity or absolute viscosity, μ Pa s that is the proportionality constant. Accordingly, μ is the proportionality constant, being independent of γ, τ is correlated to i by

$$\tau = \mu\gamma = \mu(dv/dy) \quad \text{or} \quad \mu = \tau/\gamma \tag{20.1}$$

The unit of p in SI is Pa s = (N/m^2)s = (kg m/m^2 s^2)s = kg/m s.

In cgs or MKS units, μ = (dyne/cm^2) = poise = 0.1 Pa s; 1 Pa = 10 dyne/cm^2; and 1 centipoises (cp) = 1 m Pa s. Kinematic viscosity is also expressed in stokes, S or cS and 1 S = 100 cS.

The Newtonian fluids may also be expressed in terms of kinematic viscosity as follows: Kinematic viscosity = (dynamic viscosity/density) = μ/p), m^2/s.

Fluid materials, having low molecular weights, mutually soluble components that are made up of single phase, are Newtonian.

Some Newtonian fluids found in the food and agricultural industries are water, vegetable oils, glycerin, ethanol, ethylene, gasoline, skim milk, most honeys, filtered fruit juices, and syrups. The viscosity of water and honey is about 1 and 8880 cp, respectively. The fluids that do not follow the equation of Newton's law of viscosity are called non-Newtonian fluids.

The *continuity equation* for mass balance in a stream tube fluid flow follows the principle of conservation of mass and it can be expressed by

$$\rho_1 A_1 v_1 = \rho_2 A_2 v_2 = \rho A v = w = \text{constant} \tag{20.2}$$

$A_1 v_1 = A_2 v_2 = Av$ for isothermal incompressible fluid flow as its density is constant, and so its flow through a circular cross section or pipe or tube may be presented by

$$(\pi/4)D_1^2 \rho_1 v_1 = (\pi/4)D_2^2 \rho_2 v_2 = w, \quad \text{or} \quad D_1^2 \rho_1 = D_2^2 \rho_2 V_2 \quad \text{and}$$

$$\text{when} \quad \rho_1 = \rho_2, \quad v_1 = (D_2/D_1)^2 v_2, \quad \text{and} \quad v_2 = (D_1/D_2)^2 V_1 \tag{20.3}$$

Reynolds Number

When a fluid flows over a solid surface, a velocity profile develops owing to its viscous properties. The fluid adjacent to the solid wall has a sticking tendency to the solid surface with an increasing velocity away from the wall. As a result, a boundary layer develops within the flowing fluid, which has a pronounced influence on viscous properties of fluid. There are certain boundaries that demarcate the fluid flows into three groups, namely, laminar, transitional, and turbulent flows. The exact demarcation is made by the dimensionless Reynolds number (Re), which is defined by the ratio of inertial forces, ρv^2 to viscous forces, $\mu v/D$ as follows:

$$\text{Re} = (\text{Inertial forces/viscous forces}) = \rho v^2/\mu v/D = (Dv\rho/\mu) = DG/\mu$$

$$\text{or,} \quad \text{Re} = (Dv\rho/\mu) = 4w/\mu\pi D \tag{20.4}$$

where
D is the pipe diameter, m (ft)
ρ is the density, kg/m³ (lb$_m$/ft³)
μ is the viscosity, kg m/s or Pa s (lb$_m$/ft)
A is the pipe cross-sectional area = $(\pi D^2/4)$
v is the average velocity (volumetric flow rate/pipe's cross-sectional area), m/s (ft/s)
G is vp, the mass flow rate, kg/(m² s)
w is the fluid flow in kg/s

Points 1 and 2 indicate two locations of a pipe

In practice diameter, D is considered to be the hydraulic diameter, D_H, as defined as follows:

$$D_H = 4 \,(\text{Cross-sectional area, } A/\text{wetted perimeter, } P)$$

$$D_H = 4[(\pi/4)D^2/(\pi D)] = D \text{ for a circular pipe}$$

$$D_H = (\pi/4)[(D_o^2 - D_i^2)/\pi(D_o + D_i)] = (D_o - D_i) \tag{20.5}$$

for an annular pipe, where D_o and D_i are outer and inner diameters of the annulus, respectively.

The Re gives an insight into energy dissipation due to the viscous effects. The aforementioned physical definition of Re shows that if viscous forces dominate on the effect of energy dissipation then Re is small (Re < 2100) or flow is laminar.

The large value of Re (Re > 4000) indicates turbulent flow revealing small influence of viscous forces on energy dissipation. Therefore in a straight circular pipe, laminar or viscous flow occurs at low fluid velocities without the presence of swirls or eddies for Re < 2100 and turbulent flow prevails at high fluid velocities in the presence of swirls and eddies for Re > 4000 and above except in some special cases.

The fluid velocity distribution or profile across a pipe for both laminar and turbulent flow is shown in Figure 20.1.

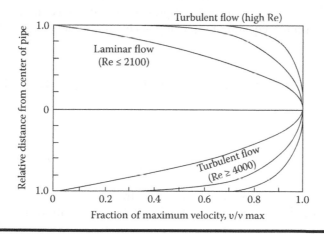

Figure 20.1   Fluid velocity distribution across a pipe.

Bernoulli Equation

On the basis of overall mechanical energy balance, the following differential form of Bernoulli equation for frictionless flow of ideal incompressible fluid may be expressed, which is applicable for either along a streamline or a stream tube:

$$d(v^2/2) + (dz)g + (dp/\rho) = 0 \quad \text{(SI)}$$

$$d(v^2/2g_c) + dz(g/g_c) + (dp/\rho) = 0 \quad \text{(English)}$$

(20.6)

The first term *kinetic energy* (KE) of the equation refers to the energy stored within the moving fluid body, which is equal to the work required to bring a body from rest to the same velocity; the second term *potential energy* (PE) gives the energy due to the relative height of the fluid; and the third one that corresponds to fluid's *pressure* and *volume* (PV) shows the energy released or needed due to change of pressure from one location to other.

Integrating the aforementioned equations within proper limits of two definite cross sections of a stream tube or between two particular points 1 and 2 of a streamline, the following equations can be derived:

$$\left(v_1^2/2\right) + z_1 g + p_1/\rho_1 = \left(v_2^2/2\right) + z_2 g + p_2/\rho_2 \quad \text{(SI)}$$

$$\left(v_1^2/2g_c\right) + (z_1 g/g_c) + p_1/\rho_1 = \left(v_2^2/2g_c\right) + (z_2 g/g_c) + p_2/\rho_2 \quad \text{(English)}$$

(20.7)

All terms of the equation must have consistent unit, J/kg (ft lb$_f$/lb).

Assuming constant fluid density, uniform equivalent pressure over the cross section, single phase, uniform material properties and neglecting effects of thermal energy, Bernoulli equation for incompressible fluid flow in a pipe line can be expressed as

$$\left(v_1^2/2\alpha_1\right) + z_1 g + p_1/\rho_1 - W_s \eta = \left(v_2^2/2\alpha_2\right) + z_2 g + p_2/\rho_2 + \Sigma F \quad \text{(SI)} \qquad (20.8)$$

In general for isothermal flow of a liquid, $\alpha_1 = \alpha_2$ and $\rho_1 = \rho_2$

$$\left(1/2\alpha\right)\left[\left(v_2^2 - v_1^2\right)\right] + g(z_2 - z_1) + (p_2 - p_1)/\rho + \Sigma F + W_s \eta = 0 \quad \text{(SI)} \qquad (20.9)$$

velocity head	elevation head	pressure head	friction losses	pump/ shaft work

$$(1/2\alpha)[(v_2^2) - (v_1^2)]/g_c + g(z_2 - z_1)/g_c + (p_2 - p_1)/\rho + \Sigma F + W_s \eta = 0 \quad \text{(English)}$$

where

p is the pressure, N/m², Pa (lb$_f$/ft²)

z is the height, m (ft)

g_c is the gravitational conversion factor = 9.8 kg m/kg f s² (32.174 ft lb/lb$_f$ s²)

$-W_s$ is the actual shaft work done by pump, ft lb$_f$

$-W_{sr}$ is the frictionless shaft work done by pump, ft lb$_f$

η is the overall pump efficiency ($-W_{sr}/W_s$)

α is the KE correction factor (dimensionless) varying from 0.5 for laminar flow to 1.0 for turbulent flow

ΣF is the sum total of all frictional losses owing to pipes of different diameters, valves, and various fittings, J/kg (ft lb$_f$/lb)

The subscripts 1 and 2 refer to two specific locations in the system.

In Bernoulli equation, the total friction losses ΣF may be identified as major friction losses during transportation of fluid in long straight pipes and the minor *friction losses* due to *expansion, contraction, valves,* and *other fittings* measured by the coefficients, K_e, K_c, K_v, and K_f, respectively, which may be expressed as follows:

$$\Sigma F = \underbrace{\left(4fL/D\right)v^2/2}_{\text{major friction losses}} + \underbrace{(K_c + K_e + K_v + K_f)v^2/2}_{\text{minor friciton losses}} \quad \text{(SI)}$$

(20.10)

$$\Sigma F = (4fL/D + K_c + K_e + K_v + K_f)v^2/2g_c \quad \text{(English)}$$

K_c, K_e, K_v, and K_f can be calculated for turbulent flow in both sections by the following:

$$h_e = (v_1 - v_2)^2/2\alpha = (1 - A_1/A_2)^2 \left(v_1^2/2\alpha\right) = K_e \left(v_1^2/2\alpha\right) \quad (20.11)$$

where

h_e is the friction loss due to expansion, J/kg

Sudden expansion loss coefficient, $K_e = (1 - A_1/A_2)^2$

v_1 is the upstream velocity in the smaller area

v_2 is the downstream velocity, m/s

$\alpha = 1$ (approximately) for turbulent

$\alpha = 1/2$ for laminar flow

$$h_c = 0.55(1 - A_2/A_1)\left(v_2^2\right)/2\alpha \quad (20.12)$$

where

h_c is the friction loss due to contraction, J/kg

Sudden contraction loss coefficient, $K_c = 0.55(1 - A_2/A_1)$

v_2 is the average velocity of smaller section in the downstream, m/s

Laminar Flow

Friction in Long Straight Pipe

The following Hagen–Poiseuille equation has been developed for calculation of pressure drop, Δp and friction loss, ΣF in laminar flow (Re < 2100) of Newtonian fluid through a long straight circular pipe of accurately bored uniform diameter with the assumption that the fluid temperature is constant and accordingly its density and viscosity remain constant, the velocity is fully developed at both up- and downstream, and the fluid adheres to the tube wall:

$$\Delta p = (p_1 - p_2) = 32\mu v(L_2 - L_1)/D^2 = 32\mu v\Delta L/D^2 \quad \text{(SI)} \qquad (20.13)$$

It can also be expressed as

$$\left[D\Delta p/4\Delta L\right] = \mu\left(8v/D\right) \qquad (20.14)$$

Thus in a circular pipe, the shear rate of Newtonian fluids can be calculated from the relationships

$$\gamma = (8v/D) = \left(32Q/\pi D^3\right) = \left(32w/\pi\rho D^3\right) \qquad (20.15a)$$

where
  Q is the volumetric flow rate, m³/s
  w is the mass flow rate, kg/s
  ρ is the density of the fluid, kg/m³

The shear rate at the wall, τ_w, in a circular tube becomes

$$\tau_w = D\Delta p/4\Delta L = \mu\left(8v/D\right) \qquad (20.15b)$$

The maximum velocity, v_{max}, at the center and the average velocity, v, in a circular pipe for a fluid flow in the laminar regime are expressed as follows:

$$v_{max} = \left[(\Delta p r^2)/(4\mu\Delta L)\right] \qquad (20.16)$$

$$v = \left[(\Delta p r^2)/(8\mu\Delta L)\right] = \left[(\Delta p D^2)/(32\mu\Delta L)\right] = \tfrac{1}{2}v_{max} \qquad (20.17)$$

where
  p_1 and p_2 are pressures, N/m² at points 1 and 2
  v is the average fluid velocity in tube, m/s
  D is the inside diameter of tube, m
  r is $(D/2)$, the radius of the tube, m
  $(L_2 - L_1)$ is ΔL, straight tube length, m

$$\text{Friction loss, } \Sigma F = \text{(N m)/kg} \quad \text{or} \quad \text{J/kg} \quad \text{(SI)}$$

$$\text{Friction loss, } \Sigma F = \text{(ft lb}_f\text{)/ lb}_m \quad \text{(English)}$$

Fanning Friction Factor

Fanning friction factor, f, for circular pipes is defined as the drag force per wetted surface area (shear stress τ_w) divided by the KE:

$$f = \left[(\tau_w)/(1/2)(\rho v^2)\right] = \left[(\Delta p r^2/2\pi r \Delta L)/(1/2)(\rho v^2)\right] = \left[(\Delta p r)/(\Delta L \rho v^2)\right]$$

(20.18)

Rearranging and rewriting in terms of diameter, D, the following Fanning equation is obtained for pressure drop in terms of Fanning friction factor, which is valid for all types of fluid flow from laminar regime Re < 2100 to turbulent regime Re > 4000 provided appropriate f is available:

$$\Delta p = \left[(2 f \Delta L \rho v^2)/D\right] = 4f\left(\Delta L/D\right)(v^2/2)\rho$$

(20.19)

Friction loss, $\Sigma F = [(p_1 - p_2)/\rho] = (\Delta p/\rho)$

$$\Sigma F = 4f\left(\Delta L/D\right)(v^2/2) \quad \text{(SI)}$$

$$\Sigma F = 4f(\Delta L/D)(v^2/2g_c) \quad \text{(English)}$$

(20.20)

The average velocity, v, in a circular pipe for laminar flow is expressed as follows:

$$v = (Q/\pi r^2) = (1/\pi r^2)[\pi(\delta P)r^4/(8L)] = [(\delta P)D^2/(32\mu\Delta L)]$$

The pressure drop per unit length can be obtained from the aforesaid equation as follows:

$$(\delta P)/L = (32\mu v)/D^2$$

(20.21)

From these equations, friction factor, f, for *laminar flow* in pipes can be expressed in a convenient form:

$$f = 16/\text{Re} = 16/(Dv\rho/\mu)$$

(20.22)

Therefore, the plot of log f vs. log Re will yield a straight line with a slope of –1.

Turbulent Flow

In turbulent fluid flow friction factor, f cannot be predicted theoretically from Reynolds number. The relationship between f and Re is available from the experimental data. Another parameter called relative roughness (K/D) also influences f

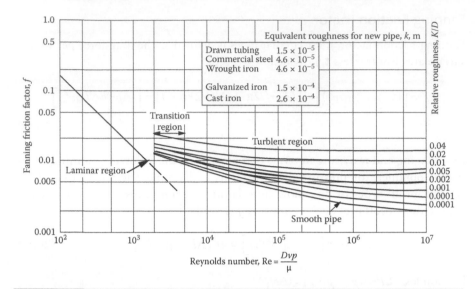

Figure 20.2   Fluid friction factors inside pipes. (From Moody, L.F., *Trans. ASME*, 66(8), 671, 1994.)

significantly in case of rough pipes, where K is the average roughness height and D is the base diameter. The copper, brass, and glass pipes are taken as smooth pipes. The log–log plot of f vs. Re is available for $10^2 \leq \text{Re} \leq 10^7$ (Figure 20.2) and is known as Moody diagram. For rough pipes, f does not change appreciably beyond Re $> 10^6$ (Tables 20.1 and 20.2).

Table 20.1   Frictional Losses for Turbulent Flow through Fittings and Valves in terms of Equivalent Resistance, Le/D and Number of Velocity Head, K_f

Fittings	Le/D	K_f	Fittings	Le/D	K_f
45° Elbows	15	0.3	Gate valves, open	7.0	0.15
90° Elbows	35	0.7	Gate valves, open	225	21
Tees (branched flow)	50	1.0	Gate valves, half open	300	6.0
Tees line flow threaded	—	0	Angle valves, open	170	2.0
Couplings and unions	2.0	0.04	Globe valves, half open	475	9.5
Union threaded	—	0.8	180 retum bend, threaded	—	0.8

Equivalent resistance, Le/D = equivalent length of straight pipe/internal pipe diameter.

Table 20.2 Dimensions of Commonly Used Steel (S) Pipe of Schedule 40

Nominal Size (in.)	S Pipe ID in (m)	S Pipe OD in (m)	Nominal Size (in.)	S Pipe ID in (m)	S Pipe OD in (m)
1/2	0.622 (0.01579)	0.840 (0.02134)	3/4	0.824 (0.01021)	1.050 (0.0127)
1	1.049 (0.02644)	1.315 (0.03340)	11/2	0.1.610 (0.04089)	1.900 (0.04826)
2.0	2.067 (0.0525)	20.375 (0.06033)	21/2	2.469 (0.06271)	2.875 (0.07302)
3.0	3.068 (0.07793)	3.50 (0.08890)	4.0	4.026 (0.10226)	4.50 (0.11430)

For estimation of f in turbulent flow, the following Von Karmon correlation is very useful:

$$(1/\sqrt{f}) = 4.0 \log_{10}(\text{Re }\sqrt{f}) - 0.4 \qquad (20.23)$$

For friction factor, f, in turbulent flow through smooth tubes and pipes, the following empirical equations, respectively, can also be used with an accuracy of ±5%:

$$f = 0.00140 + 0.125/(\text{Re})^{0.32} \text{ for tubes} \qquad (20.24)$$

$$f = 0.0035 + 0.264/(\text{Re})^{0.42} \text{ for pipes} \qquad (20.25)$$

Transition region: The friction factor, f, in the transition region $2100 < \text{Re} < 4000$ cannot be predicted accurately by the equations.

Example 20.1

A pump draws water at an atmospheric condition at a rate of 0.145 m³/min from a large open reservoir on the ground floor. The pump discharges to the overhead tank that is 6.62 m above the ground floor. The frictional losses in the piping system of inside diameter 0.0508 m are taken as 2.4 J/kg. At what height in the tank must the water level be kept if the pump can develop a power 0.0932 kW only?

Let h m be the height of water of level in the reservoir and the efficiency of the pump is 100%. Frictional losses in the pipe line, $\Sigma F = 2.4$ J/kg and water density, $\rho = 1000$ kg/m³ (approximately).

Water flow in pipe = 0.145 m³/min = 2.417×10^{-3} m³/s
Water velocity in the pipe = $Q/A = (2.417 \times 10^{-3} \text{ m}^3/\text{s}) (\pi/4) (0.0508 \text{ m})^2 = 1.2$ m/s

Mass flow $= 2.417 \times 10^{-3}\,m^3/s \times 1000\,kg/m^3 = 2.417\,kg/s$
Hence shaft work, $w_s = -93.2\,J/s \times 1/(2.417)\,kg/s = -38.56\,J/kg$
Using Bernoulli equation between two sections

$$p_1/\rho + z_1 g + \left(v_1^2\right)/2 - w_s \eta = p_2/\rho + z_2 g + \left(v_2^2\right)/2 + \Sigma F$$

According to the present conditions, $p_1 = p_2 = 1\,atm$, $v_1 = 0$, $z_1 = 0$, $\eta = 1$, and $z_2 = (4.62 - h)\,m$. We get the following:

$$(6.62 - h)g + (1.2)^2/2 + 2.4 = -(-38.56) = 38.56$$

$$\text{or} \quad (6.62 - h)g = 38.56 - 2.40 - 0.72 = 35.44$$

$$\text{or} \quad (6.62 - h) = 35.44/9.81 = 3.61\,m$$

Hence the height of water in the reservoir, $h = (6.62 - 3.62) = 3.00\,m$.

Example 20.2

Water at 15.5°C is pumped from a large reservoir at the ground level to the top of a hill through a pipe of 0.1524 m (ID) at an average velocity of 3 m/s. The pipe discharges into the atmosphere at a height of 1000 m above the level of the reservoir. The pipe line is 1540 m long. If the overall efficiency of the pump is 75%, estimate the efficiency needed to pump the water.

Pipe length = 1540 m, pump efficiency, $\eta = 75\%$, pipe diameter, $D = 0.1524\,m$.
Water viscosity, μ at $15.5°C = 1.129 \times 10^{-3}\,kg/m\ s$, water density, $\rho = 1000\,kg/m^3$ (approximately).
$Re = (Dv\rho)\mu = (0.1524 \times 3.0 \times 1000)/(1.129 \times 10^{-3}) = 4.05 \times 10^5$ (turbulent flow)
For turbulent water flow in the smooth pipe,

$$f = 0.0014 + 0.125/(Re)^{0.32} = 0.0034$$

Total friction losses in the pipe line,

$$\Sigma F_s = 4f(L/D)(v^2/2) = 4 \times 0.0034 \times (1540/0.1524) \times 3^2/2 = 618.43\ J/kg$$

Applying Bernoulli equation between two sections and taking top of the water level in the large reservoir as a datum line, $Z_1 = 0$, $v_1 = 0$, and $p_1 = p_2 =$ atmospheric pressure,

$$-W_s \eta = Z_2 g + (v_2^2/2) + \Sigma F_s$$

$$-W_s \times 0.75 = 1000 \times 9.81 + (3^2/2) + 618.43 = 10,432.93$$

$$-W_p = 13,910.57\ J/kg$$

$$\text{Power} = mW_s = \rho Av \times W_s = 1000 \times (\pi/4)(0.1524)^2 \times 3 \times 13,910.57$$

$$= 760,501.9\ W = 760.50\ kW$$

Flat Plate/Slit Flow

Applying Newton's law of viscosity, the following relationships for the velocity profile of flat plate can be derived:

the average fluid velocity

$$v = \Delta p \delta^2 / 3\mu L \tag{20.25}$$

the maximum fluid velocity

$$v_{max} = (\Delta p \delta^2)/2\mu L \tag{20.26}$$

Hence, average fluid velocity

$$v = (2/3)v_{max} \tag{20.27}$$

The hydraulic diameter for slit, $D_H = 4\delta$ and

$$Re = (D_H v \rho)/\mu = (4\delta v \rho)/\mu \tag{20.28}$$

Shear stress on the plate wall,

$$\tau_w = \mu(3v/\delta) \tag{20.29}$$

The fanning friction factor, fin terms of shear stress, τ_w on the plate wall: $f = \tau_w (1/2)\rho v^2$. The pressure difference can be written in the form $\Delta p = 2fp(L/2\delta)(v^2/2)$; $f = (2\Delta p\delta/L\rho v^2)$. Hence in the *laminar region*, the friction factor,

$$f = 24/Re \tag{20.30}$$

for slit flow.

For slit flow, the *f* value is 50% higher than that of pipe flow in the laminar region.

Isothermal Compressible Fluid Flow in Pipe

The density of compressible fluids like gases varies with the changes of both temperature and pressure. Their density may not remain constant in the direction of flow as well. The *ideal gases* follow the *ideal gas laws* and it is assumed that their molecules should neither occupy any volume nor exert forces to one another. No real gases obey these laws exactly. However, many gases at ordinary temperatures and at pressures not more than few atmospheres follow the ideal gas law within a reasonable accuracy sufficient for engineering calculations.

The ideal gas law is usually expressed in the following form:

$$pV = nRT$$

where
p is the absolute pressure, N/m^2
V is the gas volume, m^3
n is the number of kg mol of gas
T is the absolute temperature, K
R is the universal gas constant = $8314.34\,kgm^2/(kg\ mol\ s^2\ K)$
$R = 8314.34\,J/(kg\ mol\ K) = 8314.34\,m^3\,Pa/(kg\ mol\ K) = 82.057\,cm^3\,atm/(g\ mol\ K) = 1.987\,gcal/(g\ mol\ K) = 1.987\,kcal/kg\ mol$
$K = 8.314\,kJ/kg\ mol\ K = J/g\ mol\ K = Btu/lb_{mol}\ R$

Ideal gas mixtures are also governed by the simple laws.

Dalton's law for mixture states that the total pressure of a gas mixture is equal to the sum of the individual partial pressures:

$$P = p_1 + p_2 + p_3 + \cdots \qquad \text{at constant temperature and volume} \qquad (20.31)$$

where
P is the total pressure
$p_1, p_2, p_3,\ldots$ are the partial pressures of the components 1, 2, 3,...

The mole fraction of each component n_1, n_2, and n_3 can be written in the form:

$$n_1 = p_1/P = p_1/(p_1 + p_2 + p_3 + \cdots) \qquad (20.32)$$

Therefore, $n_2 = p_2/P$, $n_3 = p_3/P,\ldots$
According to *Amagat's law*,

$$V = v_1 + v_2 + v_3 + \cdots \qquad \text{at constant temperature and pressure}$$

where
V is the total volume
$v_1, v_2, v_3,\ldots$ are the partial volumes of the components 1, 2, 3,...

The mole fraction of each component n_1, n_2, $n_3,\ldots$ can be written in the following form:

$$n_1 = v_1/V = v_1/(v_1 + v_2 + v_3 + \cdots) \qquad (20.33)$$

Similarly, $n_2 = v_2/V, v_3/V, \ldots$

The mole fraction of each component can also be expressed as

$$n_1 = (m_1/M_1)/(m_1/M_1 + m_2/M_2 + m_3/M_3) = n_1/(n_1 + n_2 + n_3 + \cdots)$$

$$n_2 = (m_2/M_2)/(m_1/M_1 + m_2/M_2 + m_3/M_3) = n_2/(n_1 + n_2 + n_3 + \cdots) \text{ and so on}$$

$$(20.34)$$

The average molecular weight, M_{av}, of a mixture of substances is defined by the equation:

$$n_2 = (m_1 + m_2 + m_3 + \cdots)/m_1/M_1 = m_2/M_2 + m_3/M_3 \qquad (20.35)$$

where $m_1, m_2, m_3, \ldots$ are masses of individual pure components 1, 2, 3,... in mixture $M_1, M_2, M_3, \ldots$ are molecular weights of pure components.

Gases are compressible fluids, as density does not remain constant with the changes in pressure, volume, and temperature.

Pressure drop and friction loss for isothermal flow of compressible ideal gases through uniform straight pipes. The pressure drop and friction factor can be evaluated for isothermal flow of ideal gases in pipe as follows.

The same basic differential form of Bernoulli equation can be expressed over a short pipe length for ideal gases with suitable modifications:

$$(dp/\rho) + gdz + d(v^2/2\alpha) + dF_s = 0 \quad \text{(SI)}$$

$$(dp/\rho) + (g/g_c)dz + d(v^2/2\alpha g_c) + dF_s = 0 \quad \text{(English)}$$

For turbulent flow, $\alpha = 1$ without any shaft work, $W_s = 0$, the Bernoulli equation for a differential length, dL can be written as

$$vdv + gdZ + (dp/(4fv^2 dL)/2D) + dF_s = 0$$

and for a horizontal pipe, putting $dZ = 0$ and $dF_s = (4fv^2 dL)/2D$, the aforesaid equation becomes

$$vdv + gdZ + (dp/p) + (4fv^2 dL/2D) = 0$$

For a steady-state flow, a uniform diameter, $v = 1/p$, $G = v\rho = (v/V) = \text{constant}$, $dv = GdV$. Substituting these values in the aforementioned equation and after rearranging we get

$$G^2(dv/V) + (dp/\rho) + (2fG^2/D)dL = 0$$

Also $pV = (1/M)RT$, taking $f = $ constant, solving for V and substituting it in the aforementioned equation and integrating within proper limits.

The variation in density or specific volume is to be taken into account if the gas pressure changes more than 10%. The pressure drop and friction factor can be evaluated for isothermal flow of ideal gases. After integration of the differential form of the Bernoulli equation between two points and proper rearrangements, it yields the following equations:

$$(P_1 - p_2) = 4f(\Delta L/D)(G^2/2\rho_a) + (G^2/\rho_a)\ln(p_1/p_2) \quad \text{(SI)} \tag{20.36}$$

and

$$(p_1^2 - P_2^2) = 4f(\Delta L/D)(G^2 RT/M) + (2G^2 RT/M)\ln(p_1/p_2) \quad \text{(SI)} \tag{20.37}$$

When the changes in pressure and density are more than 10%, the variation in density or specific volume is to be taken into account. If the density and pressure changes remain within 10% limit, then the pressure drop and friction factor can be predicted for isothermal flow of ideal gases as follows:

$$(p_1 - p_2) = 4f(AL/D)(G^2/2p_a) \quad \text{(SI)} \tag{20.38}$$

and

$$(p_1^2 - p_2^2) = 4f(AL/D)(G^2 RT/M) \quad \text{(SI)} \tag{20.39}$$

where
$Re = DG/m$ G, kg/m^{2}s, a constant independent of density and velocity variation
p_a is the average density, kg/m^3
$R = 8314.3 \, J/(\text{kg mol K})$
M is the molecular weight in kg/kg mol
T is the gas temperature, K

$$RT/M = (p_a/p_a), \, p_a = (p_1 + P_2)/2, \, p_a = (p_a + P_b)/2$$

$$(L_2 - L_1) = AL = L \text{ if } L_1 = 0$$

Actual gas velocity, v, and maximum velocity, v_{max}, can be computed from ideal gas laws as

$$v = (GRT)/Mp_a \tag{20.40}$$

$$V_{max} = (RT/M)^{0.5} \tag{20.41}$$

Example 20.3

Nitrogen at an isothermal temperature of 24.9°C passes through a smooth pipe of 200 m length and 0.01 m internal diameter at a mass flow rate of 9.0 kg/s m². If the entrance pressure in the pipe is 2.0265 × 10⁵ Pa, what is the outlet pressure?

Data given: inlet pressure, $p = 2.0265 \times 10^5$ Pa, being isothermal $T_1 = T_2 = 297.9$ K, mass flow rate, $G = 9.0$ kg/s m², $D = 0.01$ m, and $L = 200$ m, air viscosity at 24.9°C, $\mu = 1.78 \times 10^{-5}$ Pa s, $M = 28$ kg/kg mol, and $R = 8314.3$ J/kg mol K.

It may be assumed that the pressure drop is less than 10%.

$$\text{Re} = (DG/\mu) = (0.01 \times 9.0)/(1.78 \times 10^{-5}) = 5056 \text{ (turbulent flow)}$$

For Re = 5056, f can be found to be 0.009 for smooth pipe. Substituting the aforementioned data in the equation,

$$\left(P_1^2 - P_2^2\right) = (4fLG^2RT)/(DM)$$

$$(2.0265 \times 10^5)^2 - p_2^2 = (4 \times 0.009 \times 200 \times 9.0^2 \times 8314.3 \times 297.9)/(0.01 \times 28)$$

$$p_2^2 = 3.591 \times 10^{10}$$

So the outlet pressure, $p_2 = 1.895 \times 10_5$ Pa.

Non-Newtonian Fluids

Introduction

Non-Newtonian fluids do not follow Newton's law of viscosity.

Most of the liquid foods such as ketchup, yogurt, creamy salad dressings, tomato puree, baby foods, and soups do not follow the behavior of Newtonian fluids when they are pumped, agitated, and subject to any processing condition. These fluids or semifluid materials in the form of pastes, suspensions, or dispersions fall under the category of non-Newtonian fluids. Unlike Newtonian fluids, the shear rate varies with shear stress nonlinearly and even not necessarily in the same direction also.

They may exhibit increasing or decreasing viscosity with increasing shear rate, a yield stress, time independency, or time dependency (viscoelastic effects). Some of these may require a yield stress, τ_o, which must be attained before the flow begins.

The properties of non-Newtonian liquids may be broadly divided into time-independent and time-dependent. The time-independent non-Newtonian liquids begin to flow as soon as small amount of shear stress is applied. Unlike Newtonian liquids, the correlation between shear stress and shear rate is nonlinear. The time-independent non-Newtonian liquids can be further subdivided into two important types, namely, shear-thinning liquids and shear-thickening liquids. Most liquid fluids follow *shear thinning or pseudo-plastic or power law Bingham plastic, Harscel–Bulkley*, and *shear thickening or dilatant*.

Based on the assumption that the non-Newtonian liquids obey Newtonian law of constant viscosity, an apparent viscosity can be evaluated by selecting a particular shear rate, a straight line may be drawn from the selected point on the curve to the origin to where its slope gives the value of apparent viscosity. Therefore, apparent viscosity is entirely based on the selected shear rate otherwise it is useless. In the case of *shear-thinning* liquids, the shear rate increases with the decrease in apparent viscosity. They are also named as pseudo-plastic or power law liquids as for example mayonnaise, condensed milk, fruit purees, and vegetable soups. Except some gels and sols, most liquid foods fall under the first group. *Pseudo-plastic liquids, Bingham plastic liquids,* and *dilatant liquids* are time-independent fluids (Figure 20.3).

Figure 20.3 shows shear rate vs. shear stress for Newtonian and non-Newtonian fluids. *Thixotropic* liquids and *rheopectic* liquids belong to the group of time-dependent fluids. Bingham plastic fluids follow the linear shear stress–shear rate relationships as follows:

$$(\tau - \tau_y) = -\mu(dv/dr) \tag{20.42}$$

These fluids generate a positive shear stress/shear rate at zero shear rate.

Milk, concentrated juice, and pulp are pseudo-plastic fluids. Margarine, chocolate mixtures, jellies are *Bingham plastic* fluids.

Most of the fluids are pseudo-plastic ($n < 1$). The following Ostwald–De Waelete equation can represent the shear stress–shear rate relationship:

$$\tau = K(-dv/dr)^n \tag{20.43}$$

where
$n < 1$
K is the consistency index, N s^n/m^2
n is the flow behavior index, dimensionless

These fluids may also be referred to *power law fluids* due to power on the shear rate.

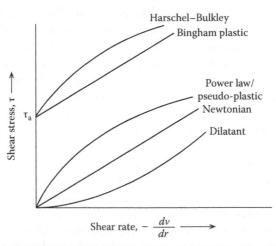

Figure 20.3   Shear stress–shear rate diagram.

Dilatant fluids also follow the same power law equation but the shear stress–shear rate diagram is concave upward as $n > 1$. Dough, cornflower, and sugar solution come under dilatant fluids but they are rare.

For non-Newtonian time-independent *pseudo-plastic* and *dilatant liquid* materials, the nonlinear functional relationship between shear stress, τ, and shear rate, γ (1/s), or velocity gradient (dv/dy) can be described by the following power law model:

$$\text{Power law} \quad \tau = K(dv/dy)^n = K\gamma^n \tag{20.44}$$

$$\text{Bingham plastic} \quad \tau = \tau_o + \mu\gamma \tag{20.45}$$

$$\text{Harschel–Bulkley} \quad \tau = \tau_o + K(dv/dy)^n = \tau_o + K\gamma^n \tag{20.46}$$

where
 τ_o (Pa) is the yield stress
 K (Pa s^n) is the consistency coefficient, and the dimension less, n is the flow behavior index of the fluid

The *power law* equation may be considered as a special case of the general *Harschel–Bulkley model*.

Thixotropic and *rheopectic* are time-dependent fluids.

Velocity Profile of a Power Law Fluid

It may be seen from the equation $\tau = \tau_o + K(-dv/dr)^n$ that the shear stress, τ, and shear rate, 1/s or velocity gradient, dv/dr, for plastic and dilatant liquid materials are related by power law model as follows:

$$\tau = K(-dv/dr)^n = K\gamma^n$$

From equations $\Delta p/L = 2\tau/r$ and $\tau = K(-dv/dr)^n$, it can be derived easily as follows:

$$K(-dv/dr)^n = \Delta pr/2L$$

Rearranging and integrating within proper limits of the center to the wall of the pipe, we get the following velocity profile equation:

$$V(r) = (\Delta p/2LK)^{1/n}(n/n+1)(R^{n/n+1} - r^{n/n+1}) \tag{20.47}$$

where
 K is the consistency coefficient, Pa s
 n is the flow behavior index (dimensionless)

For non-Newtonian fluids, the shear rate can be predicted from the following equation using the correction factor in the brackets:

$$\gamma = (8v/D)\big[(3/4)+D\,\ln(8v/D)/(4D\ln\tau_w)\big] \tag{20.48}$$

Generalized Reynolds Number

The generalized Re (Re_g) for non-Newtonian fluids is calculated from the following equation:

$$Re_g = \big[(D^n v^{2-n}\rho)/(K'8^{n-1})\big] \tag{20.49}$$

Another method of estimation of Re_g for non-Newtonian fluids is to calculate the apparent viscosity μ' from the following equation:

$$\mu' = K\gamma^{n-1} \tag{20.50}$$

Most of the liquid foods being pseudo plastic ($n<1$), their apparent viscosity decreases with the increase of the shear rate (shear-thinning materials).

Power Law Fluid

For non-Newtonian fluids, the power law fluid model is very useful among all models.

The volumetric flow rate of a power law fluid, v, and pressure drop per unit length, $(\delta P)/L$, in a circular pipe is given by

$$v = (Q/\pi r^2)\Big[\pi\big((\delta P)/2LK\big)^{1/n}\big(n/(3n+1)\big)\ rn/n+1\Big]\big[(1/\pi r^2)\big] \tag{20.51}$$

$$((\delta P)/L) = \big[(4v^n K)/(D^{1+n})\big]\big((2+6n)/n\big)^n \tag{20.52}$$

Laminar Flow

The power law friction factor, f_{pl}, can be predicted in the laminar region that is analogous to the expression for Newtonian laminar flow.

Pressure drop, Δp for flow in closed pipe

$$\Delta p = 4f\rho(L/D)(v^2/2) \quad \text{(SI)}$$

For power law, $f = (16/\text{Re}_g)$

$$\Delta p = (64/\text{Re}_g)\rho(L/D)(v^2/2) \tag{20.53}$$

$$\text{or,} \quad \Delta p = 32\gamma(L/D)(v/D)^n$$

$$f = (\Delta p/L)\left[D/(2\rho v^2)\right] = \left[(4v^n K)/(D^{1+n})(2+6n/n)^n\right](D/2\rho v^2 L) \tag{20.54}$$

and

$$f = 16/\text{Re}_g \tag{20.55}$$

The corresponding power law Re_g is the same as shown earlier:

$$\text{Re}_g = \left[(D^n v^{2-n}\rho)/\{K 8^{n-1}\}(3n+1)/4n^n\right] \tag{20.56}$$

Turbulent Flow

For turbulent flow, f can be calculated by Dodge and Metzner equation as follows:

$$\left(1/\sqrt{f}\right) = (4/n^{0.75})\log\left[(\text{Re}_g)f^{1-n/2}\right] - (0.4/n^{1.2}) \tag{20.57}$$

$$\Delta p = 4f\rho(L/D)(v^2/2) \quad \text{(SI)}$$

For turbulent flow, a modified Moody diagram (Figure 20.4) applicable for non-Newtonian fluids may be used though f for *pseudo-plastic fluids* in rough tubes is slightly smaller than the values given by the Moody diagram.

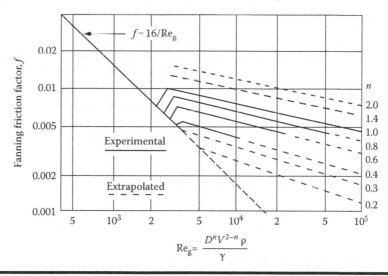

Figure 20.4 **Fanning friction factor for non-Newtonian fluid flow in pipes. (From Dodge, C.W. and A.B. Metzner, *AIChE J.*, 5(7), 189, 1959.)**

Figure 20.4 shows the modified Moody diagram for power law fluid non-Newtonian power law fluid flow occurs mostly in the laminar region:

For Newtonian fluid in laminar state $(KE/kg) = (v^2/2\alpha)$, $\alpha = (1/2)$ and

For non-Newtonian power law fluids, $\alpha = (2n+1)(5n+3)/3(3n+1)^2$

$$(20.58)$$

Coefficient of viscosity, μ', generalized coefficient of viscosity, γ, and Re_g may also be expressed by

$$\mu' = K'(8v/D)^{n-1} = K'(8v/D)^{n-1}$$

$$\gamma = K'8^{n-1} \quad \text{and}$$

$$Re_g = (Dv\rho/\mu') = \left[Dv\rho/K'(8v/D)^{n-1}\right]$$

$$Re_g = \left[(D^n V^{2-n}\rho)/\gamma\right]$$

$$(20.59)$$

In case of turbulent non-Newtonian fluid flow in tubes, modified Moody diagram (Figure 20.4) can be used to estimate fanning friction factor f, though f values for *pseudo-plastic liquids* in rough tubes are slightly smaller than the values given by the Moody diagram.

Example 20.4

A power law fluid, namely, applesauce having a density, ρ of $1030\,kg/m^3$, is being pumped at a velocity of $3\,m/s$ through a pipe of internal diameter (D) $25.4\,mm$ and length (L) $15.24\,m$. The rheological properties of the fluid are $K=0.5\,Pa\,s$ $(N\,s^n/m^2)$ and $n=0.65$.

(i) Assuming laminar flow evaluate (a) the pressure drop, Δp, and (b) the Re_g using both power law and friction factor, f.
(ii) Estimate the average velocity v and pressure drop Δp for turbulent flow, where $Re_g = 5000$.

The data given are $K=0.5\,N\,s^n/m^2$, $n=0.65$, $D=0.0254\,m$, $L=15.24\,m$, $v=0.1524\,m/s$, $\rho = 1030\,kg/m^3$, Δp and Re_g by power law
(i) (a) $\Delta p = (4KL/D)(8v/D)^n$

$$= (0.5 \times 4 \times 15.24/0.0254)(8 \times 0.1524/0.0254)^{0.65} = 14{,}858.948\,N/m^2$$

Friction loss, $\Sigma F = (\Delta p/\rho) = k'4L/D(8v/D)^n$

$$\Sigma F = (0.5 \times 4 \times 15.24/0.0254)(8 \times 0.1254/0.0254)^{0.65} = 14.423\,J/kg$$

(i) (b) $Re_g = D^n v^{2-n} \rho/(K8^{n-1}) = (0.0254)^{0.65} (0.1524)^{1.35} \times 1030/(0.5 \times (8)^{-0.35}) = 30.925$
As the flow is laminar

$$f = (16/Re_g) = 16/30.925 = 0.51$$

$$\Delta p = 4f(L/D)(v^2/2) = (4 \times 0.51 \times 15.24 \times 0.1524 \times 0.15240/(0.0254 \times 2)$$

$$= 14,640.59 \text{ N/m}^2$$

(ii) $Re_g = (D^n v^{2-n} \rho)/(K8^{n-1})$ and $\Delta p = (4KL/D)(8v/D)^n$
Putting the value of $Re_g = 5000$

$$5000 = (0.0254^{0.65} \times v^{1.35}) \times (1030/0.5 \times 8^{0.35})$$

$$v^{1.35} = 12.7757$$

$$v = 6.31 \text{ m/s and}$$

$$\Delta p = (0.5 \times 4 \times 15.24/0.0254)(8 \times 6.308/0.0254)^{0.65}$$

$$= 1,671,003.7 \text{ N/m}^2$$

Example 20.5

A ketchup is allowed to pass through a pipe using a pump. Assuming the power law fluid flow and the following properties of the ketchup as well as the necessary data for the same pipe line:
$K = 2.08 \text{ N s}^n/\text{m}^2$ (Pa s^n), $n = 0.84$, $L = 3.05 \text{ m}$, $D = 0.0125 \text{ m}$, $v = 2 \text{ m/s}$, $\rho = 1050 \text{ kg/m}^3$, calculate the pressure drop, Δp, in the aforesaid system.
Here pressure drop, $\Delta p = K4L/D (8v/D)^n = 2.08 \times 4 \times 3.05/0.0125(8 \times 2.0/0.0125)^{0.84}$
$\Delta p = 827,121.91 \text{ N/m}^2$.
Friction loss, $\Sigma F = (\Delta p/\rho) = 827,122.226/1050$

$$\Sigma F = 787.74 \text{ J/kg}$$

$$Re_g = D^n v^{2-n} \rho/(K8^{n-1}) = (0.0125)^{0.84} \times (2)^{2-0.84} \times 1050/(2.08 \times 8^{0.84-1}) = 396.48$$

So the fluid flow is laminar and accordingly

$$f = 16/Re_g = 16/396.48 = 0.04035$$

$$\Delta p = 4f(L/D)(v^2/2) = 4 \times 0.04035(3.05/0.0125)(2^2/2) = 83,936.0 \text{ N/m}^2$$

Pressure drop, $\Delta p = 83.94 \text{ kN/m}^2$

Exercises on Fluid Mechanics

Ideal Gas Laws

20.1 A natural gas with the composition CH_4—94.5%, C_2H_6—2.5%, and N_2—3% passes through a pipe from the reservoir at 24.5°C and 2.9 atm.

   If the ideal gas laws are applicable, calculate (i) volume of N_2/10 m³ of gas and partial pressure of N_2 and (ii) density of the gas.

20.2 Atmospheric air at 34°C and 89% humidity has to be conditioned to 224°C and 62% humidity by cooling a portion of the air to 12°C and mixing it with atmospheric air. The resulting final mixture is reheated to 24°C. For 20 L/s of air at 24°C and 62% humidity, estimate (a) the volume of entering air and (b) the percentage of entering air that is bypassed.

Newtonian Fluid

20.3 A liquid passes through a steel pipe of diameter, D of 0.04089 m and equivalent $K = 4.6 \times 10^{-5}$ m. Its density and viscosity are 780 kg/m³ and 4.45×10^{-5} Pa s, respectively. Estimate the pressure drop over 50.0 m length of the pipe.

20.4 Atmospheric air at 34°C and 89% humidity has to be conditioned to 224°C and 62% humidity by cooling a portion of the air to 12°C and mixing it with atmospheric air. The resulting final mixture is reheated to 24°C. For 20 L/s of air at 24°C and 62% humidity, estimate (a) the volume of entering air and (b) the percentage of entering air that is bypassed.

20.5 Water at 25°C is pumped from a storage tank through 120 m length and 3.0 cm diameter of pipe. The pipeline has two open globe valves and three 90° elbows. Water is discharged into another open tank through a spray nozzle. The discharge is at a height of 19.5 m above the level of water in the storage tank. The pressure required at the nozzle entrance is 2.5 kg/cm² gauge. Calculate (i) the energy loss due to friction, (ii) pump work required per kilogram of water, and (iii) theoretical hp required for the pump.

   Data given: Viscosity of water at 26°C = 0.975 cp, 1.0 hp = 0.746 kW, equivalent length in terms of pipe diameter, open globe value = 300 D, and 90° elbow = 30D.

20.6 Water is allowed to pass through 298 m of horizontal pipe at the rate of 0.06 m³/s. A head of 6.5 m is available. Estimate the pipe diameter if the friction factor $f = 0.0055$.

Non-Newtonian Fluid

20.7 A fruit syrup with the consistency index of 2.65 Pa s^n and the flow behavior index of 0.79 is flowing through a pipe of cm ID. If the average velocity of the syrup in the pipe is

(ii) $Re_g = (D^n v^{2-n} \rho)/(K8^{n-1})$ and $\Delta p = (4KL/D)(8v/D)^n$

Putting the value of $Re_g = 5000$

$$5000 = (0.0254^{0.65} \times v^{1.35}) \times (1030/(0.5 \times 8^{-0.35}))$$

$$v^{1.35} = 12.753$$

$$v = 6.308 \text{ m/s} \quad \text{and} \quad \Delta p = (0.5 \times 4 \times 15.24/0.0254)(8 \times 6.308/0.0254)^{0.65}$$

$$= 1{,}671{,}003.7 \text{ N/m}^2$$

Example 20.6

A vegetable ketchup has to be passed through a pipe using a pump. Here assuming the power law fluid flow and the necessary properties of the ketchup as well as other needed data for the aforementioned pipe line, calculate the pressure drop, Δp in the aforementioned system: $K = 2.08$ N s^n/m^2 (Pa s^n), $n = 0.84$, $L = 3.05$ m, $D = 0.0125$ m, $v = 2$ m/s, $\rho = 1050$ kg/m^3, calculate the pressure drop, Δp in the system.

Here pressure drop, $\Delta p = K4L/D(8v/D)^n = 2.08 \times 4 \times 3.05/0.0125(8 \times 2.0/0.0125)^{0.84}$

$$\Delta p = 827{,}121.91 \text{ N/m}^2$$

Friction loss, $\Sigma F = (\Delta p/\rho) = 827{,}122.226/1050$

$$\Sigma F = 787.74 \text{ J/kg}$$

$$Re_g = D^n \times v^{2-n} \rho/(K8^{n-1}) = (0.0125)^{0.84} \times (2)^{2-0.84} \times 1050/(2.08 \times 8^{0.84-1})$$

$$= 396.48$$

So the fluid flow is laminar and accordingly

$$f = 16/Re_g = 16/396.48 = 0.04035$$

$$\Delta p = 4f(L/D)(v^2/2) = 4 \times 0.04035(3.05/0.0125)(2^2/2) = 83{,}936.0 \text{ N/m}^2$$

Pressure drop, $\Delta p = 83.94$ kN/m^2

Heat Transfer

There are three modes of heat transfer, namely, conduction, convection, and radiation. In conduction, heat flows through a body by the transfer of momentum of individual atoms or molecules without mixing. Here more precisely, the major portion of heat is being transferred from the hotter to the cooler molecules by movement of free electrons. Because of this reason, good electrical conductors are also good thermal conductors. Generally, solids are heated by this method.

Heat may be transferred by any one or more of the aforesaid three basic mechanisms. Any of the rate transfer processes, namely, momentum transfer, mass transfer, or heat transfer can be presented by the basic equation:

$$\text{Rate of heat transfer process} = (\text{Driving force/resistance})$$

Conduction of heat, q, is expressed by Fourier's law with a similar equation as given earlier:

$$q/A = -kdT/dx = -dT/(dx/k) \qquad (20.60)$$

When heat flows by mixing of warmer portions with cooler portions of the same material, the mechanism is called convection. Actually, the hot molecules move from one place to another and they collide with cooler molecules and transfer thermal energy. Convection is restricted to the flow of heat in fluids. According to Newton's law of heating or cooling, the rate of heat transfer from a solid to a fluid or vice versa is expressed by the following equation:

$$q = hA(T_w - T_f) \qquad (20.61)$$

where h is called the film or convective heat transfer coefficient as explained subsequently. Radiation refers to the transfer of energy through space by means of electromagnetic waves almost the same way as light transmits. Radiation is the mode of heat transfer in vacuum as well as gases. The basic equation for heat transfer by thermal radiation emitted by a perfect black body with an emissivity of unity is

$$q = A\sigma T^4 \qquad (20.62)$$

where σ is known as Stefan–Boltzman constant $= 5.669 \times 10^{-8}\,\text{W/m}^2\,\text{K}^4$

Conduction

Steady-State Conduction

Under steady state where the rate of transfer process is constant, the resistance term can be mathematically defined by Fourier's law for heat conduction of solids or fluids in one direction, x, in fluids or solids as follows:

$$q = -kA(\Delta T/\Delta x)$$

Here, $(\Delta T/\Delta x)$ is negative for a positive distance Δx and the negative sign also indicates that heat always flows from higher to lower temperature.

The same equation can be written in the following differential form:

$$q = -kA(dT/dx)$$

Integrating the aforementioned equation within proper limits and rearranging

$$(q/A) = k(T_2 - T_1)/(x_2 - x_1) = (T_2 - T_1)/(x_2 - x_1)/k = \Delta T/R \quad (20.63)$$

where

Thermal resistance, R (a reciprocal of thermal conductivity) $= (x_2 - x_1)/k$
Heat transfer rate $= q$, W
Heat flux $= q/A$, W/m^2
Area $= A$, m^2
Temperature $= T$, K
Distance $= x$, m

The proportionality constant termed thermal conductivity, a basic property of the material $= k$, W/m K in SI and it can also be expressed as

$$k = -(q/A)/(\Delta T/\Delta x)$$

Therefore, thermal conductivity also refers to the rate of heat transferred per unit area per unit temperature gradient and the unit of k is then (W/m^2)/(K/m).

Thermal conductivity, k, of liquids varies linearly with temperature within a moderate range of temperature as correlated as follows.

$k = k_o + \gamma T$, where k_o and γ correspond to the thermal conductivity at 0°C (273 K) and the empirical constant (Table 20.3).

Table 20.3 Thermal Conductivities, k in W/m K of Some Solids, Liquids, and Gases

Metals	Insulating Materials	Gases	Foods, Liquids, and Others
Silver-416	Window glass—0.78	Oxygen—0.026	Water—0.5
Copper-386	Dry wood—0.16	Nitrogen—0.025	Skim milk—0.54
Mild steel-43	Loose asbestos—0.15	CO_2—0.015	Potato—0.55
Mild steel-43	Glass wool—0.05	Air—0.024	Ice—0.78

Conduction through a Hollow Cylinder

The equation for conduction through a hollow cylinder with the inner and outer radii of r_1 and r_2, respectively, and the corresponding surface temperatures of T_1 and T_2 (Figure 20.5) and a total length of L is

$$q = -kA(dT/dr) = -k(dT/dr)(2\pi rL)$$

$$\text{or} \quad (dr/r) = -(2\pi rLk/q)dT$$

Here A is the total cylinder's surface area perpendicular to heat flow $= 2\pi rL$, $T_1 > T_2$, $r_2 > r_1$.

Integrating the earlier equation within the said limits and rearranging,

$$q = k(2\pi L)\,(T_1 - T_2)/\ln(r_2/r_1) = k(2\pi L)(r_2 - r_1)(T_1 - T_2)/(r_2 - r_1)\ln(r_2/r_1) \quad \text{or}$$

$$q = kA_m(T_1 - T_2)/(r_2 - r_1) = (T_1 - T_2)/r_2 - r_1/(kA_m) = (T_2 - T_1/R) \tag{20.64}$$

where
Log mean area, $A_m = 2\pi L(r_2 - r_1)\ln(r_2/r_1) = (A_2 - A_1)/\ln(A_2/A_1)$
Log mean radius, $r_m = (r_2 - r_1)/\ln(r_2/r_1)$
k_m is the cylinder's mean thermal conductivity
Resistance, $R = (r_2 - r_1)/(kA_m) = \ln(r_2/r_1)/(2\pi kL)$

In practice, if $A_2/A_1 < 1.5$, its accuracy of arithmetic mean area, $(A_2 + A_1)/2$ is used as its accuracy is within 1.5% of log mean area.

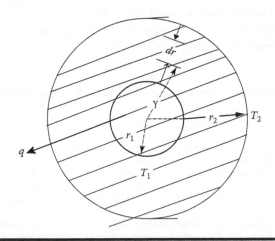

Figure 20.5 Heat conduction through cylinder.

Conduction through a Series of Slabs

The thickness, thermal conductivity, and temperature variations of three layers of different solids materials are Δx_1, k_1, ΔT_1, Δx_2, k_2, ΔT_2, and Δx_3, k_3, ΔT_3, respectively, and A is the common cross-sectional area perpendicular to the flow of heat (Figure 20.6).

The equation for each slab layer according to Fourier law

$$q = (k_1 A/\Delta x_1)/\Delta T_1 = (k_2 A/\Delta x_2)/\Delta T_2 = (k_3 A/\Delta x_3)/\Delta T_3$$

$$q/A = k_1(T_1 - T_2)/\Delta x_1 = k_2(T_2 - T_3)/\Delta x_2 = k_3(T_3 - T_4)/\Delta x_3$$

$$(20.65)$$

The Fourier equation for the combined series of layers where T_1 and T_4 are initial and final temperatures of the first and the last surfaces:

$$q = \left[(T_1 - T_4)/\{(\Delta x_1/k_1 A) + (\Delta x_2/k_2 A) + (\Delta x_3/k_3 A)\}\right]$$

$$= \left[(T_1 - T_4)/(R_1 + R_2 + R_3)\right]$$

$$(20.66)$$

$$(q/A) = (T_1 - T_4)/\Sigma R = \text{(driving force/total resistance)}$$

where
Resistances, $R_1 = \Delta x_1/(k_1 A)$, $R_2 = (\Delta x_2/k_2 A)$, and so on
Temperatures are T_1, T_2, T_3, and T_4
Overall temperature difference $= (T_1 - T_4)$

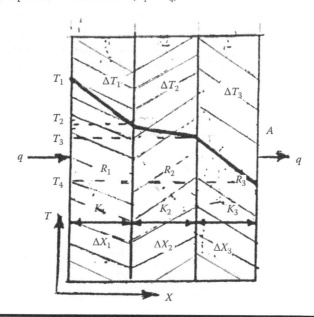

Figure 20.6 Heat conduction through composite wall.

Total resistance, ΣR

$$\Sigma R = (R_1 + R_2 + R_3)$$

Conduction through each of the different cylinders

$$q = (T_1 - T_2)/(r_2 - r_1)/(k_1 A_{1m}) = (T_2 - T_3)/(r_3 - r_2)/(k_2 A_2 m)$$

$$= (T_4 - T_3)/(r_4 - r_3)/(k_3 A_3 m)$$

$$q = \left[(T_1 - T_4)/\left\{(r_2 - r_1)/(k_1 A_1 m) + (r_3 - r_2)/(k_2 A_2 m) + (r_4 - r_3)/(k_3 A_3 m)\right\}\right]$$

$$q = \left[(T_1 - T_4)/(R_1 + R_2 + R_3)\right] = (T_1 - T_4)/\Sigma R \qquad (20.67)$$

Conduction through any number of concentric layers of a hollow cylinder

$$q = 2\pi L(T_1 - T_n)/(1/k_1)\ln(r_1/r_2) + (1/k_2)\ln(r_2/r_3)$$

$$+ (1/k_4)\ln(r_4/r_3) + \cdots + (1/k_n)\ln(r_n/r_{n-1}) \qquad (20.68)$$

Unsteady-State Heat Transfer in a Flat Slab

Unlike steady-state heat transfer, in unsteady state the temperature gradient $(\delta T/\delta x)$ varies with time and location (Figure 20.7). The driving force for heat transfer gradually decreases with time. It may be assumed that the Fourier law is applicable.

When a solid flat slab is heated by unsteady state of conduction, a very thin piece of solid of thickness dx located at a distance x from the hot side of the slab may be considered. The temperature gradient and the heat input at x in time $d\theta$ are presented by $\delta T/\delta x$ and $-kA(\delta T/\delta x)d\theta$, respectively.

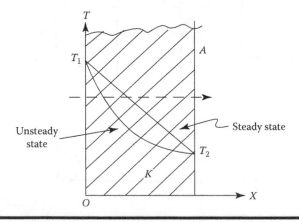

Figure 20.7   Heat conduction through wall (under steady and unsteady state).

Then the heat flow out of the solid slab at a distance of $x + dx$ is expressed by

$$-kA\{(\delta T/\delta x) + kA\{(\delta/\delta x)(\delta T/\delta x)dx\}d\theta = kA(\delta^2 T/\delta x^2)dxd\theta$$

Therefore, accumulation of heat in layer dx is

$$-kA\{(\delta T/\delta x) + kA\{(\delta/\delta x)(\delta T/\delta x)dx\}d\theta = kA(\delta^2 T/\delta x^2)dxd\theta$$

The heat accumulation in the above thin layer dx may also be presented as

$$(\rho A dx)c_p(\delta T/\delta\theta)d\theta$$

Applying heat balance,

$$kA(\delta^2 T/\delta x^2)dxd\theta = \rho c_p A dx(\delta T/\delta\theta)d\theta$$

After rearrangement, the following partial differential equation is used for one directional (x) transient or unsteady conduction heat transfer:

$$(\delta T/\delta\theta) = k/\rho c_p(\delta^2 T/\delta x^2) = \alpha(\delta^2 T/\delta x^2) \qquad (20.69)$$

where
α is the thermal diffusivity is a property of the material, m^2/s
ρ is the density, kg/m^3
k is the thermal conductivity, $W/m\ K$
c_p is the specific heat of the material, $J/kg\ K$

The heat transfer through a food material can be expressed by the two basic heat transport properties, namely, thermal conductivity, k, $W/m\ K$, and thermal diffusivity, α, m^2/s.

For integration of the aforementioned general unsteady partial differential equation for heating or cooling of an infinite slab of a certain thickness from both sides by a medium at constant surface temperature yields the following equation:

$$\left[(T_s - T_a)/(T_s - T_i)\right] = (8/\pi^2)\left[(e^{-alFo}) + (1/9)(e^{-9alFo}) + (1/25)(e^{-alFo}) + \cdots\right] \qquad (20.70)$$

where
T_s is the constant slab surface temperature, K
T_i is the initial slab temperature, K
T_a is the average slab temperature at time, θ_t

θ_t is the heating or cooling time, s or (h)
α is the thermal diffusivity, m^2/s
$S = (\frac{1}{2})$ of the slab thickness, m
$a_1 = (\pi/2)^2$
Fo is the dimensionless Fourier number (Fo) as defined, $\alpha\theta_t/S^2$

The dimensionless Fo compares the rate of thermal energy flux with the rate of thermal energy absorption causing a rise in the temperature of the material. The term Fo is defined by the following expression:

$$\text{Fo} = (k/\rho c_p)\theta_t/S_2 = \alpha\theta_t/S_2 = (\text{transient heat conducted/stored heat})$$

(20.71)

A common example of unsteady-state heat transfer is within a steam- or heat-jacketed round bottom pan that is frequently used in food processing to prepare batch recipes. Initially, the temperature difference is comparatively large, and as the temperature of the food in the pan approaches to that of the surrounding jacket, the temperature difference gradually falls. For simplicity, it may be considered that the temperature throughout the solid be uniform at any instant of time. This kind of idealized system of analysis is called lumped heat capacity of solid $(v\rho c_p)$ method. Here, if the energy of a small solid body of high thermal conductivity, k, is compared to its surface heat transfer coefficient, h, with the assumption of uniform temperature throughout the solid, then the energy of the solid at instant is a function of its temperature and total heat capacity. That is why this method is named lumped heat capacity. A billet quenching in water is its common example. Following this method, it can be derived that

$$(T/T_i) = e^{-(hL/k)\,(\alpha\theta t/L^2)} = e^{-(\text{Bi})(\text{Fo})}$$

(20.72)

It shows that T vs. θ relationship is a straight line. The characteristics dimension of the solid, $L = V/A$, for a few fixed solid geometries are as follows.

Plate: $L = A2S/2A = S$ (A and S are the surface area and half thickness, respectively)
Cube: $L = (1^3/61^2) = 1/6$ (1 is the side length)
Cylinder: $L = (\pi R^2 1/2\pi R1) = R/2$ (R and 1 are the radius and length)
Sphere: $L = (4/3\pi R^3/4\pi R^2) = R/3$ (R is the radius)

where
V is the solid volume
A is the solid surface area
Biot number, $\text{Bi} = hL/k$
Fourier number, $\text{Fo} = \alpha\theta_t/L^2$

In this case, the prediction of temperature can be made easily using Gurney–Lurie charts that consider unidirectional heat flow as it occurs in the cases of infinite slab and infinite cylinder.

Biot Number

The Bi is a dimensionless group that is defined as the ratio of the magnitudes of internal conduction resistances in the solid to the external surface convection resistances in the fluid during unsteady-state heat transfer. Mathematically, it is written as $Bi = (D/kA)/(1/hA) =$ (thermal resistance of internal conduction/thermal resistance of surface convection).

Accordingly, lumped heat capacity analysis may be applied for low Bi generally when Bi < 1.

Such a condition allows for a maximum of 5% variation in the temperature of the body from the surface to the center.

Here, k is the thermal conductivity of the material inside the object, W/m K and D is the characteristic geometric dimension of the body associated with the heat transfer system, which is the shortest distance between the body surface and the thermal center. As for example, it is the radius for a sphere and an infinite cylinder whereas half the thickness for an infinite slab. Lumped heat capacity analysis may be used for low Bi when Bi < 1. That means, it is applicable when k is very high compared to so small h.

The Bi also indicates a measure of the temperature drop in the solid body compared to the temperature difference between the surface and the bulk of the fluid.

The Bi is important as it is necessary to predict or evaluate the transient temperature response of a solid body in a fluid. Based on the properties of solids and the fluids including the conditions of fluid flow surrounding the body, the Bi may be considered under three conditions: (i) negligible internal resistance when Bi < 0.1, (ii) finite internal and surface resistance for 0.1 < Bi < 40, and (iii) the Bi > 40 suggests negligible surface resistance to heat transfer as happens with condensing steam.

In the first case, magnitude of Bi < 0.1 clearly indicates that the internal conduction resistance is negligible compared to the surface convection resistance. The cooling of a very hot small metal casting in a quenching fluid bath falls under this condition. This condition is not usually applicable for solid foods due to their relatively low thermal conductivity. The rate of heat flow is dependent not only on the food itself but also on the surface resistance of stationary layer of air or water on the food. However, the relative rates of heat flow through the surface of food and from its surface to the surface are sometimes expressed by the Bi (Heldman et al., 2003).

In case of a thoroughly stirred liquid food in a container, such a condition can be achieved because for Bi << 1, it may be assumed that there is uniform temperature across a sold at any time during transient thermal process without any spacious temperature gradient as well. However, the relative rates of heat flow through the surface of food and from its surface to the surface are sometimes expressed by the Bi.

On the basis of Newton's law of heating or cooling or lumped thermal capacity $(c_p \rho V)$ method, if a solid body at a low uniform temperature of T_o, K at time $\theta = 0$, s immersed in a hot fluid at a constant temperature T having a constant film heat transfer coefficient, h with time as well. Using a heat balance on the solid body of a surface area, A, m² for a very small time interval $d\theta$, s the heat transfer from the fluid to the object is equal to the change in heat energy of the object:

$$hA(T_* - T)d\theta = c_p \rho V dT$$

where
T is the object's average temperature, K at time, θ, s
V is the volume, m³
ρ is the object's density, kg/m³

The term $c_p \rho V$ is separating the variables and integrating within proper limits, and the following equation is obtained:

$$(T - T_*)/(T_o - T_*) = \exp(-hA/c_p \rho V)\theta$$

After further rearrangement of the earlier equation, it can be expressed in terms of meaningful dimensionless form:

$$(T - T_*)/(T_o - T_*) = \exp(-\text{Bi/Fo})$$

where
Fourier number, $\text{Fo} = (\alpha\theta/D^2)$
D is the characteristic geometric dimension, m, as for example D is half the thickness of a slab
Thermal diffusivity, $\alpha = k/c_p \rho$, m²/s

The second condition $0.1 < \text{Bi} < 40$ deals with a finite thermal resistance internally as well as at the surface of the object during heating or cooling. The temperature difference across the solid should be much larger than that between the surface and the fluid.

The high $\text{Bi} > 40$ implies negligible surface resistance to heat transfer. That indicates the magnitude of h is reasonably higher than k/D. It occurs when steam is used as a heating medium for foods.

Convection

It is known that convection is prevalent in the transfer of heat through fluids. Convective heat transfer can be classified into forced convection and natural convection. When fluids are allowed to flow mechanically by means of fans, blowers, and pumps, the convection is known as forced. In natural or free convection, the fluid flow occurs owing to the difference in density between hot and cold fluid.

Convective heat transfer may be presented by the following equation:

$$q/A = h\Delta T$$

When a fluid flow outside a solid is under forced or natural convection, the rate of heat transfer from the solid to the fluid or vice versa is expressed by

$$q = hA(T_w - T_b)$$

where
q is the heat transfer rate, W
A is the area, m^2
T_w and T_b are the temperatures of solid surface and bulk of fluid
h is the convective heat transfer coefficient or film heat transfer coefficient, W/(m^2 K) in SI or Btu/(ft^2 °F) in English system
h, also known as the film coefficient, is a function of fluid properties, temperature difference, flow velocity, and system geometry

It can be noted that $1/h$ is nothing but convective resistance.

The convective heat transfer coefficients are generally correlated by developing some empirical or semiempirical relationships between different dimensionless groups or numbers, namely, Nusselt number (Nu), Re, Prandtl number (Pr), and others. Re has already been defined and explained its importance in classifying fluid flow.

Prandtl Number

During heating or cooling of a fluid through a pipe, both hydrodynamic boundary layer and thermal boundary layer are developed, which have a major influence on the rate of heat transfer between the pipe wall and the fluid. In order to determine the convective heat transfer rate, a dimensionless number named as Prandtl number, Pr, is required.

Physically, the Pr represents the ratio of the shear component of diffusivity of momentum, μ/ρ, to the molecular diffusivity of heat, $k/\rho c_p$, and it simply compares the thickness of the hydrodynamic boundary layer to the thermal boundary layer.

Mathematically, the Pr may be expressed for Newtonian fluids.

$$Pr = (\text{Molecular diffusivity of momentum/thermal diffusivity})$$

$$\text{or,} \quad Pr = (\text{Kinematic viscosity, } \gamma/\text{thermal diffusivity, } \alpha)$$

$$= (\mu/\rho)/(k/\rho c_p) = (c_p \mu/k) \tag{20.73}$$

where
c_p is the fluid specific heat, J/(kg K)
μ is the fluid dynamic viscosity, kg/(m s)
ρ is the fluid density, kg/m³
k is the fluid thermal conductivity, W/(m K)
α is the thermal diffusivity, m²/s

The Pr unity that signifies the thickness of the hydrodynamic layer is the same as thermal boundary layer. If $Pr \ll 1$, the thermal diffusivity will be much larger than the molecular diffusivity of momentum and in that case the heat will dissipate much faster.

The Pr may range from 0.5 to 0.9 for metals; it is about 10 for water whereas it may vary from 2 to around 10^4 for different liquids.

Nusselt Number

The dimensionless Nusselt number, Nu, is a quantity that relates all the important physical factors that determine convective heat transfer rates and thus reduces the total number of variables needed to define the heat transfer coefficient. If a fluid is stationary, the heat transfer rate, q_c, will be governed by conduction, and heat transfer will be evaluated by

$$q_c = -kA(\Delta T/1)$$

If the fluid is flowing, the heat transfer will take place predominantly by convection, q_{cv}, following the Newton's law of cooling:

$$q_{cv} = hA\Delta T$$

Ratio of the aforementioned two equations will lead to a very important dimensionless number, $Nu(q_{cv}/q_c) = hA\Delta T/(kA\Delta T/1) = (h_1/k) = hDe/k = Nu$

If the thickness, 1, is replaced by a more general term, the characteristic dimension, De, then

$$Nu = h\,De/k \qquad (20.74)$$

Simply, Nu compares the heat transfer by convection through the fluid to the heat transfer by conduction. Physically, Nu is a measure of enhancement in the rate of heat transfer through a fluid layer caused by convection over the conduction mode across the same fluid layer. In practice, the Nu relates heat transfer coefficient to the fluid thermal conductivity, k, along with a dimension called characteristic length, 1, for a flat slab or a more general characteristic dimension, De, for spheres, etc. This physical dimension clearly defines their geometry of the state of heat transfer.

The Nu = 1 means that there is no improvement in heat transfer rate due to convection of immobile fluid layer whereas Nu = 3 indicates that the rate of convective heat transfer due to fluid movement is three times the rate of heat transfer if the fluid in contact with the solid surface is static.

Equivalent Diameter

In case of circular pipe, De is same as pipe diameter, D. For circular annulus where heat transfer takes place through the inner surface, De is given by

De = (cross-sectional area, m²/heat transfer perimeter, m)

$$De = \left[4 \times (\pi/4) \times \left(D_o{}^2 - D_i{}^2 \right)/(\pi D_i) \right] = \left(D_o{}^2 - D_i{}^2 \right)/D_i$$

In an annulus where the heat transfer occurs through both inside and outside the surfaces,

$$De = \left[4 \times (\pi/4) \times \left(D_o{}^2 - D_i{}^2 \right)/(\pi D_o + D_i) \right] = \left(D_o{}^2 - D_i{}^2 \right)/D_i$$

where D_i and D_o are the inside and outside diameters of the annulus.

As described earlier, three-dimensionless groups/numbers Re, Pr, and Nu incorporate most of the properties and variables that are needed for calculation of the convective heat transfer coefficient. The plot of Nu against Re on a log–log graph paper for a given fluid with a constant Pr yields a straight line and for various given values of Pr, the plots Nu vs. Re will yield a series of straight lines on log–log scale. By dimensionless analysis, the following linear relationship can be established, which will help in calculating film coefficient easily:

$$Nu = C(Re)^m (Pr)^n \qquad (20.75)$$

In *forced convection fluid flow*, the following functional relationship among the related dimensionless groups such as Nu, Re, Pr, and L/D can be derived by applying the dimensional analysis known as Buckinghum Pi theorem:

$$Nu = f(Re, Pr, L/D)$$

Film Heat Transfer Coefficient for Laminar Flow Inside a Pipe

In the laminar flow region where Re < 2100, the film coefficient h can be estimated by the following Sieder–Tate equation:

$$Nu = 1.86 \left[Re\ Pr\ (D/L) \right]^{1/3} (\mu_b / \mu_w)^{0.14} \qquad (20.76)$$

where μ_b and μ_w are dynamic fluid viscosity at bulk and wall temperatures, respectively.

Film Coefficient in the Turbulent Region Inside a Pipe

When Re > 6000, 0.7 < Pr < 1600 and (L/D) > 60, an empirical equation known as Dittus–Bolter equation can be used to evaluate h:

$$Nu = 0.027 Re^{0.8}\ Pr^{1/3} (\mu_b / \mu_w)^{0.14}$$

All physical fluid properties are calculated on the basis of bulk mean temperature and only μ_w is evaluated at the wall temperature.

For helical coils, the right-hand side of the earlier equation has to be multiplied by

$$(1 + 3.5 D/D_{coil})$$

Film Coefficient for Transitional Flow in a Pipe

There is no accurate equation to predict the film heat transfer coefficient for the transitional flow within 2100 < Re < 6000 to determine film coefficient, h.

However, recently a relationship has been developed by dimensional analysis for the fluid flow in the wide range of Re from 2320 to 10,000:

$$Nu = (1/300)\ Re\ (Pr)^{0.33} \qquad (20.77)$$

The dimensionless equations for evaluation of film coefficients under the general and the particular conditions are given as follows.

Geometry	Equation	Boundary Conditions
Immersed bodies	Average film coefficient, h, is given by $Nu = C\,Re^m\,Pr^{1/3}$	C and m are empirical constants[a] and depend upon different configurations and fluid properties that are determined at film temperature $T_f = (T_b + T_w)/2$, T_b, and T_w at fluid bulk and wall temperature
Flow passed a sphere for being heated or cooled	$Nu = 2 + 0.6\,Re^{0.5}\,Pr^{1/3}$	$1 < Re < 7000$, $0.6\,Pr < 400$ and fluid properties are evaluated at film temperature, T_f

[a] Refer to Geankoplis (1997) to obtain the different values of C, m for earlier and other physical geometries.

Example 20.7

The fruit slices are being dried (under constant drying period) in a tray dryer of length 0.76 m with the parallel flow of hot air having an average velocity of 1 m/s at 66°C. The fruit slices remain at a constant temperature of 37.4°C. The properties of drying air at the film temperature (T_f) of 51.7°C $[T_f = (66 + 37.4)/2]$ are density, $\rho = 1.09\,kg/m^3$, specific heat, $c_p = 1.0069\,kJ/kg\,K$, and viscosity, $\mu = 1.965 \times 10^{-5}\,kg/m\,s$, thermal conductivity, $k = 0.0271\,W/mK$.

Estimate the film coefficient, h

$$Re_L = (Lv\rho/\mu) = (0.76 \times 1 \times 1.09/1.965 \times 10^{-5}) = 4.216 \times 10^4$$

$$Pr = (c_p\mu/k) = (1.0069 \times 10^3 \times 1.965 \times 10^{-5})/0.0271 = 0.73$$

Nu can be evaluated with the following equation as $Re < 3 \times 10^5$ and $Pr > 0.7$:

$$Nu = (hL/k) = 0.664 Re_L^{0.5} Pr^{1/3}$$

$$Nu = 0.664 \times (4.216 \times 10^4)^{0.5} \times (0.73)^{1/3} = 22.762$$

$$h = (122.762 \times k)/L = (122.762 \times 0.0271)/0.76$$

$$h = 4.38\ W/m^2\ K$$

Example 20.8

A stream of hot air at a temperature of 170.4°C with a velocity (v) of 1.75 m/s passes over a fruit juice droplet of 120×10^{-4} cm diameter to dry it. The droplet attains a steady temperature of 44°C. All necessary properties of drying air at the film temperature (T_f), 107.2°C are as follows: density, $\rho = 0.923$ m³/kg, specific heat, $c_p = 1.0111$ kJ/kgK, thermal conductivity, $k = 0.0321$ W/mK, and viscosity, $\mu = 2.21 \times 10^{-5}$ kg/m s.

Evaluate the film heat transfer coefficient, h

$$\text{Re} = (Dv\rho/\mu) = (1.75 \times 120 \times 10^{-6} \times 0.923/2.21 \times 10^{-5}) = 8.771$$

$$\text{Pr} = (c_p\mu/k) = (1.0111 \times 10^3 \times 2.21 \times 10^{-5})/0.0321 = 0.696$$

Under the given fluid conditions of Re and Pr values, the following equation is applicable:

$$\text{Nu} = 2 + 0.6\,\text{Re}^{0.5}\text{Pr}^{1/3} \quad \text{for} \quad 1 < \text{Re} < 7000, \text{ and } 0.6\,\text{Pr} < 400$$

$$\text{Nu} = (hD/k) = 2 + 0.6\,\text{Re}^{0.5} \times \text{Pr}^{1/3} = 2 + 0.6\,(8.771)^{0.5} \times (0.696)^{1/3} = 3.57$$

$$h = 3.57 \times k/D = 3.57 \times 0.0321/(120 \times 10^{-6}) = 954.98 \text{ W/m}^2 \text{ K}$$

Film Coefficient for Non-Newtonian Flow Inside a Pipe

Food products like sugar syrup, applesauce, condensed milk, and tomato puree containing comparatively a high percentage of solid or a low percentage of water are non-Newtonian in nature.

Laminar Flow

For laminar flow of the earlier food products in tubes or pipes, the Metzner and Gluck equation can be employed:

$$\text{Nu} = 1.75\,\delta^{1/3}(\text{Gz})^{1/3}(\gamma_b/\gamma_w)^{0.14} \tag{20.78}$$

This equation is valid for the dimensionless Graetz number, Gz > 20; the dimensionless flow behavior index, $n > 1$ and natural convection is negligible owing to the high value of consistency index K, N s^n/m².

So also $\delta = (3n + 1)/4n$; γ_b and γ_w are generalized viscosity coefficients, N s^n/m² at bulk and wall temperatures, respectively, and Gz is expressed by

$$\text{Gz} = (\pi/4)\,\text{Re}\,\text{Pr}(D/L) = (\pi/4)(Dv\rho/\mu)(c_p\mu/k)(D/L)z$$

$$= (\pi/4)(D^2 v\rho c_p)/kL$$

$$= (mc_p/kL) \tag{20.79}$$

where m is the mass flow rate in kg/s.

The generalized coefficient of viscosity, γ

$$\gamma = K'8^{n-1} = K[(3n+1)/4n]^n 8^{n-1}; \quad (\gamma_b/\gamma_w) = (K_b/K_w) \text{ and}$$

the Re_g

$$\text{Re}_g = (v^{2-n} D^n \rho)/\gamma = (v^{2-n} D^n \rho)/[K 8^{n-1}\{(3n+l)/4n\}^n] \qquad (20.80)$$

where

γ_b and γ_w are the generalized coefficients of viscosity at bulk mean and wall temperatures

γ, K, and n are the flow property constants for non-Newtonian fluids

K' has the same units of N s^n/m^2

When $n = 1$ and $K' = \mu$, the flow follows Newtonian fluid flow only.

Turbulent Flow

The following equation can be used for turbulent flow in tubes or pipes:

$$\text{Nu} = 0.0041 \text{Re}_{g1} (\text{Pr}_g)^{0.4}$$

where

$$\text{Re}_g = (v^{2-n} D^n \rho)/\gamma$$

$$\text{Pr}_g = (c_p K'/k)(8v/D)^{n-1}$$

Example 20.9

A thick fruit puree (following power law) passes at a rate of 0.0333 kg/s through a tube of inside diameter of 25.4 mm and length 1.372 m. The puree enters the tube at 14°C and leaves at 61.4°C. The tube wall temperature remains at a constant temperature of 79°C. The consistency indices at 37.7°C and 79°C correspond to 5.6 and 3.3 Pa s^n whereas the flow behavior remains at constant value of 0.46 over the temperature range.

The average density, specific heat, and thermal conductivities are 1118 kg/m³, 3650 J/kg K, and 0.555 W/m K, respectively.

Estimate the film coefficient, hL

The data given are $\rho = 1180$ kg/m³, $c_p = 3650$ J/kg K, $k = 0.555$ W/m K, $K_b = 5.6$ Pa s^n, $K_w = 3.3$ Pa s^n, $n = 0.46$, $L = 1.327$ m, $m = 0.0333$ kg/s

$$\gamma = K[(3n+1)/4n]^n 8^{n-1}(\gamma_b/\gamma_w) = (K_b/K_w) \text{ and } \delta = (3n+1)/4n$$

$$\gamma = 5.6(3 \times 0.46 + 1)/(4 \times 0.46)^{0.458} \times 8^{-0.542} = 2.3 \text{ Pa } s^n$$

$$\text{Re}_g = (v^{2-n} D^n \rho)/\gamma \text{ and } v = m/(\pi D^2 \rho/4)$$

$$v = 0.0333/(3.142 \times 0.0254 \times 0.0254 \times 1118/4) = 0.0557 \text{ m/s}$$

$$\text{Re}_g = (0.0557)^{1.14} \times (0.0254)^{0.46} \times (1118/2.3) = 1.04$$

Therefore, the fluid flow is in the laminar region.

Now, $Gz = (mc_p/kL) = (0.0333 \times 3.65 \times 10^3/0.555 \times 1.372) = 159.62$

As $Gz > 20$, $Nu = (h_L k/D) = 1.75 \times \delta^{1/3} \times (Gz^{1/3}) \times (\gamma_b/\gamma_w)^{0.14}$

$(h_L k/D) = 1.75 (3n + 1/4n)^{1/3} \times (mc_p/kL)^{1/3} \times (\gamma_b/\gamma_w)^{0.14}$

$$= 1.75 (3 \times 0.458 + 1/4 \times 0.458)^{1/3} \times (159)^{1/3} \times (5.6/3.3)^{0.14} = 11.13$$

$(h_L D/k) = 11.13$, therefore $h_L = (11.13 \times 0.554/0.0254) = 242.86$

Hence film heat transfer coefficient, $h_L = 243 \, W/m^2 \, K$.

Convection Inside and Outside of a Pipe and Overall Heat Transfer Coefficient

The Pr is the ratio of shear component of diffusivity for momentum, μ/ρ, to the thermal diffusivity, $k/\rho c_p$, which is expressed as

$$Pr = (\mu/\rho)/(k/\rho c_p) = (\text{kinematic viscosity/thermal diffusivity}) = (c_p \mu/k)$$

Pr ranges from 0.5 to 1–2 to well over 10^4 for gases and liquids, respectively.

Nu relates heat transfer coefficient to the fluid thermal conductivity, k, and a characteristic dimension known as equivalent diameter, D_e:

$$Nu = hD_e/k$$

where
  c_p is the fluid specific heat, J/(kg K)
  μ is the fluid dynamic viscosity, kg/(m s)
  k is the fluid thermal conductivity, W/(m K)
  α is the thermal diffusivity, m²/s
  x_w is the wall thickness of a pipe, m

When there is a fluid on the inside and outside of a solid pipe surfaces, the inner wall is with a hot fluid of temperature T_i and a cold fluid of T_o on the outside surface (Figure 20.8). If the inside and outside convective coefficients are h_i and h_o, W/m² K, respectively, then the heat transfer rate can be expressed as

$$q = h_i A(T_i - T_1) = (kA/x_w)(T_1 - T_2) = h_o A(T_2 - T_o)$$

$$= (T_i - T_o)/\{1/(h_i A) + x_w/(kA) + (1/h_o A)\} = (T_i - T_o)/\sum R$$

(20.81)

In terms of overall heat transfer coefficient, U, by combining conduction and convection:

$$q = UA\Delta T$$

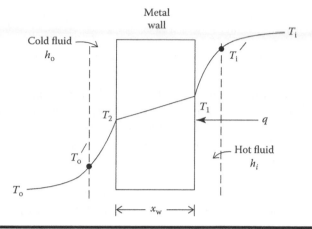

Figure 20.8 Temperature gradients for heat flow by conduction and convection from one fluid to another.

where

$$U = 1/(1/h_i + x_w/k + 1/h_o), \text{ W/m}^2\text{K (Btu/ft}^2\text{h °F)}$$
$$\Delta T = (T_i - T_o), \text{ K}$$

U can be expressed either on the basis of inside area or diameter of the pipe, U_i, or outside area or diameter, U_o. Therefore,

$$q = U_i A_i (T_i - T_o) = U_o A_o (T_i - T_o) = \Delta T/(1/U_i A_i) = \Delta T/(1/U_o A_o)$$
$$= (T_i - T_o)/\Sigma R \tag{20.82}$$

$$U_i = \left[1/\{1/h_i + (r_o - r_i)A_i/(k_w A_m) + A_i/(A_o h_o)\} \right]$$
$$\text{Also} \quad U_i = \left[1/\{1/h_i + x_w D_i/(k_w D_m) + D_i/(D_o h_o)\} \right] \tag{20.83}$$

$$U_o = \left[1/\{(A_o/A_i h_o) + (r_o - r_i)A_o/(k_w A_m) + 1/h_o\} \right]$$
$$\text{Also} \quad U_o = \left[1/\{1/h_o + x_w D_o/(k_w D_m) + D_o/(D_i h_i)\} \right] \tag{20.84}$$

In industrial practice, heat transfer surfaces cannot be kept clean due to corrosion, deposit of dirt, and various other solid deposits to form scale on either inside or both inner and outer surfaces of the pipes and include additional resistances to heat transfer reducing the overall coefficient. Assuming the deposit of scales on both inner and outer surfaces of the pipe, the earlier equations are to be corrected by adding terms for fouling resistances as follows:

$$U_i = \left[1 / \left\{ 1/h_i + 1/h_{fi} + x_w D_i/(k_w D_{lm}) + D_i/(D_o h_o) + D_i/(D_o h_{fo}) \right\} \right] \qquad (20.85)$$

$$U_o = \left[1 / \left\{ 1/h_o + 1/h_{fo} + x_w D_o/(k_w D_{lm}) + D_o/(D_i h_i) \right\} + D_o/(D_i h_{fi}) \right] \qquad (20.86)$$

where h_{fi} and h_{fo} are the fouling factors for the scale deposition on the inner and outer pipe surfaces, respectively, and fouling resistance, FR is $1/h_f$, $m^2 K/W$.

The values of overall heat transfer coefficients, U ($W/m^2 K$) for some exchangers are (i) for steam condenser, $U = 1100–5700$; (ii) steam-to-water exchanger, $U = 2400$ (approximately); (iii) water-to-water exchanger, $U = 850–1700$; (iv) steam-to-air exchanger, $U = 100$ (approximately); (v) water-to-air exchanger, $U = 95$ (approximately); and (vi) air-to-air exchanger, $U = 30$ (approximately).

The typical values of fouling resistances, FR ($m^2 K/W$) for some fluids are as follows: FR for distilled water = 0.0001, FR for sea water = 0.0002, and FR for pure steam = 0.00005.

Overall Coefficient through a Series of Multiple Cylinders

Metallic pipes are often insulated in multiple layers to minimize heat loss. If the fluid temperatures inside and outside the pipe are known, the overall coefficient, U can be predicted.

If D_i, D_2, and D_o are diameters of two concentric hollow cylinders corresponding to the interfacial temperatures of T_i, T_2, and T_3 with the innermost and outermost media temperatures of T_i and T_o, respectively, then the thermal conductivities of two cylinders are k_1 and k_2. Film coefficients of inside and outside fluids are h_i and h_o and overall coefficients based on inside and outside areas or diameters are U_i and U_o, respectively.

Then for steady state of heat transfer

$$q = h_i A_i (T_i - T_o) = h_o A_o (T_i - T_o) = U_i A_i (T_i - T_o) = U_o A_o (T_i - T_o) \qquad (20.87)$$

where

$$U_i = 1/h_i + D_i \ln(D_2/D_i)/2k_1 + D_i \ln(D_o/D_2)/2k_2 + D_i/(h_o D_o)$$

$$U_o = 1/h_o + D_o \ln(D_2/D_i)/2k_1 + D_o \ln(D_o/D_2)/2k_2 + D_o/(h_i D_i)$$

Forced Convection Heat Transfer Inside Pipes

Heat exchangers and other heat transfer equipment are designed on the basis of film heat transfer coefficient, h, $W/m^2 K$ and overall heat transfer coefficient, U, $W/m^2 K$ that can be mathematically expressed by

$$q = hA\Delta T \text{ and } q = UA\Delta T \quad \text{or,} \quad q/A = h\Delta T \text{ and } q/A = U\Delta T$$

$$1/U_i = 1/h_i + (D_i/D_m)\,(x_w/k_w) + (D_i/D_o)\,(1/h_o)$$

$$1/U_o = 1/h_o + (D_o/D_m)\,(x_w/k_w) + (D_o/D_i)\,(1/h_o)$$

Wall Temperature, T_w

To estimate the viscosity of the fluid, μ_w, at the wall temperature, T_w, a trial error method is to be used as follows:

$$(1/U_o) = (D_o/D_i h_i) + (x_w/k_w)\,(D_o/D_m) + 1/h_o$$

Each term of the right-hand side of the equation represents the individual resistance for each of the two fluids and the metal wall in between. Hence, on the basis of overall and individual resistances:

$$\Delta T/(1/U_o) = \Delta T_i/(D_o/D_i h_i) = \Delta T_w/(X_w/k_w)(D_o/D_m) = \Delta T_o/h_o \qquad (20.88)$$

where
   ΔT is the overall temperature drop
   ΔT_o is the temperature drop through outside fluid
   ΔT_w is the temperature drop through metal wall
   ΔT_i is the temperature drop through inside fluid
   ΔT is the temperature difference between bulk of the fluid and heat transfer surface
   k_w is the thermal conductivity of the metal, W/(m K)
   h_i and h_o are the film heat transfer coefficients based on inner and outer diam-
      eters, D_i and D_o, respectively
   D_m is the log mean diameter defined as

$$D_m = (D_o - D_i)/\ln(D_o/D_i) \text{ and thickness of the heating surface } x_w,$$

$$m = (D_o - D_i)/2$$

Natural Convection

Grashof Number

In natural or free convection dimensionless group, called Grashof number (Gr), is used to evaluate film heat transfer coefficient. When a fluid is heated or cooled, a natural or free convection sets in due to the upward buoyancy force in presence of the gravity without any other external driving force. The induced flow is opposed by viscous drag at the solid wall. The Gr compares buoyancy force with viscous drag. Natural convection means buoyancy-driven or self-driven flow, whereas forced convection is always driven by some external agencies.

The Gr is useful to describe natural or free convection for both laminar and turbulent flows in heat as well as mass transfer. In heat transfer, the Gr is mathematically expressed as

$$Gr = (L^3 \rho^2 g \beta \Delta T)/\mu^2 = (L^3 g \beta \Delta T)/\nu^2$$

where
L and D are characteristic length and diameter, m
ρ is the fluid density, kg/m^3
β is the fluid volumetric thermal expansion coefficient
$1/K$ ($\beta = 1/K$ for ideal gases)
ΔT = (+ve) temperature difference between wall and bulk fluid, K
μ is the fluid dynamic viscosity, Pa s
γ is the kinematic viscosity = (μ/ρ)

For horizontal plates, L may be the side of a square or the mean of the two sides of a rectangle and for horizontal cylinders, outside diameter, D_o, is used in place of L. All temperatures are measured at film temperature, T_f, K. The thermal expansion coefficient, P, is a measure of the rate at which the fluid volume changes with temperature at a particular pressure, p. The parameter $g\beta/\gamma^2$ signifies the value of the Gr for a particular length L and temperature difference ΔT.

In case of natural convection or buoyancy-driven flow, the magnitude of Gr indicates when the flow undergoes to the transition state from laminar to turbulent, whereas in forced convection the value of Re involving characteristic velocity demarcates this transition and others. In both cases, the Pr is expressed as

$$Pr = (\text{kinematic viscosity/thermal diffusivity}) = \gamma/\alpha$$

where α has an additional influence on the fluid flow. The relative values of Gr and Re reveal whether natural or forced convection is predominant. When Gr $\gg$ Re^2, the influence of forced convection is negligible with respect to that of natural convection. If the values of Gr and Re^2 are more or less same, the combined effects of natural and forced (mixed) convection will exist.

Film Coefficient

By dimensional analysis, an empirical correlation can be deducted to estimate film coefficient for natural convection:

$$Nu = (hL/k) = a(L^3\rho^2 g\beta\Delta T/\mu^2 \, c_p\mu/k)^m = a(Gr \, Pr)^m$$

where a and m are empirical constants. The values of a and m with the earlier equation for the geometry of vertical planes and cylinders when the vertical height $L < 1$ m are as follows (Geankoplis, 1993).

$Gr \cdot Pr$	a	M		$Gr \cdot Pr$	a	m
$<10^4$	1.36	0.2		10^4–10^9	0.59	0.5

In mass transfer concentration difference, Δc is the driving force, the Grashof number Gr_m is formulated in terms of $\beta \Delta c$ in place of $\beta \Delta T$, and Gr_m is expressed as $Gr_m = (L^3 \rho^2 g \beta \Delta c)/\mu^2$ where β is the concentration coefficient of volumetric expansion.

Boiling

Boiling is generally employed for industrial evaporation and distillation operations. In food industry, it is frequently used for evaporation and concentration of milk, fruit, and sugarcane juice for various end products. Figure 20.9 shows the different stages of boiling where heat flux, q/A in m² is plotted against the temperature difference between the heat transfer surface and the boiling temperature of water, ΔT in K on a full log–log graph paper. The entire plot may be divided into four regions A, B, C, and D.

In the first low ΔT region A, heat transfer occurs by natural convection as very few bubbles released from the heating surface do not disturb this free convection and the heat flux varies from 10^4 to 5×10^4 W/m².

In the second nucleate boiling region B, the circulation increases with the increase of bubble formation and the heat flux ranges from 5×10^4 to 10^6 W/m² and h varies from 5700 to 57,000 W/m² K. Heat transfer mainly depends upon the temperature difference between the heating surface and the boiling water as well as the specific thermal stress that influences the rate of formation of bubbles. The orientation of heating plates such as horizontal or vertical position also has an effect.

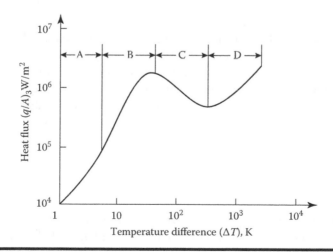

Figure 20.9 Heat flux–temperature difference diagram in boiling. Different stages of boiling.

Table 20.4   Typical Values of Film Coefficient, *h* for a Few Common Heating and Cooling Media

5–25 W/m²K for free convection of air
10–500 W/m²K for forced convection of air
100–15,000 W/m²K for forced convection of water
(5000–2100 Btu/h ft² °F) for drop-wise condensing steam
(1000–3000 Btu/h ft² °F) for film-type condensation of steam
(5–20 Btu/h ft² °F) for superheating of steam
2000–6000 W/m²K (350–1000 Btu/h ft² °F) for boiling water
(50–3000 Btu/h ft² °F) for heating or cooling of water
500–2000 W/m²K (80–350 Btu/h ft² °F) for a moving stream of water
(0.2–10 Btu/h ft² °F) for heating or cooling of air

The following relationships between film heat transfer coefficient *h* and heat flux *q* are useful for horizontal plates (Saravacos and Kostaropoulos, 2002):

$$h = 170 \, q^{0.26} \text{ W/m}^2 \text{ K} \quad \text{for } q < 17{,}500 \text{ W/m}^2 \tag{20.90}$$

and

$$h = 1.54 \, q^{0.26} \text{W/m}^2 \text{ K} \quad \text{for } 17{,}500 < q < 200{,}000 \text{ W/m}^2 \tag{20.91}$$

For evaporation at vertical surfaces, these values can be multiplied by a factor of 1.25. The next region *C* is called transition boiling region where rapid production of bubbles takes place with the formation of a layer of insulating vapor. As ΔT increases, this vapor layer also increases and consequently heat flux drops from 10^6 to 7×10^5 W/m² in this region. In the final film-boiling region D, bubbles detach themselves that rise upward and help increase both heat flux and film coefficient, *h* (Table 20.4).

Condensation

Film-Type Condensation

Laminar Flow

In the film-type condensation on a vertical surface, an average convective heat transfer coefficient, *h*, for laminar flow can be predicted by the following equations:

$$h = 1.13 \left[\left\{ k_1^3 \lambda_c \rho_1 (\rho_1 - \rho_v) g \right\} / \left\{ \mu_1 (T_s - T_w) L \right\} \right] \quad \text{for Re} < 1800 \tag{20.92}$$

This equation can be expressed for Nu as

$$Nu = (hL/k) = 1.13\left[\lambda_c \rho_1 (\rho_1 - \rho_v) g L^3 / k_1 \mu_1 (T_s - T_w)\right]^{1/4}$$

$$Re = (4m/\pi D\mu_1) \quad \text{for vertical tubes and} \quad Re = (4m/w\ \mu_1)$$

(20.93)

Nu is expressed in the following for Re > 1800 or turbulent flow over vertical plates

$$Nu = 0.0077 \left[(\rho_1^3 L^3 g)/\mu_1\right] (Re)^{0.4}$$

(20.94)

The film coefficient, h, for horizontal pipe is given by

$$Nu = 0.725[\lambda_c \rho_1 (\rho_1 - \rho_v) g L^3 / k_1 \mu_1 (T_s - T_w)]^{1/4}$$

(20.95)

where
All liquid properties are measured at film temperature, $T_f = (T_s + T_w)/2$ in K
T_s and T_w are condensing vapor and wall temperature, respectively, K
ρ_1 and ρ_v are density of liquid and saturated vapor, respectively, kg/m³
λ_c is the latent heat of condensation, kJ/kg
k_1 is the liquid thermal conductivity, W/m K
μ_1 is the liquid viscosity, Pa s
g is the acceleration due to gravity, 9.81 m/s²
m is the liquid mass flow rate, kg/s
D is the pipe diameter, m
w is the horizontal plate width, m

Mass Transfer and Heat Transfer

The dimensionless Schmidt number (Sc) is mathematically expressed as $Sc = (D\rho/\mu)$. The Sc is physically interpreted as the ratio of momentum diffusivity to mass diffusivity. The dimensionless Sherwood number (Sh) is defined by the ratio $Sh = (k_G d/D)$ By dimensional analysis, the following functional relationships can be shown:

$$Nu = f(Re,\ Pr) \quad \text{and} \quad Sh = f(Re,\ Sc)$$

To find an analogy between mass transfer and heat transfer, the Sc is referred to mass transfer whereas Pr is relevant to heat transfer.
For free convection, Schmidt also provided analogous relationships as follows:

$$Nu = f(Gr,\ Pr) \quad \text{and} \quad Sh = f(Gr,\ Sc)$$

The Lewis number (Le) is the ratio of heat diffusivity to mass diffusivity, $Le = (\alpha/D)$.

For an ideal fluid, v (kinematic viscosity) $= \alpha$ (thermal diffusivity) $= D$ (mass diffusivity), and $Pr = Sc = Le = 1$.

The aforementioned dimensionless groups including others[a] are summarized as follows for convenience.

Dimensionless Group	Definition	Significance/Uses
Biot number, Bi	hD/k	Heat transferred by convection/heat conducted in solid
Fanning friction factor, f	$-\Delta pD/2Lv$	Fluid flow
Fick number, Fi	$(D_f\theta_r)/D^2$	Corresponds to Fourier number
Fourier number, Fo	$(k\theta_r/\rho c_p)/S^2 = (\alpha\theta_r/S^2)$	Transient conduction/stored heat
Heat transfer J'_H factor, J'_H	$(h/c_pG)(c_p\mu/k)^{2/3}$ $(\mu_w/\mu)^{0.14} = St\ Pr^{2/3}(\mu_w/\mu)$	Heat transfer
Grashof number, Gr	$L^3\rho^2\beta g\ \Delta t/\mu^2$	Hydrostatic buoyancy/viscosity
Graetz, Gz	GAc_p/kD	Heat transfer
Lewis number, Le	$(k/c_p\rho)/D_r = \alpha/D_f = Sc/Pr$	Relates heat and mass transfer
Mass transfer factor, J_H	$St\ Sc^{2/3}$	Mass transfer
Nusselt number, Nu	hD/k	Convectional heat/conduction heat in medium
Peclet number, Pe(Re Pr)	$(vD/k/c_p\rho l) = LGc_p/k$	Convectional heat-transferred/heat conducted
Prandtle number, Pr	$c_p\mu/k$	Heat produced by friction/conduction heat
Reynolds number, Re	DG/μ	Inertial forces/frictional forces

(continued)

Stanton number, St (Nu/ Re Pr)	h/Gc_p	Heat transferred from outside/convectional heat transported
Schmidt number, Sc	v/D	Mass transfer (corresponds to Prandtl number)

where

A is the interfacial area, m²

D is the diameter, m

α is the thermal diffusivity, m²/s

c_p is the specific heat, J/kg K

g is the acceleration due to gravity, 8.91 m²/s

$G = v\rho$, mass flow rate, kg/m² s

h is the film heat transfer coefficient, W/m² K

L is the length, m

Δp is pressure drop, Pa

S is one-half slab thickness, m

v is velocity, m/s

α is thermal diffusivity

θ_T is the heating or cooling time, s

μ is the viscosity, Pa s

ρ is the density, kg/m³

v is the kinematic viscosity, m²/s

v is the velocity, m/s

k_G is mass transfer coefficient, m/s

d is characteristic length, m

D_f is the mass diffusivity, m²/s

[a] Kessler, H.G., *Food Engineering and Dairy Technology*, Verlag, Germany, 1981.

Logarithmic Mean Temperature Difference

The temperature differences characterizing the heat exchangers' operation are based on the temperatures at the two ends as shown in Figures 20.10 and 20.11.

For parallel or cocurrent flow heat exchanger: $\Delta T_1 = T_{h1} - T_{c1}$ and $\Delta T_2 = T_{h2} - T_{c2}$

For countercurrent flow heat exchanger: $\Delta T_1 = T_{h2} - T_{c1}$ and $\Delta T_2 = T_{h1} - T_{c2}$

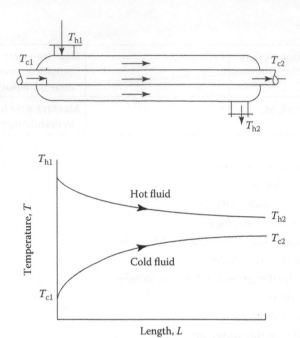

Figure 20.10   Cocurrent.

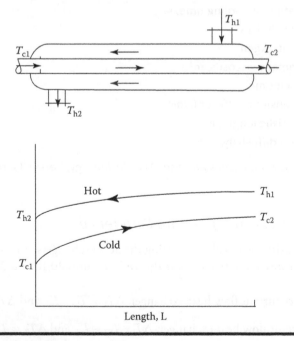

Figure 20.11   Countercurrent.

When ΔT_1 and ΔT_2 are neither same nor very close to each other, the overall temperature difference, ΔT, of heat transfer equation is taken as the following log mean temperature difference, ΔT_m:

$$\Delta T_m = (\Delta T_1 - \Delta T_2)/\ln(\Delta T_1/\Delta T_2)$$

When ΔT_1 and ΔT_2 are equal or close to each other, the arithmetic mean temperature difference, $\Delta T_{am} = (\Delta T_1 + \Delta T_2)/2$ are generally used.

Example 20.10

Water flowing at a rate of 1 kg/min is heated from 15°C to 80°C with the flue gas at a rate of 3 kg/min having a temperature of 180°C and a specific heat (c_h) of 1.03 kJ/kg K. If the overall heat transfer coefficient (U) is taken as 120 W/m^{2}K, calculate the heating surface area (A) required for (i) parallel current and (ii) countercurrent flow.

Heat lost by the hot flue gas = heat gained by the cold water

$$q = m_h c_h \, \Delta T_h = m_c c_c \, \Delta T_c$$

$$[3 \times 1.05 \times 1000 \times (180 - T_{h2}/60)] = [1 \times 4.2 \times 1000 \times (80 - 15)/60]$$

$$\text{or } (180 - T_{h2}) = 86.67$$

Outlet temperature of the flue gas, $T_{h2} = 180 - 86.67 = 93.33°C$

For parallel flow

$$q = m_c c_c \, \Delta T_c = (1 \times 4.2 \times 1000 \times 65)/60 = 4550 \text{ W}$$

$$\Delta T_m = \{(180 - 15) - (93.33 - 80)\}/\ln(165/13.33) = 60.2°C$$

Heating surface area

$$A = (q/U \, \Delta T_m) = (4550/120 \times 60.26) = 0.629 \text{ m}^2$$

For countercurrent flow

$$\Delta T_m = \{(180 - 15) - (93.33 - 80)\}/\ln(165/13.33) = 60.2°C$$

$$= (100 - 78.33)/0.245 = 88.5°C \quad \text{and} \quad A = (4550/120 \times 88.5) = 0.428 \text{ m}^2$$

Heat Exchangers

There are various types of heat exchangers that are available in food processing industries depending on the different requirements of heating, cooling, evaporation, thermal processing, baking, and so on. As a whole, the heat exchangers accomplish these heating and cooling operations. The mechanisms of heat exchange are mainly conduction and convection. The general principles of heat exchangers are outlined later (Figures 20.12 and 20.13).

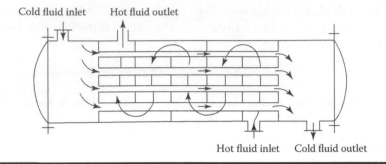

Figure 20.12   One-shell pass and one tube (1-1 heat exchanger).

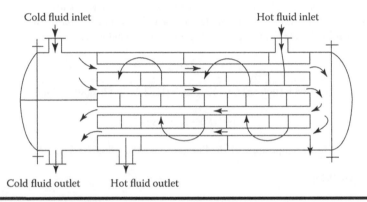

Figure 20.13   One-shell and two-tube passes (1-2 heat exchanger).

Types of Heat Exchangers

Double Pipe Exchangers

The concentric double pipe heat exchangers are very common and simple in design. Both double and triple pipe heat exchangers are available. The double pipe type consists of two concentric pipes with the cold stream usually flowing through the inner pipe and the heating medium through the annulus. The cold and hot fluids may flow in either cocurrent or countercurrent direction. In case of triple tube, the working fluid usually flows through the annulus of the middle tube. Other types of concentric tube heat exchangers may be of helical or spiral design. Both double tube and triple tube heat exchangers can be designed in any of the earlier forms for the purpose of space saving.

Single-Pass 1-1 Exchanger

The double pipe exchangers cannot handle large fluid rates in few tubes. Here one shell can serve for many tubes, which is economical as well. This type of exchanger, because of its one shell side pass and one tube side pass, is called single pass 1-1 exchanger. The exchanger in whom the shell side fluid flows in one shell pass the tube side fluid in two or more passes is the 1-2 exchangers.

From a practical point of view, it is difficult to obtain a high velocity when one of the fluids flows through all the tubes in a single pass. This can be circumvented by introducing a system that tube fluid is carried through fractions of the tubes consequently as in the case of a two pass tube exchanger, where all the tube fluid flows through the two halves of the tubes successively. In these, the velocity and turbulence of the shell side fluid are as important as those of the tube side fluid. If a fluid releases a large amount of heat due to phase change or any other reason and other fluid undergoes a small rise in temperature owing to a large quantity of flow, it may be taken as a good condition for this type of exchanger. The smaller fluid flow occurs through a bundle of tubes housed in a large cylindrical vessel, which allows the large quantity of fluid to pass. In this system, it is possible to provide two, four, or six tube passes. The one shell–two tubes passes and two shells–four tubes passes are most common.

The baffles can be installed in the shell to decrease the cross section of the shell side fluid and force the fluid to flow across the tube bank rather than parallel with it and guide the fluid flow from one end to the other end of the shell. The turbulence created in this type of flow further increases the shell side coefficient.

In cases of shell and tube, plate, and cross flow heat exchangers, the mean temperature difference is to be used with a correction factor, F, due to the flow directions in shell and tube and plate heat exchanger may change from parallel to countercurrent flow along the whole length of the exchangers. The F for shell and tube exchangers depends on two dimensionless temperature ratios:

$$P = (T_{to} - T_{ti})/(T_{si} - T_{ti}) \quad \text{and} \quad R = (T_{si} - T_{so})/(T_{to} - T_{ti})$$

where

T_{ti} and T_{to} are tube side fluid inlet and outlet temperatures, respectively, K
T_{si} and T_{so} are shell side fluid inlet and outlet temperatures, respectively, K

The F values can be found from the graphics corresponding one shell–two tubes passes and two shells–four tubes passes heat exchangers that are available in Geankoplis (1997). It may be noted that the factor F can be obtained by locating the earlier values of P on the X-axis, and corresponding R from the curved lines.

Plate Heat Exchangers

The plate heat exchangers are widely used in food, pharmaceutical, and other industries due to their high thermal efficiency, compactness, and hygienic design. These exchangers are most suitable to heat or cool low viscosity, particle-free liquids like milk.

A common type of plate heat exchanger employed for liquid foods comprises of steel plates separated from each other by sealing gaskets on the edges of their surfaces and the whole assembly is rigidly fixed in a frame. Generally, large bolts are used to fix the plates rigidly kept to the support base. The narrow space between the plates is filled with either the liquid food or the heating or cooling fluid/medium. At each end of every plate, there are two holes and on the face of the plates around each of these holes a gasket may be fitted.

The liquid products as well as the heating or cooling fluid flow alternately in cocurrent or countercurrent direction through the slit between the two plates (Figure 20.14). The plates can be organized in such a way so that the four wholes correspond to supply of fluid A, the supply of fluid B, the exit of fluid A, and the exit of fluid B. Their important feature is that they can be disassembled easily for thorough cleaning. Usually, the plates are having various patterns of ribs or corrugated ridges to increase the surface area of plate and improve heat transfer by enhancing turbulence in the liquids.

Placing polymer gaskets in between these plates seals the narrow gaps. These gaskets seal the thin plates around their edges and these are made of some elastomers for temperatures as high as 135°C.

The length and width of the plates may range from 1 to 2.5 m and 0.25 to 1 m or even higher according to the requirement of heat transfer area. The gaps between plates may be kept between 2 and 6 mm; the flow rates may vary from 0.5 to 5000 m³/h; operating temperatures from –35°C to 200°C; and pressure up to 25 bars.

The *film heat transfer coefficient* is correlated as

$$\mathrm{Nu} = c\, \mathrm{Re}^n\, \mathrm{Pr}^{1/3} \quad \text{for Re} > 5,\ c = 0.352;\ n = 0.539$$

where Re is based on equivalent diameter of the flow channel, which is four times the hydraulic radius (Saravacos and Kostaropoulos, 2003).

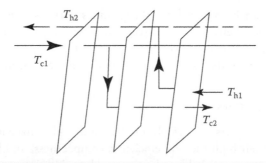

Figure 20.14 **A line diagram of a plate heat exchanger system.**

Example 20.11

In a double tube water heater, water flows through the inner steel tube of 35.05 mm ID and 42.16 mm OD with an average velocity of 1.5 m/s. The outer-jacketed steam at a temperature of 170°C heats the inner tube. Thermal conductivity of the steel tube, k_m, is 45 W/m K. The film coefficient on the steam side is 1 1500 W/m² K.

(i) Evaluate the film coefficient on the water side in the inner tube and (ii) estimate the length of the steel tube required to heat the water from 42°C to 88°C

(i) Thickness of the pipe wall $= (42.16 - 35.05)/2 = 3.56\,mm = 0.0356\,m$

Average water temperature $= (42 + 88)/2 = 65°C$
Properties of water at 65°C

$$c_{pw} = 4.19 \text{ kJ/kg K}; \quad k_w = 0.654 \text{ W/m K}; \quad \mu_w = 0.432 \times 10^{-3} \text{kg/ms};$$

$$\rho_w = 981.6 \text{kg/m}^3; \quad \text{and} \quad \text{Pr} = 2.72$$

For water side

$$\text{Re} = (D v \rho_w / \mu_w) = (0.03505 \times 1.5 \times 981.6/0.432 \times 10^{-3}) = 119,462.08 \text{(turbulent)}$$

Neglecting viscosity correction factor,

$$\text{Nu} = (h_i D/k_w) = 0.023 \, (\text{Re})^{08} \, (\text{Pr})^{1/3} = 0.023 \, (119,462.1)^{0.8} (2.72)^{1/3} = 369.90$$

Water side film coefficient, $h_i = (369.90 \times 0.6629/0.03505) - 6901.997 \text{ W/m}^2\text{K}$

(ii) Mass flow rate, $m = A v \rho_w = 3.142 \, (0.03505)^2 \times 1.5 \times 981.6/4 = 1.421 \text{ kg/s}$

Overall resistance $(1/U) = (1/h_o A_o) + \ln(D_o/D_i)/(2\pi k L) + (1/h_i A_i)$

$(1/U) = R = (1/L)[\{1/(h_o \pi D_o)\} + \{\ln(D_o/D_i)/(2\pi k)\} + 1/(h_i D_i)]$

$= (1/L)[\{1/(11,500 \times 3.142 \times 0.04216)\} + \{\ln(0.04216/0.03505)/(2 \times 3.142 \times 45)\}$

$\quad + 1/(6901.99 \times 3.142 \times 0.03505)] = (2.53 \times 10^{-3})/L$

Now log mean temperature difference, $\Delta T_m = (\Delta T_2 - \Delta T_1)/\ln(\Delta T_2/\Delta T_1)$

$\Delta T_m = (170 - 42) - (170 - 88)/\ln(170 - 42)/(170 - 88) = 104.5°C$

Heat flow $= q = m c_{pw} \Delta T = \Delta T_m/(1/U)$

$m c_{pw} \Delta T = \Delta T_m/(1/L)[\{1/(h_o \times 3.142 D_o)\} + \{\ln(D_o/D_i)/(2 \times 3.142 k)\}$

$\quad + \{1/(h_i \times 3.142 D_i)\}]$

$q = 1.421 \times 4190 \times 46 = (103.3 \times L)/(2.53 \times 10^{-3})$

Length of the steel tube, $L = 6.71$ m.

Example 20.12

Water flowing through tubes in a number of passes at the rate of 4.1 kg/s is heated from 36°C to 56°C in a shell and tube heat exchanger with hot water (heating fluid). The hot water enters at 96°C in the shell side in a single pass at a rate of 2.25 kg/s.

The hot water velocity in 19.05 mm internal diameter is 30.5 cm/s.

The tube length should be restricted 2 m due to space limitation.

Assuming overall heat transfer coefficient as 900 W/m^2 K, calculate (i) the number of tube passes, (ii) the number tubes, and (iii) the length of tube.

As a first step, let us assume one tube pass only.

Heat lost by hot water = heat gained by cold water

$$q = m_h \times c_h \times \Delta T_h = m_c \times c_c \times \Delta T_c$$

$$\Delta T_h = (m_c \times c_c \times \Delta T_c)/m_h \times c_h) = (4.1 \times 4.2 \times 10^3 \times 20)/(2.25 \times 4.2 \times 10^3) = 36.4°C$$

Hot water outlet temperature, $T_{h2} = (96 - 36.4) = 59.6°C$

$$q = 4.1 \times 4.2 \times 20 \times 10^3 = 344.4 \text{ kW}$$

The log mean temperature difference, $\Delta T_m = \{(96 - 56) - (59.6 - 36)/\ln(40/23.6)\}$

$$\Delta T_m = (96 - 56) - (59.6 - 36)/1 = (16.4/0.528) = 31.1°C$$

$$A = (q/U \times \Delta T_m) = (344.4 \times 10^3/900 \times 31.1) = 12.3 \text{ m}^2$$

Total cross-sectional flow area of tubes, $A_c = (mc/v\rho) = (4.1/0.305 \times 1000) = 0.013 \text{ m}^2$
If n be the number of tubes,

$$(n\pi D^2/4) = 0.013 \text{ m}^2$$

Therefore, $n = (4 \times 0.013/3.142 \times 0.01905 \times 0.01905) = 45.6 = 46$ tubes

Heat exchanger surface area required, $n\pi DL = A$

Tube length, $L = (A/n\pi D) = (12.3/46 \times 3.142 \times 0.0305) = 2.8$ m

As the length of the tube, L should not exceed the permissible limit of 2 m. Let us try two-tube pass system.

$$R = (T_{h1} - T_{h2})/(T_{c2} - T_{c1}) = (96.0 - 59.6/56 - 36) = 1.82$$

$$P = (T_{c2} - T_{c1})/(T_{h1} - T_{ci}) = (56 - 36)/(96 - 36) = 1/3$$

For $P = 0.33$ and $R = 1.82$ the correction factor, $F = 0.82$ (Geankoplis, 1997)

$$A = (q/U\Delta T_m F) = (344.4 \times 10^3/900 \times 31.1 \times 0.8) = 15.4 \text{ m}^2$$

$$A = 2n\pi DL \text{ or, } L = (A/2n\pi D) = (15.4/2 \times 46 \times 3.142 \times 0.0305) = 1.75 \text{ m}$$

Hence

Number of passes, $n = 2$
Number of tubes = 46
Length of the tube, $L = 1.75$ m

Radiation

All materials at temperatures above absolute zero emit radiations of electromagnetic waves that are independent of external agencies. Radiations that are the result of temperature only are known as thermal radiations. The electromagnetic spectrum that is of importance in heat transfer lies in the wavelength ranging from 0.5 to 50 μm. The fraction of the radiation falling on a body that is reflected is named *reflectivity*, ρ; the fraction that is absorbed by the body is known as *absorptivity*, α; and the transmitted part of it is called *transmitivity*, τ. When radiation falls on different bodies, it is ultimately transformed into heat energy. As radiation is a form of energy according to conservation of energy:

$$\alpha + \rho + \tau = 1$$

Kirchoff's Law

Kirchoff's law states that at equilibrium temperature, the ratio of the total radiating power of any body, W_1, to the absorptivity of that body, α_1, depends only on the temperature of the body. Accordingly, for any two bodies at equilibrium temperature with the surroundings:

$$(W_1/\alpha_1) = (W_2/\alpha_2)$$

where W_2 and α_2 are the total radiating power and absorptivity of the other body.

A *black body* absorbs all radiations incident upon it and reflects or transmits none and therefore according to Kirchoff's law it attains a maximum absorptivity of unity at any temperature.

Emissivity

The emissivity of a body is defined as the ratio of the total radiating power of the body to that of a black body at the same temperature.

If the first of the earlier Kirchoff's equation is a black body where $\alpha_1 = 1$ and

$$W_1 = W_b = W_2/\alpha_2$$

the emissivity of the second body, ε_2 is

$$\varepsilon_2 = W_2/W_b = \alpha_2$$

Thus, emissivity and absorptivity of a body are same when it is at equilibrium temperature with its surroundings. For other materials, they are not equal. There is no perfect black body. However, some carbon blacks approach blackness.

Generally, emissivity increases with temperature. The emissivities of the polished metals are of low values varying from 0.03 to 0.08, those of oxidized metals vary from 0.6 to 0.9, those of refractories, building materials range from 0.65 to 0.9, and those of common paints vary between 0.8 and 0.95.

Stefan–Boltzman Equation

The following basic equation for black body radiation actually follows the Stefan–Boltzman law:

$$W_b = \sigma T^4 \quad \text{and} \quad q/A = \sigma T^4$$

where
q is the heat flow, W
A is the surface area of the body, m^2

The maximum thermal radiation for the black body is a constant, $\sigma = 5.669 \times 10^{-8}$ W/(m^2 K^4) = 0.1714 × 10^{-8} Btu/h ft^2 R^4). All material bodies other than perfect black body radiate less energy at any temperature and their values of a vary with the material as follows:

$\sigma = 0.1$–0.2 for highly polished surface of metals; $\sigma = 0.2$–0.4 for matted surface of metals; $\sigma = 4$–5.5 for most of the liquids, paints, ice foods, and packaging materials.

$W_b = 5.676 \ (T/100)^4$ (W/m^2) and $W_b = 0.1714 \ (T/100)^4$(Btu/h ft^2)

Any real body other than black body called *gray body* that has an emissivity <1 and the emissive power is reduced by 8, or

$$q = A\varepsilon\sigma T^4 \tag{20.96}$$

Hence, the total radiation from a unit area of an opaque or gray body, q/A is

$$q/A = \varepsilon\sigma T^4$$

Radiation between two parallel large plane black surfaces where each black surface views only the other and consequently all radiations from each of the surfaces fall on the other surface and are completely absorbed. Therefore, the total loss of energy by the first plane and the total gain of energy by the second one is

$$\sigma\left(T_1^4 - T_2^4\right) \ \text{W/m}^2 \quad \text{(SI)} \quad \text{or}$$

$$\sigma\left(T_1^4 - T_2^4\right) \ \text{(Btu/ft}^2\text{h)} \quad \text{(English)}$$

Heat exchange between two radiating parallel black planes of areas, A_1 and A_2 with the geometric factors or view factors F_{12} and F_{21} at T_1 and T_2, respectively, is given by

$$q_{1\ 2} = \sigma A_1 F_{12}\left(T_1^4 - T_2^4\right) \quad \text{for } A_1 \tag{20.97}$$

$$q_{2-1} = \sigma A_2 F_{21}\left(T_1^4 - T_2^4\right) \quad \text{for } A_2 \tag{20.98}$$

Comparing the earlier two equations $A_1F_{12}=A_{21}$ and if A_1 can see only the surface A_2, the view factor, $F_{12}=1$.

For an equal area of two surfaces,

$$q_{2-1} = \sigma AF\left(T_1^4 - T_2^4\right) \tag{20.99}$$

where

q_{2-1} is net radiation between two surfaces, W (Btu/h)
A is the same area of the two surfaces, m²
F is the dimensionless geometric or view factor, generally symbolized by F_{ij}

Radiation view factor may be defined as the fraction of energy reaching a particular surface compared to the total energy emitted from another surface and its general symbol is F_{ij}.

$F12$ is the fraction of energy leaving surface 1 that reaches surface 2
$F21$ is the fraction of energy leaving surface 2 that reaches surface 1

A *gray body* is defined as a surface for which the monochromatic properties are constant over all wave lengths. Applying Kirchoff's law, it may be shown that the total absorptivity and emissivity are equal for a gray body, even if the body is not in thermal equilibrium with its surroundings. In practice, exact gray body does not exist as it is an ideal concept.

When *radiation occurs to a small gray object* from large surroundings in an enclosure, the net heat of absorption is expressed by the Stefan–Boltzman equation:

$$q = A_1\sigma\varepsilon_1 T_1^4 - A_1\alpha_{12}\sigma T_2^4 = A_1\sigma\left(\sigma\varepsilon_1 T_1^4 - \alpha_{12}\sigma T_2^4\right) \tag{20.100}$$

The earlier equation can be simplified with the assumption of one emissivity of the small body at T_2

$$q = A_1\varepsilon\sigma\left(T_1^4 - T_2^4\right)$$

where the small body emits radiation to the surroundings in the enclosure $= A_1\varepsilon_1\sigma T_1^4$ with its emissivity being ε at T_1 and it also absorbs energy from the surroundings $= A_1\alpha_{12}\sigma T_2^4$ at T_2, α_{12} being the absorptivity of the body 1 due to radiation from the enclosure at T_2 and approximately $\alpha_{12}=\varepsilon_2$ at T_2.

Heat Exchange between Two Infinite Parallel Gray Planes

Where there is heat exchange between two gray planes only, the net heat flow between them is obtained from

$$q_{1-2} = (Eb_1 - Eb_2)/(1-\varepsilon_1/\varepsilon_1 A_1)+(1/A_1 F_{12})+(1-\varepsilon_2/\varepsilon_2 A_2) \tag{20.101}$$

Heat Exchange between Infinite Parallel Gray Planes

Let the discussion be confined to three specific cases.

1. *Between two equal gray surfaces A_1 and A_2:*
   When surfaces $A_1 = A_2 = A$, view factors $F_{12} = F_{21} = 1$, and also their emissivities and absorptivities of $\varepsilon_1 = \alpha_1$ and $\varepsilon_2 = \alpha_2$, in general the net exchange between the two parallel planes 1 and 2 is given by

   $$q_{1-2} = A_1\sigma\left(T_1^4 - T_2^4\right) / \left[1/(1/\varepsilon_1) + (1/\varepsilon_2) - 1\right] \tag{20.102}$$

   Similarly,

   $$q_{2-1} = A_2\sigma\left(T_1^4 - T_2^4\right) / [1/(1/\varepsilon_1) + (1/\varepsilon_2) - 1]$$

   When surfaces $A_1 = A_2 = A$, view factors $F_{12} = F_{21} = 1$, and also their emissivities and absorptivities of $\varepsilon_1 = \alpha_1$ and $\varepsilon_2 = \alpha_2$,

   $$q = q/A = \sigma\left(T_1^4 - T_2^4\right) / \left[1/(1/\varepsilon_1) + (1/\varepsilon_2) - 1\right] \tag{20.103}$$

2. *Between two long parallel concentric cylinders*
   The heat exchange is between cylinders outside surface A_1 and inside surface A_2. Since all the energy leaving A_1 reaches A_2, then $F_{12} = 1$, $F_{21} = A_1/A_2$
   Therefore, the earlier equation is simplified to

   $$q = \left[\sigma A_1\left(T_1^4 - T_2^4\right) / \left\{1/(1/\varepsilon_1) + (A_1/A_2)(1/\varepsilon_2) - 1\right\}\right] \tag{20.104}$$

Example 20.13

Liquid nitrogen having the boiling point −195.7°C (77.3 K) at normal pressure is stored in a spherical aluminum container of 0.305 m OD. The container is placed in another concentric spherical aluminum container of 0.455 m ID with annular space being evacuated for insulation. If the outer sphere with an emissivity of 0.31 is subjected to 38°C, evaluate (i) the rate of heat flow by radiation to the liquid nitrogen in the container and (ii) what would be the reduction in heat gain if polished aluminum with an emissivity of 0.04 is used for the container?
As the outer and inner surfaces are parallel, the following equation is applicable:

(i) $q = \left[\sigma A_1\left(T_1^4 - T_2^4\right) / \left\{1/(1/\varepsilon_1) + (A_1/A_2)(1/\varepsilon_2) - 1\right\}\right]$

$= [(5.669 \times 10^{-8})(3.142(0.305)^2(77.3)^4 - (311)^4] / [(1/0.31)$

$\quad + \{(0.305/0.455)^2(1/0.31) - 1)\}]$

$= -36.5$ W

(ii) where emissivity, $\varepsilon = 0.04$

$$q = [(5.669 \times 10^{-8})(3.142(0.305)^2(77.3)^4 - (311)^4]/[(1/0.04)$$

$$+ \{(0.305/0.455)^2(1/0.04) - 1\}]$$

$$= -4.3 \text{ W}$$

The reduction in heat gain $= [\{(36.5 - 4.3)/36.5\}(10 = 88\%$.

Combined Heat Losses by Conduction, Convection, and Black Body Radiation

The heat losses from a hot pipeline by conduction–convection and radiation to its surroundings are appreciable and total or combined heat loss can be calculated as follows:

$$q_t = A_1(q_{cv} + q_r) = h_{cv} A_1 (T_1 - T_2) + h_r A_1 5.676 \varepsilon \left[(T_1/100)^4 - (T_2/100)^4 \right]$$

$$q_t = A_1(q_{cv} + q_r) = h_{cv} A_1$$

$$(T_1 - T_2) + h_r A_1 5.676 \varepsilon \left[(T_1/100)^4 - (T_2/100)^4 \right]$$

In the case of combined radiation and convection, the radiation heat transfer can be expressed: $q_r = A_1 \varepsilon \sigma \left(T_1^4 - T_2^4 \right)$.

Convective heat transfer, $q_{cv} = h_{cv} A_1 (T_1 - T_2)$ and radiative heat transfer, $q_r - h_r A_1 (T_1 - T_2) = A_1 \varepsilon \sigma \left(T_1^4 - T_2^4 \right)$.

Then the combined heat transfer rate by convection, q_c, and radiation, q_r, is given by

$$q_t = q_{cv} + q_r = h_{cv} A_1 (T_1 - T_2) + h_r A_1 (T_1 - T_2) = (h_{cv} + h_r) A_1 (T_1 - T_2)$$

For h_r, equating $q_r = q_{cv}$,

$$A_1 \varepsilon \sigma \left(T_1^4 - T_2^4 \right) = h_r A_1 (T_1 - T_2) = (h_{cv} + h_r) A_1 (T_1 - T_2)$$

$$= \varepsilon (5.676) \left[\{ (T_1/100)^4 - (T_2/100) \}/(T_1 - T_2) \right] \quad \text{(SI)}$$

(20.105)

and

$$h_r = \varepsilon (0.1714) \left[\{ (T_1/100)^4 - (T_2/100)^4 \}/(T_1 - T_2) \right] \quad \text{(English)}$$

A plot of temperature of one surface, °F in linear scale vs. h_r, Btu/(hft^2 °F) in semilog scale for various temperatures of another surface is available, which can be calculated from the aforementioned equation where ε is unity. T_o use these values from the said graph; the value obtained from the plot is to be multiplied by 6 to give the actual of h_r to use in the aforementioned combined heat transfer equation.

When air temperature is not at T_2 of the enclosure, the equations for q_{cv} and q_r are to be used separately and cannot be combined together as done earlier.

Exercises on Heat Transfer

20.1 A pipe of 77.9 mm ID and 88.90 mm OD is insulated with a first layer of 50.80 mm thickness of an insulating material of average thermal conductivity 0.05 W/m² K and second layer of another material of 31.75 mm thickness having 0.15 W/m² K thermal conductivity. If the temperatures of outer surface of the pipe and the outer surface of second layer of insulation are 640 and 324 K, respectively, evaluate the heat loss from the system.

20.2 A furnace wall is made of 63.5 mm (Δx_1) fire brick ($k_1 = 1.13$ W/m K), 127.0 mm (Δx_2) block insulation (k_2), and 0.0635 mm (Δx_3) steel plate (k_3). The hot gas inside the furnace and the inside surface of fire brick wall temperatures are at 770°C (T_i) and 768°C (T_1), respectively. The fire brick and block insulation interface is at 737°C (T_2) and block insulation and steel plate interface is at 78.5°C (T_3), the steel plate outer surface of wall is at 78.4°C (T_4) whereas outside air surrounding the furnace is at 36.5°C (T_o).

Estimate (i) loss of heat per unit area of furnace wall, q/A under steady state, (ii) thermal conductivities of block insulation (k_2) and steel plate (k_3), and (iii) combined convective and radiative heat transfer coefficient for the outside surface of the furnace wall, h.

20.3 The outer walls of a cubical furnace are made of 100 mm thick brick having $k = 0.82$ W/m K and 35 mm asbestos plaster having $k = 0.47$ W/m K.

Evaluate the thickness of loosely packed glass wool insulation of $k = 0.05$ W/m K, which is to be added to minimize the heat transfer through the wall by 75%.

20.4 Two concentric gray surface tubes having diameters 30 and 25 cm, respectively, are at 542°C and 220°C and their corresponding emissivities are 0.86 and 0.25. Determine the net heat transfer by radiation between the surfaces of the tubes in kcal/h m² for the second tube. Given: $\sigma = 5.669 \times 10^{-8}$ W/m² K⁴ and the equation:

Heat Transfer Coefficient

20.5 An organic liquid passes through the inner pipe of a double tube heat exchanger to cool the water flowing through its annular space between the two tubes. The inside diameter (D_i), outside diameter (D_o), and wall thickness (x_w) of the inner steel tube are 0.03341 m, 0.02664 m, and 0.000338 m², respectively, and its thermal conductivity corresponds to 50 W/m K. If the following are the inside and outside fouling factors, $h_{di} = 5700$ and $h_{do} = 2850$, film coefficients of the solvent and water $h_1 = 1020$ and $h_o = 1700$ in W/m² K, respectively.

Evaluate the overall coefficient U_o based on outside diameter of the inner tube.

20.6 The milk while flowing at a rate of 75 L/h through a pipe of diameter 1.25 cm and length 5 m is heated from 28°C to 82°C. The wall temperature is at 118°C. The properties of milk at 55°C are as follows: Density = 1020 kg/m³, viscosity = 8.8 × 10⁻⁴ Pa s, specific heat = 3.8 kJ/kg K, and thermal conductivity = 0.56 W/m K. If the viscosity of milk at 120°C = 0.48 × 10⁻³ Pa s, estimate the heat transfer coefficient.

Heat Transfer for Non-Newtonian Fluid

20.7 After thermal processing of a food can of 0.116 m diameter kept at 79.6°C, it is being cooled vertically in a room under ambient conditions at 31°C by convection. Estimate of the film heat transfer coefficient for the given properties of air at about 55°C are: density = 1.078 kg/m³, $c_p = 1.007$ kJ/kg K, $k = 0.0279$ W/m K, $\mu = 1.977 \times 10^{-5}$, Pa s, and $\beta = 3 \times 10^{-3}$.

20.8 A fruit puree is flowing at 170 kg/min through a pipe of 1.25 cm diameter, and length 3.0 m where the puree is heated from 13°C to 62°C. For the wall temperature of 80°C and bulk temperature of 38°C, the flow behavior index remains constant at 0.46. The consistency indices at 38°C and 81°C are 5.6 and 3.4 Pa s″, respectively. Find out the film heat transfer coefficient.

Heat Exchanger

20.9 A liquid food is being cooled from 70°C to 40°C in a double pipe heat exchanger by using chilled liquid food as a cooling media. The liquid inlet and outlet temperatures are 6°C and 36°C and the flow rate of hot as well as cold liquid is 135 kg/min. If the specific heat of the liquid is 4.01 kJ/kg K and the overall heat transfer coefficient for the countercurrent exchanger is 600 W/m² K, calculate the surface area of the exchanger.

20.10 In a double tube heat exchanger, a fruit juice is be pasteurized by using hot water as a heating medium. The inlet temperatures of the juice and water correspond to 20°C and 85°C whereas the outer temperature is kept at 55°C. The specific heats of juice and water are 4.01 and 4.18 kJ/kg K while the flow rates are maintained at 540 and 720 kg/h, respectively. If the overall heat transfer coefficient is estimated to be 900 W/m² K, evaluate area of the heat exchanger.

20.11 Water heater made of an iron pipe having $k = 60$ W/m K with an ID of 31.75 mm and a wall thickness of 3.2 mm. The pipe is being heated by steam at 180°C. The water passes through the pipe at a velocity of 120 cm/s. The film heat transfer coefficient on the steam side is 11,200 W/m² K. What should be the length of the pipe to heat the water from 25°C to 95°C?

Food Drying and Dryers

A Few Drying Equations and Some Common Problems

Drying Rate for Constant Drying Conditions

Drying theories in detail have been discussed in Chapter 3 where it is shown that the constant drying-rate period, Re can be expressed as kg H_2O/m^2 h and the drying time, θ_c h are expressed in the form

$$R_c = (W_d/A\theta_c)(X_1 - X_c) \qquad (20.107)$$

$$\text{Or,} \quad \theta_c = (W_d/A)(X_1 - X_c)/R_c \qquad (20.108)$$

R_c and θ_c may also be expressed by

$$R_c = q/A\lambda_w = (h/\lambda_w)(T - T_w)(3600) \quad \text{(SI)}$$

$$\theta_c = W_d\lambda_w(X_1 - X_c)/Ah(T - T_w) \qquad (20.109)$$

When airflow is parallel to the surface and the temperature varies from 45°C to 150°C and the mass velocity, G ranges from 2450 to 29,300 kg/h m² or a linear velocity of 0.61–7.6 m/s

$$h = 0.0204G^{0.8} \qquad (20.110)$$

When air flows perpendicular to the surface for the G of 3900–19,500 kg/h m² or a linear velocity of 0.9–4.6 m/s

$$h = 1.17G^{0.37} \qquad (20.111)$$

where
  q is the convective heat transfer or J/s
  A is the surface area, m²
  λ_w is the latent heat, J/kg
  h is the film heat transfer coefficient, W/m² K
  T and T_w are dry bulb and wet bulb temperatures, °C
  $G = v\rho$ = mass flow rate, kg/h m²
  θ_c is the drying time, h
  W_d is the bone dry solid, kg
  X_1 and X_c are initial and final critical moisture contents (d.b.), kg H_2O/kg dry
    solid

Discussion on drying rate for falling-rate period and other related topics are available in grain drying section.

Example 20.14

In a tray dryer, small fruit pieces placed are to be partially dried under constant drying-rate period. The sides and the bottom of the drying tray are insulated. The dimensions of the tray are 0.75 mm × 1.2 m with a depth of 50 mm. Necessary convective heat for drying the fruit pieces is supplied by passing a parallel flow of heated air at 60°C and 0.0125 (kg water/kg dry air) at a velocity of 5.5 m/s.

Estimate (i) the drying rate in the constant drying-rate period and (ii) the total evaporation rate for the surface area of the tray.

The constant drying rate, $Re = (h/\lambda_w)(T - T_w)(3600)$

where

The film coefficient, $h = 0.0204(G)^{0.8}$ for parallel flow of air at 45°C–150°C

Mass flow rate, G of 2450–29,300 kg/h m^2, or velocity of 0.61–7.6 m/s

h is in W/m^{2}K, latent heat of vaporization λ_w at wet bulb temperature, T_w

Using the psychometric chart for a humidity $H = 0.0125$ and a dry bulb temperature 60°C, $T_w = 30$°C and the humid volume v_H at the same conditions

$$v_H = (2.83 \times 10^{-3} + 4.56 \times 10^{-3} H)T \ \text{m}^3/\text{kg dry air}$$

$$= (2.83 \times 10^{-3} + 4.56 \times 10^{-3} \times 0.0125)(273 + 60) = 0.9614 \ \text{m}^3/\text{kg dry air}$$

The air density, ρ_A at 60°C $= 1/(22.41 \times 333/28.997 \times 273) = (1/0.944) = 1.059 \ \text{m}^3/\text{kg dry air}$.

The mass flow rate of air $G = v \times \rho_A = 5.5 \times 1.059 \times 3600 = 20{,}968.2 \ \text{kg/m}^2\text{h}$.

To estimate h the equation for parallel airflow for drying in constant rate period,

$$h = 0.0204(G')^{0.8} = 0.0204(20{,}968.2)^{0.8} = 58.46 \ \text{W/m}^2\text{K}$$

Using steam table for λ_w at wet bulb temperature 30°C = 24,034.5 kJ/kg. The constant drying rate R_c

$$R_c = (h/\lambda_w)(T - T_w)(3600)$$

$$= 58.46 \times (60 - 30) \times 3600/(2434.5 \times 1000) = 2.59 \ \text{kg/hm}^2$$

The total water evaporation rate from the entire surface area, A

$$Re \times A = 2.59 \times 0.75 \times 1.2 = 2.33 \ \text{kg/h}$$

Example 20.15

A biomaterial has to be dried from an initial 1.61 to a final 0.10 moisture content, kg water/kg dry matter (d.b.) in a dryer with the supply of 51.45 kg dry air/kg bone dry biomaterial through the dryer, leaving at an absolute humidity of 0.051 kg water vapor/kg dry air. The fresh air is supplied at the inlet at a humidity of 0.0150.

Evaluate the fraction of air recycled.

On the basis of 1 kg bone dry biomaterial:

Initial and final moisture contents are $X_1 = 1.612$ and $X_2 = 0.10$ (d.b.), respectively. Moisture removed from the biomaterial in the dryer, $(X_1 - X_2) = (1.61 - 0.10) = 1.51$ kg water/kg dry matter.

Moisture removed by 51.45 kg dry air, in the dryer, $G_1 = 1.51$ kg.

Therefore, change in humidity of air $= (1.51/51.45) = 0.029$ kg water vapor/kg dry air.

Inlet air humidity, $H_1 = (0.051 - 0.029) = 0.022$ (given exit air humidity $= 0.051$ given).

If G_2 kg dry exit air is recycled, then fresh dry air $= (51.45 - G_2)$ kg.

By making a humidity balance at a point A, it follows:

$$0.051 G_2 + (51.45 - G_2) \times 0.015 = 0.022 \times 51.45$$

$G_2 = (0.360/0.036) = 10.0$, the ratio of air recycled $(G_2/G_1) = (10/51.45) = 0.19$.

Food Dehydration

Food dehydration is usually accomplished by sun drying and mechanical drying based on mainly convection and conduction heating mechanism. In some special dryers, heat energy may be transmitted through infrared, microwave, dielectric, or any other thermal radiation. Among nonthermal processes, osmotic dehydration has been gaining importance. Though much higher energy is required for vacuum as well as freeze-drying and their combined system, these have been growing fast due to their superior quality products. The different types of drying operations and dryers used in commercial food industries are shown as follows.

Sun Drying

Sun drying of some food produce is a common practice in many developing and developed countries where solar energy is abundantly available.

Huge quantities of cereal grains, pulses, coffee, cocoa beans, and fish are dried by the direct exposure to the solar radiations under the open atmosphere in most of the tropical and subtropical countries. Sun-dried food products also include grapes (raisins), apricots, plumes, dates, figs, and others. Typical steps for sun drying of grapes, figs, and apricots are described in brief.

Generally, seedless grapes are pretreated with alkaline vegetable oil or ethyl oleate solutions to increase the moisture permeability of the grape skin. The grape brunches spread on trays are dried directly by the sun. They may also be dried by hanging the grape bunches from a string covering them with a transparent plastic sheet to protect the product from adverse conditions. Sun drying may take 10–20 days or more depending on solar insulation and weather as well. The sun-dried raisins may be mechanically separated from the stems and stored in bulk for further processing and packaging.

Usually, figs are allowed to dry in the sun after ripening on the trees and fall down to the ground and they may be further sun-dried to a moisture level of about 15%–20%.

The ripe apricots are generally cut into halves before sun drying on trays, kept on the ground. Some dried fruits such as figs and apricots may be fumigated with sulfur dioxide during storage before packaging.

Large requirement of labor is the main cost involved in sun drying. Both qualitative and quantitative losses may occur in the dried products due to sun drying in the open yard under unhygienic conditions. The higher yield and better quality of mechanically dried product may offset the equipment and other costs. That is why there is no alternative to large mechanical drying system when a huge quantity of freshly harvested wet produce has to be dried at the earliest for safe storage.

Mechanical Drying

There is no consistent system of classifying driers. Sometimes, dryers are classified on the basis of mode of heat transfer or materials to be handled.

However, the following *classification of dryers*, based on the form in which the material is handled through the drying process, may be used for convenience:

I. Materials may be carried easily on conveyors or trays
   A. Batch dryers: (i) atmospheric tray/compartment and (ii) vacuum tray
   B. Continuous dryers: tunnel dryers
II. Granular or loose materials
   A. (l) Standard rotary and (ii) roto-louver
   B. Conveyor dryers
III. Materials in continuous sheets
   A. Cylinder dryers
   B. Festoon dryers
IV. Sludge and pastes
   A. Agitator dryers: (i) atmospheric and (ii) vacuum
V. Materials in solution
   A. Drum dryers: (i) atmospheric and (ii) vacuum
   B. Spray dryers
VI. Special dryers
   A. Infrared radiations
   B. Dielectric heating
   C. Microwave heating
   D. Freeze-drying
   E. Osmotic drying and others

Some of the common dryers used for drying of fruits and vegetables and few special dryers are narrated here.

Cabinet Tray/Compartment Dryers

If the raw materials as well as dried products are consistent and easy to handle on trays, any type of cabinet tray/compartment dryer can be used. These include batches of food gains, fruits, and vegetables, as these are easy to handle in both loading and unloading without losses. A compartment dryer consists of a rectangular chamber whose walls contain heat-insulating materials. Inside the chamber, there is a set of racks made of light angles on which trays may be placed to slide freely (Figure 20.15).

In large cabinet dryers, a number of trays may also be placed on more than one mobile truck/trolley moving on rails for easy loading and unloading. The trucks are pushed into the drying chamber and the doors are closed behind it.

Air may be circulated in parallel direction over or both over and under thin layer of foods on the trays. However, airflow across/through the foods is superior to the earlier system to ensure uniform product quality and maximum drying capacity.

Dryers are equipped with heaters for heating air inside the dryer. There may be provision for circulating certain amount of air and adding a controlled amount of fresh air. Hot air is exhausted to remove evaporated moisture and maintain a relatively low relative humidity in the cabinet. Usually, the drying air temperatures between 45°C and 65°C may be used for fresh fruits and vegetables. To have good distribution of air through the drying chamber, the air velocities may be 300 ft/min or above though the contact with any tray would be short. In consequence, about 60% of the air in single pass should be circulated and only 40% of the fresh air may

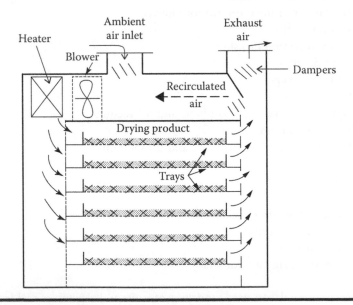

Figure 20.15 A line diagram of a cabinet tray dryer.

be introduced into the drying system for heat utilization. Cabinet dryers are flexible, simple in design, and easy to build. They are generally used for small batches 100–2000 kg of dried food per day.

A truck containing 30 trays of dimensions 80 × 80 × 5 cm each loaded with 5 kg would produce about 150 kg.

These dryers are necessarily intermittent and each unit has relatively small capacity.

Tunnel Dryer

If a large amount of material of uniform moisture content and physical properties is to be dried, a continuous system is desirable as it is always convenient compared to any intermittent batch system.

In that case, tunnel dryer with the provision of a tunnel to convey the material through it on cars/trucks either continuously or by having the tunnel is so arranged that a car/truck leaves the discharge end when a fresh car is put into the entrance. As many as 50 trays can be incorporated in each truck. Usually, it takes around 6 h for drying vegetables. The flow of heated air in a tunnel dryer may be parallel-, counter-, cross-current, or a combination through the trucks (Figure 20.16).

The heated may be supplied by heat exchangers using steam at a pressure of about 7 bars. Recirculation of air to the extent of around 50% may be practiced for heat economy. Drying with any type of tunnel or cabinet dryer is a slow process and it is desirable to use a lower drying temperature to prevent scorching/case hardening.

Throughput capacity of the tunnel dryer is mainly dependent upon the number as well as size of trays, size of the tunnel, air temperature, and airflow rates among many other factors.

Tunnel dryers are similar to cabinet dryers except that trays placed on mobile shelves of trucks move along the tunnel. The operation is mostly semicontinuous where fresh produce is added at one end and the dried product is removed from the other end of the tunnel at a predetermined interval.

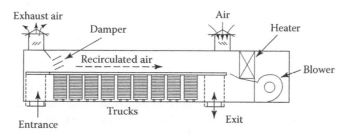

Figure 20.16   A counterflow tunnel dryer with recirculated air.

Tunnel dryers should be used for various seasonal fresh produce of different fruits and vegetables to maximize their operational time and minimize the production cost. These dryers can be made of low cost materials; they are easy to operate, and are suitable for economical drying of fruits and vegetables. That is why tunnel dryers are widely used for drying of fruits, vegetables, and lumbers as well.

Conveyor Dryer

The chamber of a typical conveyor dryer consists of two or more separate sections, preferably each with its own blower and air-heating unit to provide a maximum exposure of drying material to the air. Both downward and upward airflows are used. At the feeding end of the dryer, the air usually passes upward through the screen and near discharge end, where the material in dry air passes downward through the screen.

There may be variation in air temperature and humidity in the different sections; the exhaust is also controlled in each section to maximize drying rate and optimize energy efficiency at each section.

The continuous circulation screen conveyor dryers are suitable for drying of cut vegetables, coarse granular, flaky, or fibrous materials and used widely in the food industries.

In operation, wet materials on perforated metal or plastic belts are conveyed from one end to the other. A layer of 5–10 cm wet material to be dried slowly is carried on traveling screen through the long drying chamber. The food products may be dehydrated from 80% to 90% moisture to below 5% (w.b.) moisture either in a single dryer, or initially to 20% in a dryer and then the final part of drying may be accomplished in a separate bin or tunnel dryer for economy. Screen conveyor drying system may be relatively expensive compared to cabinet or tunnel dryer, but it provides higher throughputs with more consistent product quality.

A conveyor vegetable dryer chamber may range from 30 to 60 m in length and 2.5 to 3 m width with the throughput drying capacity varying from about 500 to 2500 kg/h depending on the type of product. Usually, the depth of vegetable loading on belt is kept between 5 and 12 cm. The air velocities as high as 3–6 m/s may be used for fast drying of vegetables.

Two or more stages/passes are needed for some food materials. A three-belt convey dryer suitable for fruits and vegetables dehydration is shown in Figure 18.12. The width and length of each belt are about 1–3 m and 10–30 m, respectively. The drying air temperatures of about 70°C–100°C and air velocities of 0.5–1.5 m/s are generally used and drying rate may vary from 2 to 15 kg/h m².

Belt drying of some vegetables may be stopped at about 12%–15% moisture and the raining part of drying to about 5% can be carried in a bin dryer for economy. Necessary heat energy can be supplied either by indirect heat exchanger using steam-or oil-fired burner or by direct combustion of LPG or CNG.

It can be used round the clock and the year round with some rest periods for cleaning, repairing, and maintaining.

Rotary Dryer

Rotary dryer consists of a slightly inclined long cylindrical shell mounted on rollers so that it can be rotated slowly. The granular material to be dried is fed to the high end. The materials go forward with a tumbling action as assisted by the rotation of internal flights and gradually move toward the discharge end by the rotation of the whole dryer. The heated air is usually supplied by a suitable blower and heat exchanger from outside to the dryer shell. Rotary dryer has been covered in grain drying chapter earlier.

Drum Dryer

The typical conventional single roll and double roll drum dryers are operated under atmospheric pressure. The dryer consists of one or two large hollow cylindrical high-grade caste iron or stainless steel rolls with smooth external surface. The diameter and length of the drum dryers may vary from about 0.5 to 3 m and 1 to 3.5 m, respectively. These rolls are horizontally mounted and equipped with a feeding system, a scraper, an applicator, and other accessories. The two rolls rotate slowly toward each other at a speed of 1–10 rpm and the feeding liquid from the top is fed into the v-shaped space between the rolls. Of the five types of feeding systems applied in commercial large drum dryers, namely, spraying, roll feeding, nip feeding, dipping, and splashing, selection of a feeding system depends on the drum arrangement, solid concentration, viscosity, and wetting ability of the product (Figures 20.17 and 20.18).

Pipes for introducing steam and removing condensate from the interior of the drum pass through the trunnions on which the rolls are supported. In cases of large dryers, special feeding devices are used for uniform feeding. The thickness of the coating on the rolls is adjusted by the space between them. Doctor knives/scrapers are placed near the top of the rolls on the outside and the dry product falls into the conveyors. In case of a single drum dryer, the bottom of the roll is usually placed into the liquid trough to feed it and a doctor knife is set at the lower part of the drum to remove the dry product. The diagrams of single drum and double drying systems are shown later.

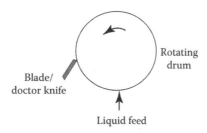

Figure 20.17   A single drum dryer.

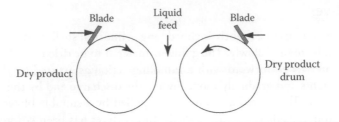

Figure 20.18 A double drum dryer.

During operation, steam at a temperature of about 150°C–200°C or higher is introduced to heat the inner surface of the drum. The steam pressure of around 2.5–4 bars may be used and the evaporation capacity 10–60 kg water/h m² or even higher with the efficiencies of 70%–85% may be achieved. Comparatively low heat energy of about 3 MJ/(kg water evaporated) is required for drum drying due to efficient conductive heat transmission through the drum wall. Surface temperature and liquid concentration as well should be controlled during drum drying. A thin layer of about 0.5–2.5 mm thick wet material is uniformly applied on the outside surface of the drum. The maximum amount of moisture is evaporated at the boiling point. The residence time may vary from few seconds to about 30 s or more for drying the product to a level of 5% (w.b.) with the production rate varying between 5 and 50 kg/(h m²), which depends upon the initial and final moisture content, type of food, surface temperature, and other conditions. Drum drying is very effective for concentrated, thick liquid food containing too large particles for spray drying.

To take care of the product quality, sometimes it may be necessary to add other materials to change its surface tension and viscosity. The costly vacuum-drying system may be preferred to reduce the drying temperature for a heat-sensitive material.

Spray Dryer

In spray dryers, liquid food materials in solutions, slurries, and pastes can be dried in very short periods of 5–30 s by spraying as fine droplets of 5–500 μm diameters into a stream of hot air or gas at temperatures ranging from 150°C to 250°C in its large cylindrical drying chamber. The atomized particles of liquid evaporate rapidly and dry before they are swept to the sides of the chamber. As a result, the liquid particles are transformed into a dry powdered product, which falls to the conical bottom of the chamber for its collection. The major part of the exit gas may be led to a cyclone-type dust collector before being discharged into the atmosphere (Figure 20.19).

Since heat is so rapidly transmitted to the particles from the hot gas, the entrapped liquid portion of the drop vaporizes *and* expands to 3–10 times the original size,

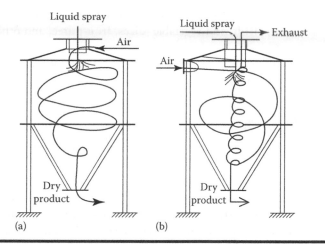

Figure 20.19   (a and b) Different types of spray dryers.

eventually exploding a small blowhole in the wall and escaping to leave hollow dried shell of solid product as hollow beads of low density.

Three types of atomizers, namely, pressure spray atomizers operating at 50–100 bars, centrifugal disks rapidly rotating at 5000–25,000 rpm with peripheral velocities of 100–200 m/s, and two-fluid/pneumatic nozzles operating at an air pressure of 3 bars are used in spray dryers. The atomizers/high-pressure nozzles are to produce nearly uniform droplets of narrow size distribution and their capacity is also restricted to about 1 T/h liquid. Therefore, multiple nozzles would be necessary for large spray dryers. The centrifugal disks having higher capacities than the pressure nozzles produce a wider range of droplet size. Nonuniform size distribution leads to overheating of the smaller droplets. Despite all, centrifugal disk atomizers are preferable in some food products due to ready solubility of their larger particles produced by this system, and the pneumatic nozzles may be used in small-scale spray drying systems.

During operation, the liquid food may be preheated to reduce viscosity and improve the drying economics. In spray drying, moisture evaporates so rapidly that a major part of drying takes place under constant drying-rate period keeping the droplets at the wet bulb temperature (40°C–50°C) of the heated air. Though the dried product remains in the drying chamber at about 90°C–100°C, the very short residence time of about 1–20 s helps to prevent heat damage. As the energy consumption in spray dryers is comparatively high about 6 MJ/kg water mainly due to their high drying temperature, preheating of drying air would improve the thermal efficiency.

Spray dryers are available from a small-scale to a large commercial scale plant of about 50 T/day or higher capacity. Though spray drying is mainly applied to the large-scale commercial production of milk and whey products in the dairy industry, it is also being used in other food industries to produce heat-sensitive products instant coffee, fruit, and vegetable juices, fruit puree, and other heat-sensitive products.

Liquid foods may be concentrated in multiple effect evaporators before their spray drying for heat economy.

Fluidized Bed Drying

During World War II, the large-scale fluidized bed dryers were introduced for commercial exploitation, as this drying system was suitable to produce uniformly dried products at constant temperature.

In fluidized bed convection drying, the solid particles are elevated in an upward moving hot air/gas stream where solid particles are so mobilized that each solid particle is subjected to the hot carrier gas/air uniformly. At a certain gas velocity/flow rate, the solid particles behave like a fluid and ensure intimate contact between them. Fluidization depends mainly on size distribution, shape, density, and viscosity.

A typical industrial fluidized bed dryer consists of a perforated bed, which is fixed in a cylindrical drying chamber where hot gas/air at a proper velocity is blown through the drying particle bed from the bottom and the exhaust gas/air is escaped from the top (Figure 20.20). A vibratory mechanism may be added to have a better contact of the drying product with the hot gas. The fluidized bed batch dryers are relatively simple and easy to use for small, uniform discrete food products such as peas, diced vegetables, etc. Fluidized beds are widely utilized in dairy processing industry in combination with spray drying to produce agglomerated milk powder and dried whey as well.

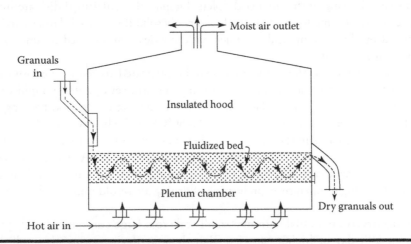

Figure 20.20 Fluidized bed dryer.

Freeze-Drying

The basic principle behind freeze-drying or lyophilization is that under high vacuum ranging from about 0.5 to 4.5 torr, the frozen water in food sublimes into vapor without passing through liquid state resulting in freeze-dried food characteristics.

In practice, prefrozen foods on trays are brought into an airtight vacuum-drying chamber in a specially designed tunnel freezer. A vacuum under the reduced pressure of about 0.1–2.0 torr is maintained in the drying chamber. There are heating plates in close contact with the bottom of the food trays to supply thermal energy just needed for sublimation of ice into vapor. Water vapor is then carried away from the food product onto the condensing plates or coils, usually kept at a low temperature of about –35°C to –50°C.

Retention of colors, flavors, and nutrients without any remarkable loss of sensory quality is the main feature of freeze-dried food products and these products also rehydrate rapidly owing to their porous structure.

The simple batch as well as the complex continuous dryer is commercially used for the freeze-drying of many aromatic spices like pepper, cardamom, etc., vegetables like peas, broccoli, green beans, etc., delicious fruits, dairy products, and others.

As freeze-drying is an energy-intensive and lengthy process, it comes out to be expensive for commercial use. The freeze-dried products being susceptible to oxidation need packaging materials with sufficient oxidation barrier for storage.

Using microwaves in place of conduction heating can augment the freeze-drying rate substantially and thus increasing the drying efficiency appreciably may offset the higher initial costs of this combined equipment.

Osmotic Drying/Dehydration

In osmotic dehydration, moist food products are dipped into concentrated sugar or salt solutions. It initiates osmosis between solute of the solution and water of the food in opposite or countercurrent directions leading to an increase in solid content of the food product. Thus, tasty solute-infused dried products like pineapples, cherries, blue berries, and others are produced by osmotic dehydration.

Osmotically dehydrated solid-infused foods may be finally dried by the controlled conveyor air-drying method.

The osmotic drying may pose a disposal problem of large amount residual fluid if it is not recycled and further processed into products like jam, jelly, and puree.

Combining osmotic drying with vacuum drying a very good-quality intermediate-moisture food product may be produced.

A combined method has been developed to minimize the disposal problem. Initially, food materials are dried to a level of 20%–50% of their original weight by convective air-drying and then mixed with precalculated solutes for further concentration of solids in food by infusion. The product can be compressed and rehydrated easily keeping its quality.

Another potential method of combining osmotic drying with vacuum drying may be very useful to produce high-quality intermediate-moisture food.

Exercises on Drying and Dryers

20.1 A continuous countercurrent dryer of 250 kg/h capacity is employed to dry potato slices from 4 to 0.087 kg moisture/kg dry solid. The inlet and outlet temperatures of the product are 40°C and 75°C, respectively. Ambient air at 25°C and 0.007 kg H_2O vapor/kg dry air humidity is heated to 90°C before entering into the dryer.

Assuming the specific heat, c_{ps}, of the dry potato 1.81 kJ/kg °C, estimate (a) airflow rate kg dry air/h required to dry the aforesaid product and (b) properties of the exhaust air.

20.2 Fifty percent of the exhaust air at 58°C and 0.02 kg/kg dry air from the drying system mentioned before.

20.3 In a spray dryer, a liquid food to be dried is sprayed into a stream of hot inlet air at 290°C having a humidity of 0.0045, and 800 mm (Hg) pressure. The inlet air stream is blown at 40 m³/min (NTP). The other data available are (i) inlet liquid food sprayed at 140 kg/h and 24.9°C with 15% solid content, (ii) outlet solid product at 19.9°C, and (iii) outlet air at 92°C, 1 atm pressure with a humidity of 0.04.

Assuming c_p of dry solid = 0.19 cal/g °C and c_p of air = 0.24 cal/g °C, calculate (a) composition outlet solid product and (b) heat utilized for drying liquid food by vaporization of air at 60°C and 750 mmHg with 10% humidity passes through a dryer at the rate of 20 m³/min. The water evaporated in the dryer is at a rate of 0.4 kg/min. The air leaves the dryer at 35°C and 740 mmHg. Estimate (a) percent humidity present in the air at the exit of the dryer and (b) volumetric flow rate of wet air at the dryer exit. Data available for the dryer: water vapor pressures are 150 and 42 mmHg at 60°C and 35°C, respectively. In a spray dryer, a liquid food solution to be dried is sprayed into a stream of hot gas. Inlet air: 60 m³/min (measured at NTP) at 300°C, 780 mm (Hg) and $\dfrac{0.005 \text{ kg } H_2O}{\text{kg dry air}}$.

Inlet solution: 2.5 kg/min, 15% solids, at 20°C.

Outlet solid: 40°C; and outlet air: 90°C, 0.035 (kg H_2O/kg dry air), 760 mm (Hg).

Evaluate (1) the composition of the outlet solids and (2) the radiation loss, if the c_p of dry solids = 0.2 cal/g °C and the c_p of air = 0.24 cal/g °C.

Appendix

Table 20.A.1 Conversion Factors

Quantity	Dimension	Conversion Factor
Length[a]	L	1 in. = 2.54 cm
		1 ft = 30.48 cm = 0.3048 m
		1 m = 3.28 ft = 39.37 in.
		1 micron = 10^{-6} m = 1 μm
Mass[a]	M	1 lb_m = 453.59 g = 0.4536 kg = 16 oz
		1 kg = 1000 g = 2.2046 lb_m
		1 ton (metric) = 1000 kg
		1 U.S. gal = 3.785 × 10^{-6} m^3
Area	L^2	1 ft^2 = 0.0929 m^2
		1 $in.^2$ = 6.451 × 10^{-4} m^2
Volume	L^3	1 L (liter) = 1000 cm^3
		1 $in.^3$ = 16.387 cm^3
		1 ft^3 = 0.02832 m^3 = 7.481 U.S. gal
		1 U.S. gal = 3.785 × 10^{-3} m^3
Density	M/L^3	1 lb_m/ft^3 = 16.0185 kg/m^3
		1 lb_m/gal = 1.19826 × 10^2 kg/m^3
		1 g/cm^3 = 62.43 lbm/ft^3 = 100 kg/m^3
		Air density at 0°C, 760 mmHg = 1.2929 g/L
		1 kg mol ideal gas at 0°C, 760 mmHg = 1.22.414 m^3
Gravitational acceleration (g)	L/θ^2	9.80665 m/s^2 = 980.665 cm/s^2 = 32.174 ft/s^2
Gravitational conversion factor (g_c)		32.174 ft $lb_m/lb_f.s^2$ = 980.665 g.cm/g_f.s^2
Force	ML/θ^2	1 lb_f = 4.4482 N = 1.35582 J, 1 N = 1 kg m/s^2 = 10^{-5} N

(continued)

Table 20.A.1 (continued) Conversion Factors

Quantity	Dimension	Conversion Factor
Power	ML^2/θ^3	1 hb = 0.7457 kW = 550 ft lb$_f$/s
		1 hb 1 Btu/h = 0.29307 W = 70681 Btu/s
		1 W (watt) = 1 J/s (Joule/s) = 14.340 cal/min
		1 dyne = 7.233 × 10^{-5} lb$_m$.ft/s^2 (poundal)
Pressure	$M/L\theta^2$	1 atm = 14.696 psia = 1.0133 × 10^5 N/m^2
		= 760 mmHg at 0°C = 29.921 in Hg
		= 33.90 ft H$_2$O ft at 4°C
		1 bar = 1 × 10^5 Pa = 1.0 × 10^5 N/m^2
		1 psia = 1 lb$_f$/in.^{2} = 2.036 in Hg at 0°C = 51.715 mmHg at 0°C
		1 lb$_f$/ft^2 = 4.788 × 10^2 dyne/cm^2 = 47.880 N/m^2 = 47.880 N/m^2
Heat, energy, and work	$M\ L^2/\theta^2$	1 Btu = 1.055 kJ = 1055 J = 252.16 cal
		1 kcal = 4.1840 kJ and 1 ft lb$_f$ = 1.35582 N m
		1 lbf = 4.4482 N = 1.35582 J, 1 ft lb$_f$ = 1.3558 N m
		1 Btu = 778.17 ft lbf = 1.3558 J
		1 J = 1 N m = kg/m^2 s^2 and 1 kWh = 3.6 × 10^3 kWh
		1 ft lb$_f$/lb$_m$ = 2.9890 J/kg
Enthalpy		1 Btu/lb$_m$ = 2.3258 J/kg
Diffusivity	L^2/θ	1 m^2/s = 3.875 ft^2/h
		1 m^2/h = 10.764 ft^2/h
		1 ft^2/h = 2.581 × 10^{-5} m^2/s
Viscosity	$M/L\theta$	1 cp = 10^{-2} g/cm s = 2.4191 bm/ft h
		= 6.7197 × 10^{-4} lbm/ft
		1 cp = 10^{-3} Pa s = 10–3 kg/m s
		1 Pa s = 1 N s/m^2 = 1 kg/m s = 1000 cp

Table 20.A.1 (continued)   Conversion Factors

Quantity	Dimension	Conversion Factor
Enthalpy	L^2/θ^2	1 Btu/lbm = 2326.0 J/kg
Specific heat	$ML/T\theta^2$	1 Btu/hft°F = 1.7307 W/m K
		= 1 cal/g°C
Thermal conductivity	$ML/T\theta^3$	1 Btu/hft°F = 1.7307 W/m K
		= 1.731 W/m K = 5.678 cal/Scm°C
Heat transfer coefficient	$M/T\theta^3$	1 Btu/hft²°F = 5.6783 W/m²
		= 5.6783 × 10⁻⁴ W/cm²°C
		= 1.357 × 10⁻⁴ cal/Scm²°C
		1 kcal/hm²°F = 0.2048 Btu/hft²°F
Heat flow	$M\ L^2/\theta^3$	1 Btu/h = 0.29307 W, 1 Btu/min = 17.58 W
		1 cal/h = 1.622 × 10–3 W, 1 kJ/h = 2.778 × 10⁻⁴ kW
Heat flux	M/θ^3	1 Btu/hft² = 3.1546 W/m²

[a] The three fundamental quantities used in SI system are length, m; time, s; and mass, kg. Other SI units are temperature, K, and mass of an element, kg mol. Some of the derived units in SI are force—Newton, 1 N = 1 kg m/s²; work/energy/heat Joule, 1 J = 1 N m = kg m²/s²; power—Watt, 1 W = 1 J/s; pressure—Pascal, 1 Pa = 1 N/m²; acceleration due to gravity, g = 9.80665 m/s². In fps system, four units are considered fundamental units such as mass, lb$_m$; length, ft; time, s; as well as force, lb$_f$ in equation $F = kmf$, where k is not unity; rather $k = 1/g_c$, g_c being arbitrarily fixed at 32.174 ft-lb$_m$/lbf-s².

Table 20.A.2 Summary of Food Dryers and Their Remarkable Feature/Performance

Dryer	Feature	Production Temperature (°C)	Dry Time	Advantages	Disadvantages
Sun grains, fruits		Atms.	7–20 days	Low cost	Weather-dependent slow, large space
Tray	Grains, fruits, vegetables, fish, meat	40–60	3–10 h	Batch, flexible	Small cap, poor quality control
Tunnel	Grains, fruits, vegetables, fish	50–80	1–3 h	Semicontainers	1 ton/day cap large space
Conveyor	Diced fruits and vegetables	50–80	1–3 h	Contains, large cap	Large space good quality
Rotary	Grains, lose materials	60–100	0.5–1 h	Containers	Large space good quality
Fluidized	Grains	—	—	Uniform drying	Size restriction
	Vegetable pieces			Uniform production	Usually batch
Drum	Liquid foods, fruits, and vegetable purees	80–110	10–30 s	Containers, high energy efficient	Liquid modification for adhesion
Freeze-valued productions, fruit, vegetables, instant coffee		Low temps	—	Containers, excellent quality production	Slow, expensive good packaging required
Osmotic	Sugar-infused fruit productions	—		High-quality production	Multiple steps

Appendix

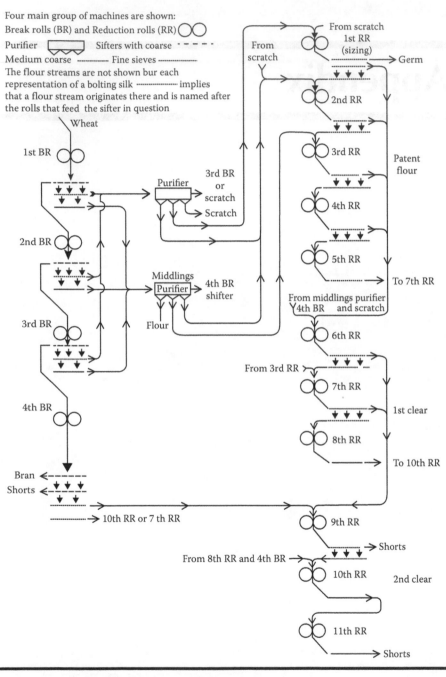

Figure A.1   Flow diagram of a wheat flour mill.

Table A.1   Grain Equilibrium Moisture Content, Percent, Wet Basis

Grain	Temperature, °C	Relative Humidity, %									
		10	*20*	*30*	*40*	*50*	*60*	*70*	*80*	*90*	*100*
Paddy	23	4.9	7.3	8.7	9.8	10.9	12.4	13.5	15.9	19.0	—
	30	—	7.1	8.5	10.0	10.9	11.9	13.1	14.7	17.1	—
	44	—	—	—	—	—	10.3	12.3	14.3	16.5	—
Wheat white	25	5.2	7.5	8.6	9.4	10.5	11.8	13.7	16.0	19.7	26.3
Wheat	32	—	5.3	7.0	8.6	10.3	11.5	12.9	14.3	—	—[a]
	49	—	—	6.2	7.4	9.6	10.4	11.9	13.6	—	
Shelled com (WD)	25	5.1	7.2	8.5	9.8	11.2	12.3	13.9	15.5	18.9	24.6
Shelled com (YD)	32	—	—	5.3	6.6	8.3	10.2	12.1	13.9	—	—[a]
	49	—	—	—	5.3	6.5	7.8	9.3	10.7	—	—[a]
	70	3.9	6.2	7.6	9.1	10.4	11.9	13.9	15.2	17.9	—[a]
Sorghum	25	4.4	7.3	8.6	9.8	11.0	12.0	13.8	15.8	18.8	21.9
	32	—	7.0	8.7	10.2	11.8	12.2	13.1	14.8	—	—[a]
	70	—	6.6	8.0	9.4	10.7	11.6	12.7	14.3	—	—[a]
Oats	25	4.1	6.6	8.1	9.1	10.3	11.8	13.0	14.9	18.5	24.1
Barley	25	4.4	7.0	8.5	9.5	10.8	12.1	13.5	15.8	19.5	26.8
Rye	25	5.2	7.6	8.7	9.9	10.9	12.2	13.5	15.7	20.6	—

[a] Haynes (1961).

Table A.2 Constants in Henderson's Equations

Material	c 1/k	n
Paddy	1.22×10^{-5}	1.35
Shelled corn	1.98×10^{-5}	1.90
Wheat	10.6×10^{-7}	3.03
Sorghum	6.1×10^{-6}	2.31

Table A.3 Bulk Densities of Grain at Different Moisture Contents

Grain	Moisture Content, % (w.b.)	Density, kg/m³
Paddy	14.0	587.9
	18.0	615.2
Wheat	11.0	789.8
	14.1	756.1
Corn (shelled)	13.0	736.9
	16.2	720.9
Barley	16.8	592.7
	10.8	576.7
Sorghum	12.0	752.9
	14.3	752.9

Table A.4   Specific Gravity of Cereal Grains

Grain	Moisture Content, % (w.b.)	Specific Gravity of Kernel
Rice	8.6	1.36
Wheat	8.5	1.41
Corn	6.7	1.29
Barley	7.5	1.42
Millets	9.4	1.11
Oats	10.33	0.99

Table A.5 Thermal Properties of Cereal Grains

Grain	Moisture Content, % (w.b.)	Temperature Range, °C	Specific Heat, kcal/ kg°C	Thermal Conductivity, kcal/(m h°C)	Thermal Diffusivity, m²/h	Reference
Paddy	12	—	0.3934	—	—	Haswell (1954)
	15	—	0.4255	—	—	
	17	—	0.4469	—	—	
Wheat	9.2	—	0.370	0.1198	0.000414	Babbit (1945)
	11.7	26.50–31.0	—	0.128	—	Oxley (1944)
Wheat hard, white	12	—	0.367	—	—	Babbit (1945)
	15	—	0.391	—	—	
Wheat soft, white	14.4	9.0–23.0	0.5	0.116	0.000295	Kazarian (1965)
Corn, yellow dent	9.8	8.3–23.2	0.438	0.1308	0.000338	Kazarian (1965)
	13.2	26.6–31.1	—	0.102	—	Oxley (1944)
Oats	12	—	0.380	—	—	Haswell (1954)
	15	—	0.415	—	—	
	17	—	0.439	—		

Table A.6 Average Composition of Cereals and Legumes

	Wheat	Brown Rice	Maize	Sorghum Lentils	Millets	Legume
Moisture, %	11.2	12.5	13.6	11.2	11.6	12.20
Calories/100 g	360	330	349	320	326	350
Protein, %	7.5	12.5	8.8	11.2	9.8	23.6
Fat, %	1.9	1.8	3.8	3.4	2.8	1.2
N–flee extract, %	775	71.6	72.1	73.1	72.8	57.5
Fiber, %	0.8	2.4	2.1	1.6	3.3	3.3
Ash, %	1.2	1.7	1.5	1.6	2.4	2.2
Thiamin, mg/100 g	0.35	0.51	0.36	039	0.71	—
Niacin, mg/100 g	4.6	4.2	23	3.8	2.4	—
Riboflavin, mg/100 g	0.06	0.14	0.12	0.16	0.36	—

Table A.7 Recommendations of Grain Drying Conditions

	Raw Paddy	Wheat	Ear Corn	Shelled Corn	Barley	Sorghum	Oats
1. Maximum harvest moisture content of grain for drying	25	20	30	25	20	20	20
a. With natural air, %	25	25	35	35	25	25	25
b. With heated air, %							
2. Maximum moisture content of grain for safe storage in a leak proof structure, for 1 year, %	12	13 (12)[a]	13	13	13	12	13 (12)[a]
3. Maximum safe temperature of drying air for heated air drying for dried grain to be used for							
a. Seed, °C	43	43	43	43	40	43	43
b. Food, °C	43	60	54	54	40	60	60
c. Feed, °C	—	82	82	82	82	82	82

Description	23–45			40–50			152–610			40–60			40–60			40–60			40–60		
4. Depth of grain bed preferable for static drying with heated air, cm	25																				
5. Maximum depth of grain at various moisture levels for drying in bin with blower capacities as listed in (6) below moisture content, %	25	20	18	20	18	16	30	25	20	30	25	20	20	18	16	25	20	18	25	20	16
Depth of grain, m	1.2	1.8	2.4	1.2	1.8	2.4	4.6	6.1	6.1	1.2	1.5	1.8	1.2	1.8	2.4	1.2	1.2	1.8	1.2	1.8	2.4
6. Minimum airflow required at different moisture contents and depths as listed in (5), $m^3/min/m^3$ — Natural air	3.2	2.4	1.6	2.4	1.6	0.3	4.0	4.0	2.4	–	4.0	2.4	2.4	1.6	0.8	–	3.2	2.4			
With supplemental heat (max. 9°C rise)	3.2	2.4	3.2	2.4	1.6	0.8	4.0	4.0	2.4	4.8	4.0	2.4	2.4	1.6	0.8	4.0	3.2	2.4	3.2	1.6	1.2

Source: Crop Dryer Manufactures, *Assn. Rev.,* Feb. 9, 1956.

[a] For seed.

Bibliography

Acharya, H.N. et al., 1982, Low temperature preparation of polycrystalline silicon from silicon tetrachloride, *Mater. Lett.*, (2).

Adams, J.B. and H.M. Brown, 2007, Discoloration in raw and processed fruits and vegetables, *Crit. Rev. Food Sci. Nutr.*, 47, 319–333.

Ajlouni, S. et al., 2001, Lycopene contents in two different tomato cultivars, *Food Aust.*, 53(5), 195.0.

Alary, R. et al., 1977, Effects of amylose content on some characteristics of parboiled rice, *Food Sci.*, 25, 261.

Ali, S.Z. and K.R. Bhattacharya, 1976, Starch retrogradation and starch damage in parboiled and flaked rice, *Starke*, 28, 233.

Ali, S.Z. and K.R. Bhattacharya, 1982, Studies on pressure parboiling of rice, *J. Food Sci. Technol.*, 19(6), 236.

Ali, N. and T.P. Ojha, 1976, Parboiling technology, *Rice: Postharvest Technology*, Araullo, E.V., D.B. de Padua, and M. Graham, (Eds.), International Development Research Centre, Ottawa, Ontario, Canada, p. 163.

Allen, J.R., 1960, Application of grain drying theory to the drying of maize and rice, *J. Agric. Eng. Res.*, 5(4), 363–386.

Almosnino, A.M. et al., 1996, Unsaturated fatty acids bioconversion by apple pomace enzyme system: Factors influencing the production of aroma compounds, *Food Chem.*, 55(4), 327.

Anderson, R.A., 1962, A note on wet-milling of high amylose corn containing 75 per cent—Amylose starch, *Cereal Chem.*, 39, 406–408.

Anderson, R.A., 1963, Wet-milling properties of grains: Bench scale study, *Cereal Sci. Today*, 8(6), 190–192, 195, 221.

Angladette, A., Rice drying principles and techniques, Informal Working Bulletin 23, p. 52, Plate PI. 6, Food and Agriculture Organization of the United Nations, Rome, Italy.

Antonio, A.A. and B.O. Juliano, 1973, Amylose content and puffed volume of parboiled rice, *J. Food Sci.*, 38, 915.

Arnold, J.H., 1933, The theory of the psychrometer, *Physics*, 4, 334–340.

Arthey, D. and P.R.A. Ashurst (Eds.), 2001, *Fruit Processing, Nutrition, Products & Quality Management*, 2nd edn., Springer.

ASHRAE, 1990, American Society of Heating, *Refrigerating and Air Conditioning Engineers (ASHRAE) Guide and Databook: Refrigeration*, ASHRAE, Atlanta, GA.

Babbit, E.A., 1945, The thermal properties of grain in bulk, *Can. J. Res.*, 23, 388–401.

Baikov, V.N., 1978, *Reinforced Concrete Structures*, MIR Pub., Moscow, Russia.

Bailey, J.E., 1992, Whole grain storage, *Storage of Cereal Grains and Tin Products*, Sauer, D.B. (Ed.), AACC, St. Paul, MN, pp. 157–182.

Baker, C.G.J. (Ed.), 1997, *Industrial Dying of Foods*, Blackie Academic & Professional, London, U.K.

Bakshi, A.S. and R.P. Singh, 1980, Drying characteristics of parboiled paddy, *Drying 80*, Mujumdar, A.S. (Ed.), Hemisphere Publishing Company, New York.

Bala, B.K., 1947, *Drying and Storage*, Sci. Pub. In, NH.

Bala, B.K., 1997, *Drying and Storage of Cereal Grains*, Science Pub., Enfield, CT.

Balasubramanian, M. et al., 1982, Paper boards from agricultural wastes—A promising agro-based industry for rural development, Seminar on Agro-industries for Rural Development, July 30–31.

Ball, C.O., 1923, Thermal process time for canned food, Bulletin 37, Vol. 7, Part I, National Research Council, Washington, DC.

Ball, C.O. and F.C.W. Olson, 1957, *Sterilization in Food Technology*, McGraw-Hill, New York.

Bandopadhyay, B. and T.K. Ghosh, 1965, Studies on the hydration of Indian paddy. Part 1—A rate equation for the soaking of the soaking equation, *Ind. J. Technol.*, 3(11), 360.

Bandopadhyay, S. and N.C. Roy, 1976, Kinetics of absorption of liquid water by paddy grains during parboiling, *Ind. J. Technol.*, 14(1), 27.

Bandopadhyay, S. and N.C. Roy, 1977, Studies on swelling and hydration of paddy by hot soaking, *J. Food Sci. Technol.*, 14(3), 95.

Bandopadhyay, S. and N.C. Roy, 1978, A semi-empirical correlation for prediction of hydration characteristics of paddy during parboiling, *J. Food Technol.*, 13(2), 91.

Barthakur, B., 1970, Improved process for soaking and parboiling of paddy, Ind. Patent 121562.

Bast, A. et al., 1998, Antioxidant effects of carotenoids, *Int. J. Vitamin Nutr. Res.*, 68, 399–403.

Beagle, E.C., 1978, Rice-husk conversion to energy, FAO Services Bulletin 31, Rome, Italy.

Becker, H.A. and H.R. Saltans, 1956, A study of the desorption isotherms of wheat at 25°C and 50°C, *Cereal Chem.*, 33, 79–90.

Bee, M., 2004, Dehydrated tomatoes in HB, *Vegetable Preservation & Processing*, Hui, Y.H. et al. (Eds.), Marcel Dekker, Inc., New York.

Beiqun, M., 2005, Genetic variation of β-carotene and lutein contents in lettuce, *J. Am. Soc. Hort. Sci.*, 130, 870–876.

Belitz, H.D. and W. Grosch, 1987, *Food Chemistry*, Springer-Verlag, Heidelberg, Germany.

Bertram, J.S. and A.L. Vine, 2005, Cancer prevention by retinoids and carotenoids: Independent action on a common target, *Biochem. Biophys. Acta*, 1740, 170–178.

Bhattacharya, K.R., 1969, Breakage of rice during milling and the effect of parboiling, *Cereal Chem.*, 37, 478–485.

Bhattacharya, K.R., 1985, Parboiling of rice, *Rice Chemistry and Technology*, Juliano, B.O. (Ed.), AACC, St. Paul, MN.

Bhattacharya, K.R. and Y.M. Indudharaswamy, 1967, Conditions of drying parboiled paddy for optimum milling quality, *Cereal Chem.*, 44(6), 592–600.

Bhattacharya, K.R. and V.P. Subba Rao, 1966a, Processing conditions and milling yield in parboiling of rice, *J. Agric. Food Chem.*, 14(5), 475.

Bhattacharya, K.R. and V.P. Subba Rao, 1966b, Effect of processing conditions on quality of parboiled rice, *J. Agric. Chem.*, 14, 476.

Bigelow, W.D. et al., 1920, Heat penetration in processing canned food, National Canners Association Bulletin 16L, Washington, DC.

Bird, R.B. et al., 1960, *Transport Phenomena*, John Wiley & Sons, Inc., New York.

Biswas, S.K. and B.O. Juliano, 1988, Laboratory parboiling procedures parboiled rice differing in starch properties, *Cereal Chem.*, 65, 417–420.

Bond, E.J., 1973, Chemical control of stored grain insects and mites, *Grain Storage: Part of a System*, Sinha, R.N. and W.E. Muir (Eds.), *Proceedings Symposium of Grain Storage*, Winnipeg, Manitoba, Canada, June 6–9, 1971, AVI Publishing Co., Westport, CT.

Bond, E.J. and H.A.U. Monro, 1961, The toxicity of various fumigants to the Cadelle *Tenebroides mauritanicus, J. Econ. Entomol.*, 54, 451–454.

Borasio, L. and F. Gariboldi, 1957, *Illustrated Glossary of Rice Processing Machines*, Food and Agricultural Organization of the United Nations, Rome, Italy.

Borasio, L. and F. Gariboldi, 1965, Parboiled rice production and use; Part I, *Prod. Rice J.*, 68(5), 32–35.

Boumans, G., 1985, Grain handlings and storage, development, *Agricultural Engineering*, Vol. 4, Elsevier, Tokyo, Japan.

Brekke, O.L., 1967, Corn dry milling: Pretempering low-moisture corn, *Cereal Chem.*, 44, 521–531.

Brekke, O.L., 1968, Corn dry milling: Stress crack formation in tempering low-moisture corn, and effect on degerminator performance, *Cereal Chem.*, 45, 291–303.

Brekke, O.L. and W.F. Kwolek, 1969, Corn dry milling: Cold tempering and degermination of corn of various initial moisture contents, *Cereal Chem.*, 46, 545–549.

Brekke, O.L. and L.A. Weinecke, 1964, Corn dry-milling. A comparative evaluation of commercial degermer samples, *Cereal Chem*, 41, 321–328.

Brekke, O.L., L.A. Weinecke, J.N. Boyd, and E.L. Griffin, Jr., 1963, Corn dry milling: Effects of first-temper moisture, screen perforations, and rotor speed on Beall degerminator throughput and products, *Cereal Chem.*, 40, 42–29.

Brennan, J.C. et al., 1990, *Food Engineering Operations*, 3rd edn., Applied Science Publications, London, U.K.

Britton, G., 1988, Biosynthesis of carotenoids, *Plant Pigments*, Goodwin, T.R. (Ed.), Academic Press, London, U.K., pp. 133–182.

Brooker, D.B. et al., 1974, *Drying and Storage of Agricultural Crops*, AVI Publishing Co., Westport, CT.

Brooker, D.B. et al., 1992, *Drying and Storage of Grains and Oilseeds*, AVI, New York.

Brown, W.B., 1959, Fumigation with methyl bromide under gas-proof sheets, Pest. Infest. Res. Bull., Vol. 1, 2nd edn., Department of Scientific and Industrial Research, London, U.K., 44pp.

Brown, A.W.A., 1961, *Insect Control by Chemicals*, John Wiley & Sons, New York.

Brunndett, G.W., 1987, *HB of Dehydration Technology*, Butts Worths, London, U.K.

Chakraverty, A. (Ed.), *HB of Postharvest Technology: Cereals, Pulses. Fruits, Vegetables, Tea & Spices*, Marcel Dekker, Inc., New York.

Chakraverty, A., 1975, Some aspects of intermittent drying of paddy, *J. Agric. Eng.*, 13(1), 15–18.

Chakraverty, A., 1976, Effects of tempering on drying characteristics of paddy, *J. Agric. Eng.*, 13(3), 130–133.

Chakraverty, A., 1978a, Analytical approach to thin layer drying of paddy, *RPEC Reporter*, 4(2).

Chakraverty, A., 1978b, Intermittent drying of paddy in thin layer, *J. Agric. Eng.*, 15(1), 33–36.

Chakraverty, A., 1978c, Effects of continuous and intermittent drying on drying characteristics of paddy, *RPEC Reporter*, 4(1).

Chakraverty, A. (Princ. Inv.), 1982, Stabilization of rice bran, Report on the ICAR Project, IIT, Kharagpur, India.

Chakraverty, A., 1987, Development of a continuous rice bran stabilizer for wet heat treatment, *Handbook on Rice Bran*, The Solvent Extraction Association of India, Bombay, India, pp. 373–376.

Chakraverty, A., 1988, *Bulletin of Paddy and Other Grain Drying Systems*, IIT, Kharagpur, India.

Chakraverty, A., 1989, *Biotechnology and Other Alternative Technologies for Utilization of Biomass/Agricultural Wastes*, Oxford and IBH Pub. Co., New Delhi, India.

Chakraverty, A., H.D. Banerjee, and S.K. Pandey, 1987, Studies on thermal decomposition of rice straw, *Thermochim. Acta*, 120, 241–255.

Chakraverty, A. and S.K. Das, 1991, Development of a solar paddy dryer, *Energy Conversion Manage.*, 33(3), 183–190.

Chakraverty, A. and D.S.K. Devadattam, 1985, Conduction drying of raw and steamed rice bran, *Drying Technol. J.*, 3(4), 567–583.

Chakraverty, A. and S. Kaleemullah, 1991a, Conversion of rice husk into amorphous silica and combustible gas, *Energy Conversion Manage.*, 32(6), 565–570.

Chakraverty, A. and S. Kaleemullah, 1991b, Production of amorphous silica and combustible gas from rice straw, *J. Mater. Sci.*, 26, 4554–4560.

Chakraverty, A. and R.T. Kausal, 1982, Determination of drying conditions for drying of parboiled wheat, *AMA*, Winter.

Chakraverty, A., P. Mishra, and H.D. Banerjee, 1985, Investigation of thermal decomposition of rice husk, *Thermochim. Acta*, 94, 267–275.

Chakraverty, A., Mujumdar, A.S. et al. (Eds.), *Handbook of Postharvest Technology: Cereals, Pulses, Fruits, Vegetables, Tea and Spices*, Marcel Dekker, Inc., New York.

Chakraverty, A. and T.P. Ojha, 1975, Effects of various air temperatures and exposure times on milling quality of rice, *J. Agric. Eng.*, 13(2), 1–6.

Chakraverty, A. and R.P. Singh, 2002, *Postharvest Technology of Cereals, Pulses, Fruits & Vegetables*, Science Pub., Enfield, CT.

Chakraverty, A. et al., 1976, Studies on thin layer drying of paddy, Rice Report, Spain.

Chakraverty, A. et al., 1979, Thin layer drying characteristics of soybean, *The Harvester*, 21.

Chakraverty, A. et al., 1983a, Design development and testing of a simple baffle type of grain dryer, *AMA*, 14(1), 41–44.

Chakraverty, A. et al., 1983b, Stabilization of rice bran by conduction and humid heat treatments, *AMA*, 14(2), 72–76.

Chakraverty, A. et al., 1984, Thin layer drying characteristics of cashew nuts, *Drying 84*, Hemisphere Publishing Company.

Chakraverty, A. et al., 1987a, Design and testing of a solar cum-husk-fired paddy dryer of IT/day cap, *Drying 87*, Hemisphere Publishing Corporation, New York.

Chakraverty, A. et al., 1987b, Sorption and desorption characteristics of raw and heat treated rice bran, *Drying Technol. J.*, 5(25), 205–212.

Chakraverty, A. et al., 1987c, Stabilization of rice bran by conduction heating in a PHTC-continuous rice bran stabilizer, *AMA*, 38(1), 41–44.

Chakraverty, A. et al., 1988, Stabilization of rice bran by steaming and conduction drying in a continuous rice bran stabilizer, *The Harvester*, (23), 5–10.

Chakraverty, A. et al. (Eds.), 2003, *Handbook of Postharvest Technology*, Marcel Dekker, Inc., New York.

Champagne, E.T. et al., 1997, Effects of drying conditions, final moisture content and degree of milling on rice flavour, *Cereal Chem.*, 74(5), 556–570.

Charley, H., 1982, *Food Science*, 2nd edn., McMillan, New York.

Charm, S.E., 1963, *Fundamental of Food Engineering*, AVI Publishing Co., Westport, CT.

Chikubu, S., 1970, Storage condition and storage method, *Training Manual for Training in Storage and Preservation of Food Grains*, APO Project, Tokyo, Japan.

Christensen, C.M. (Ed.), 1974, *Storage of Cereal Grains and Their Products*, AACC, St. Paul, MN, 549pp.

Christine, M.O. et al., 2006, Creating proteins with novel functionality via the Maillard reaction: A review, *Crit. Rev. Food Sci. Nutr.*, 46, 337–350.

Chung, D.S. and H.B. Pfost, 1967, Adsorption and desorption of water vapour by cereal grains and their products, *Trans. ASAE*, 10, 552–575.

Cornell Hugh, J. and W.H. Albert (Eds.), 1998, *Wheat Chemistry and Technology*, Technomic Pub. Co., Inc., Lancaster, U.K.

Cotton, R.T., 1963, *Pests of Stored Grain and Grain Products*, Burgess Publishing Co., Minneapolis, MN, 318pp.

Crop Dryer Manufactures Assn. Rev., February 9, 1956.

Dale, A.C. and H.K. Johnson, 1956, Heat required to vaporise moisture in wheat and shelled com, *Purdue Eng. Res. Bull.*, 131.

Dandekar Machinery Works, Catalogue for rice and pulse milling machinery, M/s. G.G. Dandekar Machine Works Ltd., Bhiwandi, Thana, India.

Das, S.K. and A. Chakraverty, 2003, Grain-drying system, *Handbook of Postharvest Technology*, Chakraverty, A. et al. (Eds.), Marcel Dekker, Inc., New York.

Desikachar, H.S.R. et al., 1967, Relative yields of total and head rice from raw and parboiled paddy, *J. Food Sci. Technol.*, 4, 156.

Desikachar, H.S.R. et al., 1969, Steaming of paddy for improved culinary, milling and storage properties, *J. Food Sci. Technol.*, 6(2), 117–121.

Devadattam, D.S.K. and A. Chakraverty, 1986, Some physico-thermal properties of raw bran related to drying technology, *Drying Technol. J.*, 4(1), 145–154.

Dexter, J.E. and A.K. Sarkar, 1993, Roller milling operation, flour, *Encyclopedia of Food Sciences*, Food Technology and Nutrition, Academic Press, New York.

Disney, R.W., 1954, The specific heat of some cereal grains, *Cereal Chem.*, 31, 229–334.

Dodge, C.W. and A.B. Metzner, 1959, Turbulent flow of non-Newtonian systems, *AIChE J.*, 5(7), 189–204.

Downing, D.L. (Ed.), 1996, *A Complete Course in Canning*, Vols. I, II, & III, 13th edn., CTI Publ., Timonium, MD.

Doymaz, I., 2004, Drying kinetics of white mulberry, *J. Food Eng.*, 61(3), 341–346.

Dubey, O.M., 1984, Optimum design of RCC silos for bulk storage of paddy, MTech thesis, IIT, Kharagpur, India.

Dutta, S.K. et al., 1988, Physical properties of grain, *J. Agric. Eng. Res.*, 39, 259–268.

Dykstra, W.W., 1968, The economic importance of rodents, *Proceedings of Conference on Rodents as Factors in Disease and Economic Loss*, Institute for Technical Interchange, Honolulu, HI, pp. 47–52.

Earle, R.L., 1983, *Unit Operations in Food Engineering*, AVI Publishing Co., Westport, CT.

Emmons, C.L. and D.M. Peterson, 1999, Antioxidant activity and phenolic contents of oat groats and hulls, *Cereal Chem.*, 76(6), 902–906.

Emrich, W., 1985, Energy from biomass, *Handbook of Charcoal Making*, Series E, Vol. 7, D. Riedil Pub. Co., Dordrecht, Holland.

Ezaki, H., 1973, Paddy husker, Group training course, Fiscal Institute of Agricultural Machinery, Japan.

FAO, 1991, *FAO Production Year Book*, Vol. 51, Rome, Italy.

Fellows, P., 1988, *Food Processing Technology: Principles & Practices*, Ellis Horswood, Chichester, England.

Fennema, O.R. (Ed.), 1991, *Food Chemistry*, 2nd edn., Marcel Dekker, Inc., New York.

Food Industries, 1979, *Infestation Control*, Chem. Eng. Ed. Dev. Centre, IIT, Madras, India.

Foster, G.H., 1953, Minimum air flow requirements for drying grain with unheated air, *Agric. Eng.*, 34(10), 681–684.

Foster, G.H., 1964, Dryeration—A corn drying process, Agr. Marketing Service Bull., USDA, Washington, DC, 532pp.

Foster, G.H., 1973, Heated air grain drying, *Grain Storage: Part of a System*, Sinha, R.H. and W.E. Muir (Eds.), AVI Publishing Co., Westport, CT.

Gallahar, G.L., 1951, A method of determining the latent heat of agricultural crops, *Agric. Eng.*, 32, 34–38.

Garibodi, F., 1974a, Parboiled rice, *Rice Chemistry*, Houston, D.F. (Ed.), AACC, St. Paul, MN.

Gariboldi, F., 1974b, Rice milling equipment operation and maintenance, Agricultural Service Bulletin No. 22, FAO, Rome, Italy.

Gariboldi, F., 1974c, *Rice Parboiling*, FAO, Rome, Italy.

Geankoplis, C.J., 1993, *Transfer Process and Unit Operations of Chemical Engineering*, 3rd edn., Prentice Hall, New York.

Geankoplis, C. J. 1997. *Transfer process & Unit Operations of Chemical Engineering*, 3rd edn., Prentice Hall, New York.

Geoffrey, C.-P. (Ed.), 2009, *Food Science and Technology*, Wiley-Blackwell, Chichester, U.K.

Gerzhoi, A.P. and V.F. Samochetov, 1958, *Grain Drying and Grain Dryers*, 3rd review and enl. edn., Ginzburg, A.S. (Ed.), Khieboizdat, Moscow, Russia.

Ghose, T.K., 1963, *Development of an Improved and Fully Mechanised Method for the Production of Parboiled Rice*, Jadavpur University, Calcutta, India.

Ghosh, S.K. and M.K. Sinha, 1977, *Indian Text J.*, 88(2), 111.

Goff, J.A., 1949, Standardization of thermodynamic properties of moist air, *Trans. ASHVE*, 55, 463–464.

Goff, J.A. and S. Gratch, 1945, Thermodynamic properties of moist air, *Trans. ASHVE*, 51, 125–164.

Gordon, G.L., 2006, *Food Packaging: Principles and Practice*, 2nd edn., CRC Press, New York.

Gordon, L.R. (Ed.), 2010, *Food Packaging and Shelf Life (A Practical Guide)*, CRC Press, Taylor & Francis Group, New York.

Goyal, 1977, Ceramic materials from rice husk, MTech thesis, IIT, Kharagpur, India.

Grabber, E. and Grigull, 1963, Grundgesetze der Warmeubertragung, Springer-Verlag, Berlin Gottingen Heidelberg.

Green, W.D. and R.R. Perry (Eds.), 2008, *Chemical Engineering Handbook*, McGraw-Hill, New York.

Haghighi, K. and Agnirre, C.G., 1999, Adaptive and stochastic finite element analysis in drying, *Drying Technol.*, 17(10), 2037–2053.

Hall, C.W., 1957a, *Drying Farm Crops*, Agricultural Consulting Association, Reynoldsburg, OH, 336pp.

Hall, C.W., 1957b, *Drying Farm Crops*, AVI Publishing Co., Westport, CT.

Hall, C.W., 1979, *Dictionary of Drying*, Marcel Dekker, Inc., New York, 300pp.

Hall, C.W., 1980, *Drying and Storage of Agricultural Crops*, AVI Publishing Co. Inc., Westport, CT.

Hang, T.D., 1987, Production of fuels and chemicals from apple pomace, *Food Technol.*, 41(3), 115.

Harkins, W.D. and G. Jura, 1944, A vapour adsorption method for the determination of the area of solid, *J. Am. Chem. Soc.*, 66, 1366–1371.

Harris, R.V., 1971, Rice bran oil and wax, *Interregional Seminar on the Industrial Processing of Rice*, UNIDO, Madras, India, February 25, 1971, pp. 4–7.

Haswell, G.A., 1954, A note on the specific heat of rice, oats and their products, *Cereal Chem.*, 31, 341–343.

Haynes, B.C., 1961, Vapour pressure determination of seed hygroscopicity, Technical Bulletin, 1929, ARS, USDA, Washington, DC.

Heldman, D.R. and D.B. Lund (Eds.), 2004, Marcel Dekker, Inc., New York.

Heldman, D.R. et al. (Eds.), 2003, *Encyclopedia of Agricultural, Food & Biological Engineering*, Marcel Dekker, Inc., New York.

Henderson, S.M., 1952, A basic concept of equilibrium moisture, *Agric. Eng.*, 33, 29–31.

Henderson, S.M. and S. Pabis, 1961, Grain drying theory, I. Temperature effect on drying coefficient, *J. Agric. Eng. Res.*, 6(3), 169–174.

Henderson, S.M. and S. Pabis, 1962, Grain drying theory, IV. The effect of airflow rate on the drying index, *J. Agric. Eng. Res.*, 7(2), 85–89.

Henderson, S.M. and R.L. Perry, 1955, *Agricultural Process Engineering*, John Wiley & Sons, New York.

Heseltine, H.K. and R.H. Thompson, 1957, The use of aluminium phosphide tablets for the fumigation of grain milling, 129, 778–783.

Holman, L.E., 1948, Adapting cribs for corn drying, *Agric. Eng.*, 29, 149–151.

Holman, J.P., 1984, *Heat Transfer*, 6th edn., McGraw-Hill International Book Co., Singapore.

Houston, D.J. (Ed.), 1972, *Rice Chemistry and Technology*, AACC, St. Paul, MN.

Hubbard, J.E. et al., 1950, Composition of the component of sorghum kernel, *Cereal Chem.*, 2, 415–420.

Hukill, W.V., 1954, Grain drying with unheated air, *Agric. Eng.*, 35, 393–395.

Hukill, W.V. and J.L. Schmidt, 1966, Drying rate of fully exposed grain kernels, *Trans. ASAE*, 3(2), 71–77, 80.

Hulse, J.H. et al., 1980, *Sorghum and the Millets: Composition and Nutritive Value*, Academic Press, New York.

Iglesias, H.A. and J. Chirfe, 1983, *HB of Isotherms*, Academic Press, New York.

IS: 604, 1969, Code of practice of practice for construction of food grain storage structure suitable for trade and Government purpose for the southern region, ISI, New Delhi, India.

IS: 607, 1965 (Revised), Bagged grain storage structure, ISI, New Delhi, India.

IS: 631, 1961, Specification for aluminium-food grain storage bins, ISI, New Delhi, India.

IS: 6940, 1973, Methods of test for pesticide and their formulations, ISI, New Delhi, India.

Jayas, D.S. et al., 1995, *Stored Grain Ecosystems*, Marcel Dekker, Inc., New York.

Jones, J.W. and J.W. Taylor, 1933, Effect of parboiling rough rice on milling quality, USDA, Circular No. 340.

Jones, J.W. et al., 1946, Effect of parboiling and related treatments on the milling, nutritional & cooking quality of rice, USDA, Circular No. 752.

Juliano, B.O., 1984, A simplified assay for milled rice amylase, *AACC*, 16, 334–340.

Juliano, B.P. (Ed.), 1985, *Rice Chemistry and Technology*, 2nd edn., AACC, St. Paul, MN.

Kachrew, R.P., T.P. Ojha, and G.T. Kurup, 1971, Equilibrium moisture content of Indian paddy varieties, *Bull. Grain Technol.*, 3(9), 186–196.

Kadam, S.S. et al., 1982, Seed structure and composition, *Handbook of World Food Language*, Vol. I, Salukkde, D.K. et al. (Eds.), CRC Press, Boca Raton, FL.

Kader, A.A., 2002, *Postharvest Technology of Horticultural Crops*, Publication 3311, Agriculture and Natural Resources, University of California, Davis, CA.

Kader, A.A., R.F. Kasmire, F.G. Mitchell, M.S. Reid, N.F. Sommer, and J.F. Thompson, 1985, *Postharvest Technology of Horticultural Crops*, Cooperative Extension, University of California, Davis, CA.

Kaupp, A., 1984, *Gasification of Rice Hulls*, Friedr, Vieweg and Soln, Braunschweig, Germany.

Kaupp, A. and J.R. Goss, 1981, *State of the Art for Small Scale (to 50 KW) Gas Producer– Engine Systems, VS*, Agency for International Development, Washington, DC.

Kausal, R.T., Drying characteristics of wheat bulgur and soybean, MTech thesis, IIT, Kharagpur, India.

Kazarian, E.A. and C.W. Hall, 1965, Thermal properties of grains, *Trans. ASAE*, 8(1), 33–37, 48.

Keey, R.B., 1973, *Drying: Principles and Practices*, Pergamon Press, Oxford, U.K., p. 355.

Kemp, I.C., 1999, Progress in dryer selection techniques, *Drying Technol.*, 17(7–8), 1667–1680.

Kent-Jones, D.W. and A.J. Amos, 1957, *Modern Cereal Chemistry*, 5th edn., Northern Publishing Co., Liverpool, U.K.

Kern, D.Q., 1984, *Process Heat Transfer*, McGraw-Hill International Book Co., Singapore.

Kerr, R.W., 1950, *Chemistry and Industry of Starch*, 2nd edn., Academic Press, New York.

Kessler, H.G., 1981, *Food Engineering and Dairy Technology*, Verlag, Freising, Germany.

Kester, E.B. and S.A. Matz, 1970, Rice processing, *Cereal Technology*, Matz, S.A. (Ed.), AVI Publishing Co., Westport, CT.

Klingspohn, U.J. et al., 1993, Utilization of potato pulp from potato starch processing, *Process Biochem.*, 28, 91.

Koga, Y., 1969, Drying, husking and milling in Japan IV and V, *Farming*, Japan.

Kranzier, G.A. and D.C. Davis, 1981, Energy potential of fruit juice processing residue, ASAE Paper 81.6006.

Kreyger, J., 1972, Drying and storing grains, seeds and pulses in temperature climates, Bulletin 205, Institute for Storage and Processing of Agricultural Produce, Wageningen, the Netherlands.

Krishna, N., 1991, Consorted report of the entitled: Pilot scale studies on the methane generate from fit and veg. processing waste (No. 5/2/29/88-BE) from March 1989 to September 1989, GOI, New Delhi, India.

Kulkarni, S.D. et al., 1986, Saturation moisture content of paddy grains and its dependence on initial moisture content and temperature of soaking, *J. Agric. Eng.*, (23), 99–104.

Kulp, K. and J.G. Ponte, 2000, *Handbook of Equal Science and Technology*, 2nd edn., Marcel Dekker, Inc., New York.

Kuprits, Y. (Ed.), 1965, Technology of grain processing and provender milling, Tekhnologiya perer atbotki zernai Kombikormovoe, Izd. 'Kolos,' Moscow. (Translation by Israel Program for Scientific Translation, Jerusalem, Israel, 1967.)

Kurien, P.P., 1979, *Pulses Milling in Food Industries*, CFTRI, Mysore, India, pp. 3.1–3.20.

Kurien, P.P. and H.A.B. Parpia, 1968, Pulse milling in India, I, Processing and milling of tur and arhar (*Cajonus cajan*), *Food Sci. Technol.*, 5(6), 203–207.

Kurien, P.P. et al., 1964, Effect of parboiling of paddy on the swelling quality of rice, *Cereal Chem.*, 41(1), 16–22.

Lahssasni, S. et al., 2004, Drying kinetics of prickly pear fruit, *J. Food Eng.*, 61(2), 173–179.

Lan, Y. et al., 1999, Mathematical model of the stress within a kernel from moisture absorption, *J. Agric. Eng. Res.*, 72(3), 247–257.

Lay, W.A., 1972, Parboiled rice, Patent 3660109.

Lee, D.S. et al., 2008, *Food Packaging Science and Technology*, CRC Press, New York.

Liu, Q. and F.W. Bakker-Arkema, 1997, Stochastic modeling of grain drying: Part 2. Model development, *J. Agric. Eng. Res.*, 66, 275–280.

Liu, W. et al., 1997, Digital image analysis method for rapid measurement of degree of milling of rice, *ISAE Annual International Meeting*, Minneapolis, MN.

Lockwood, J.F., 1960, *Flour Milling*, Northern Publishing Co., Liverpool, U.K.

Loeb, J.R. et al., 1949, Rice bran oil. IV. Storage of bran as it affects hydrolysis of the oil. *J. Am. Oil Chem. Soc.*, 26(12), 738–743.

Longstaff, B.C. and P. Corninsh, 1994, Pest man—A decision support system for post management in the Australian grain-industry, *AI Appl. Nat. Resource Mgmt.*

Lorenz, J.K. and K. Karel (Eds.), 1991, *Handbook of Cereal Science and Technology*, Marcel Dekker, Inc., New York.

Luikov, A.V., 1966, *Heat and Mass Transfer in Capillary—Porous Bodies*, Pergamon Press, London, U.K.

Lyon, B.G. et al., 1999, Effects of degree of milling, drying conditions and final moisture content on sensory texture of cooked rice, *Cereal Chem.*, 76(1), 356–362.

Mahadevappa, M. and H.S.R. Desikachar, 1968, Expansion and swelling of raw and parboiled rice during cooking, *J. Food Sci. Technol.*, 5(2), 59–62.

Majumdar, A.S. (Ed.), 1995, *Handbook of Industrial Drying*, Vols. I and II, 2nd edn., Marcel Dekker, Inc., New York.

Majumdar, A.C. et al., 1960, I—Soaking and gelatinization; II—Dehydration; III—Effect of hot soaking and mechanical drying on nutritive value of parboiled rice, *J. Biochem. Microbiol. Tech. Eng.*, 15(11), 431.

Majumdar, S.K. et al., 1973, Control of microflora on moist grain, *Ann. Technol. Agric.*, 22(3), 483.

Mallik, S.K. and A.R. Gupta, 1983, *Plain Reinforced Concrete*, Oxford & IBH Pub. Co. Pvt. Ltd., Calcutta, India, p. 1.

Mannapperuma, J.D. and R.P. Singh, 1994, Modeling of gas exchange in polymeric packages of fresh fruits and vegetables, Chapter 29, *Minimal Processing of Foods and Process Optimization*, Singh, R.P. and F.A.R. Oliveira (Eds.), CRC Press, Boca Raton, FL.

Matz, S.A., 1959, *The Chemistry and Technology of Cereals as Food and Feed*, AVI Publishing Co., Inc., Westport, CT.

McCabe, W.L. and J.C. Smith, 1976, *Unit Operations in Chemical Engineering*, 3rd edn., McGraw-Hill, New York.

Me Farlance, J.A., 1988, Storage methods in relation to post harvest losses in cereal grains, *Insect Sci. Appl.*, 9, 747–754.

Merson, R.L. et al., 1978, An evaluation of Ball's formula method of thermal process calculations, *Food Technol.*, 32(3), 66.

Midwest Plan Service, 1987, *Grain Drying, Handling and Storage Handbook*, Iowa State University, Ames, IA, p. 44.

Miller, F.C., 1993, Composting as a process based on the control of ecologically selective factors, *Soil Microbial Ecology*, Metting, F.B. (Ed.), Marcel Dekker, Inc., New York.

Milner, M. and W.F. Geddes, 1954, Respiration and heating, Chapter IV, *Storage of Cereal Grains and Their Products*, American Association of Cereal Chemists, St. Paul, MN, pp. 152–220.

Mishra, P., 1986, Investigation on physico-thermal properties and thermal decomposition of rice husk, production of pure amorphous white ash and its conversion to pure silicon, PhD thesis, IIT, Kharagpur, India.

Mishra, P., A. Chakraverty, and H.D. Banerjee, 1985, Production and purification of silicon by calcium reduction of rice husk white ash, *J. Mater. Sci.*, 20, 4387–4391.

Mitsui, E., 1970, Stored product pests and their control, *Training Manual for Training in Storage and Preservation of Food Grains*, APO Project, Tokyo, Japan.

Mohsenin, N.N., 1980, *Thermal Properties of Food and Agricultural Materials*, Gordan & Breach Sci. Pub., New York, p. 407.

Mohsenin, N.N., 1988, *Physical Properties of Plant and Animal Materials and Products*, Hemisphere Publication Company, Washington, DC, 1244pp.

Monro, H.A.U., 1969, *Manual of Fumigation for Insect Control*, FAO, United Nations, Rome, Italy, Manual 79, 381pp.

Moody, L.F., 1944, *Trans. ASME*, 66(8), 671.

Morton, I.S. and S. Wilie (Eds.), 1998, *World Grain*, 16(11), Kansas City, MI.

Mujumdar, A.S. et al., 1995, *HB of Industrial Drying*, Vols. 1 and 2, 2nd edn., Marcel Dekker, Inc., New York.

Mukanor, K., 1980, *Design of Metal Structures*, MIR Pub., Moscow, Russia.

Nagashima, T. et al., 1983, Study on parboiling of brown rice (Japonica), *J. Agric. Sci.*, 28(2), 238.

Nanda, S.K., 1993, Development of starch-hosed plastics films, PhD thesis, IIT, Kharagpur, India.

Nanda, S.K., A. Chakraverty, and S. Maiti, 1990, Starch-based plastics films. Part 1. Preparation of the film, *J. Polym. Mater.*, 7, 331–333.

Oil Technological Research Institute, 1974, Edible rice bran oil, Project Report.

Oliver, C.M. et al., 2006, Creating proteins with novel functionality via the Maillard reaction: A review, *Crit. Rev. Food Sci. Nutr.*, 46, 337–350.

Othmer, D.F., 1940, Correlating vapour pressure and latent heat data, *Ind. Eng. Chem.*, 32, 841–846.

Oxley, T.A. 1944, The properties of grain in bulk, *Soc. Chem. Indus. J. Trans.* 63.

Pabis, S. and S.M. Henderson, 1961, Grain drying theory, II. A critical analysis of the drying curve for shelled maize, *J. Agric. Eng. Res.*, 6(4), 272–277.

Pabis, S. and S.M. Henderson, 1962, Grain drying theory, III. The air–grain temperature relationship, 7(1), 21–26.

Pabis, S. et al., 1998, *Grain Drying: Theory and Practice*, John Wiley & Sons, New York.

Pal, U.S. and A. Chakraverty, 1997, Thin layer convection drying of mushrooms, *Energy Conversion Manage.*, 38(2), 107–113.

Pal, W., J. Coombs, and D.O. Hall (Eds.), 1985, Energy from biomass, *Third E.C. Conference*, Venice, Italy, Elsevier Appl. Sci. Pub., London, U.K.

Pan, Y.K. and X. Wang, 1998, *Modern Drying Technology* (in Chinese), Chemical Industry Pub., Beijing, China, 1251pp.

Pandeya, A., 1997, Drying and milling characteristics of various types of parboiled paddy and wheat, MTech thesis, IIT, Kharagpur, India.

Parkin, E.A., 1963, The protection of stored seeds from insects and rodents, *Proc. Mem. Seed. Test Assoc.*, 28, 893–909.

Patrau, J.M., 1982, *By-Products of the Cane Sugar Industry*, 2nd edn., Sugar Series 3, Elsevier Scientific Pub. Co., New York.

Peplinski, A.J. and V.F. Pfiefer, 1970, Gelatinization of corn and sorghum grits by steam-cooking, *Cereal Sci. Today*, 15, 144, 149–151.

Perry, R.H. and D.W. Green, 1997, *Perry's Chemical Engineering Handbook*, 7th edn., McGraw-Hill, New York.

Pillaiyar, P., 1988, *Rice Post Production Manual*, Wiley Eastern Ltd., New Delhi, India.

Pomeranz, Y., 1971, *Wheat Chemistry and Technology*, AACC, St. Paul, MN.

Potter, N.N., 1986, *Food Science*, AVI, New York.

Prasanna, T.N. et al., 2007, Fruit ripening phenomenon—An overview, *Crit. Rev. Food Sci. Nutr.*, 47, 319–333.

Priestly, R.J., 1976, Studies on parboiled rice, I—Comparison of the characteristics of raw & parboiled rice, *Food Chem.*, (1), 5.

Primo, E., S. Barber, E. Tortosa, J. Camacho, J. Ulldemolins, A. Jimenez, and R. Vega, 1970, Chemical composition of the byproducts obtained in the different steps of the rice milling process (in Spanish), *Rev. Agroquim Technol.*, 10, 244.

Raghavendra Rao, S.N. and B.O. Juliano, 1970, Effect of parboiling on some physio-chemical properties of rice, *Food Chem.*, 18(20), 289.

Raghavendra Rao, S.N. et al., 1967, Studies on some comparative milling qualities of raw and parboiled rice, *J. Food Sci. Technol.*, 4(40), 150–155.

Rahaman, S., 1995, *Food Properties Handbook*, CRC Press, Boca Raton, FL.

Rahaman, S.M., 1999a, Glass transition and other structural changes in foods, *HB of Food Preservation*, Marcel Dekker, Inc., New York.

Rahaman, M.S. (Ed.), 1999b, *Handbook of Food Preservation*, Marcel Dekker, Inc., New York.

Rajkondawar, R.R., 1984, Development and testing of soaking and steaming system for the production of crystal rice, MTech thesis, IIT, Kharagpur, India.

Ramaswamy, H.S. and R. Paul Singh, 1997, Sterilization process engineering, *HB of Food Engineering Practice*, Valentas, K.J. et al. (Eds.), CRC Press, New York.

Rao, J.V.P.K., S. Bat, and A. Chakraverty, 1997, Use of pneumatic pressure in parboiling paddy, *AMA*, 28(20), 69–72.

Rao, M.A. and S.S. Rizvi (Eds.), 1986, *Engineering Properties of Food*, Marcel Dekker, Inc., New York.

Reznick, R. (Ed.), 1998, *Physical Properties of Agricultural Materials and Products*, Hemisphere Publishing Company, Washington, DC, 1244pp.

Richey, C.B., P. Jacobson, and W.C. Hall, *Agricultural Engineers' Handbook*, McGraw-Hill, New York.

Ripp, B.E. (Ed.), 1984, *Controlled Atmosphere and Fumigation in Grain Storages*, Elsevier, New York.

Rockland, L.B. and L.R. Beuchat (Eds.), 1987, *Water Activity: Theory and Application to Food*, Marcel Dekker, Inc., New York, p. 404.

Roger, D. (Ed.), 1970, *Rice and Bulgur Quick Cooking Processes*, Food Proceedings Review No. 16, NJ.

Roy Chowdhury, P.K. et al., 1980, Simultaneous saccharification and fermentation of cellulose to ethanol, *Proceedings of Bioconv. and Biochem. Eng. Symp.*, Vol. II, Ghose, T.K. (Ed.), IIT, Delhi, India.

Rynk, R. (Ed.), 1992, Composting methods, *On-Farm Composting Handbook*, NER AGE Service, Cooperative Extension, Ithaca, NY.

Sakar, S.R., 2003, Rice milling and processing, *Handbook of PHT*, Chakraverty, A. et al. (Eds.), Marcel Dekker, New York.

Salunkhe, D.K. et al. (Eds.), 1984, *Nutritional and Processing Quality of Sorghum*, Oxford & I.B.H. Pub. Co. Pvt. Ltd., New Delhi, India.

Saravacos, D. and A.E. Kostaropoulos, 2002, *HB of Food Processing Equipment*, Plenum Pub., New York.

Sarkar, A.K., 1993, Flour milling, *Grain and Oil & uds—Handling, Marketing and Processing*, Vol. 2, 4th edn., Canadian International Grains Institute, Winnipeg, Manitoba, Canada.

Sarkar, A.K., 2003, Grain milling operations, *Handbook of PHT*, Chakraverty, A. et al. (Eds.), Marcel Dekker, New York.

Satake, T., 1972, Parboiled rice, US Patent 3674514.

Satake, T., 1990, *Modern Rice-Milling Technology*, University of Tokyo Press, Tokyo, Japan.

Satake Engineering Co. Ltd., 1973, Rice milling machinery, Technical Note No. 601, Extension and Training Institute, Satake Engineering Co. Ltd., Tokyo, Japan.

Satake Owner Manual, Type-03, Satake Engineering Co., Ltd., Japan.

Sazhin, B.S. and V.B. Sazhin, 1997, *Scientific Principles of Drying Techniques* (in Russian), Nanka, Moscow, Russia, p. 448.

Scott, J.H., 1951, *Flour Milling Processes*, 2nd edn., Chapman & Hall, London, U.K.

Shah, G.H. and F.A. Masoodi, 1994, *Studies on the Utilization of Waste from Apple Processing Plants*, Indian Food Packer, AFSTI, India.

Shedd, C.K., 1945, Resistance of ear corn to air flow, *Agric. Eng.*, 26, 19–20, 23.

Shedd, C.K., 1953, Resistance of grains and seeds to air flow, *Agric. Eng.*, 34, 616–619.

Shibano, M., 1973a, Construction and function of abrasive roll type rice whitening machine, Technical Note, JRMA, Japan.

Shibano, M., 1973b, Construction and function of friction type rice whitening machine, Technical Note, JRMA, Japan.

Shing-Jy, J.T. et al., 2004, Food preservation and processing: Vegetables: Types and biology, *HB of Vegetables Preservation, & Processing*, Marcel Dekker, Inc., New York.

Shivanna, C.S., 1971a, Traditional and modern methods of parboiling and drying, *Food Ind. J.*, 4(5), 1–12, 24–25.

Shivanna, C.S., 1971b, Traditional and modern methods of storage of paddy, *Food Ind. J.*, 4(1).

Shvete, I. et al., 1975, *Heat Engineering*, 2nd edn., MIR Pub., Moscow, Russia.

Shyamal, D.K., A. Chakraverty, and H.D. Banerjee, 1994, Thermal properties of wheat and wheat bulgur, *Energy Conversion Manage.*, 35(9), 801–804.

Shyamanuj, D., R.K. Mukherjee, and A.C. Chakraverty, 1989, Production technique for crystalline parboiled rice, *J. Agric. Eng.*, 26(1), 55–58.

Siebenmorgen, T.J. and T.R. Archer, 1994, Absorption of water in long-grain rough rice during soaking, *J. Food Proc. Eng.*, (17), 141–154.

Siebenmorgen, T.J. et al., 1998, Milled rice breakage due to environmental conditions, *Cereal Chem.*, 75(1), 149–157.

Simmonds, W.H.C., G.T. Ward, and E. McEwen, 1947, Drying of wheat grain, Part II, Through-drying of deep beds, *Trans. Inst. Chem. Eng.*, 31, 279–288.

Simmonds, W.H.C., G.T. Ward, and E. McEwen, 1953, The drying of wheat grain, Part I, The mechanism of drying, *Trans. Inst. Chem. Eng.*, 31, 265–278.

Singh, R.P. and D.R. Heldman, 2006, *Introduction to Food Engineering*, 4th edn., Academic Press, New York.

Sinha, R.N., 1973, Interrelations of physical, chemical and biological variables in the deterioration of stored grains, *Grain Storage: Part of a System* (*Proceedings Symposium on Grain Storage*, Winnipeg, Manitoba, Canada, June 6–9, 1971), Sinha, R.N. and W.B. Muir (Eds.), AVI, Westport, CT.

Smith, L., 1944, *Flour Milling Technology*, 3rd edn., Northern Publishing Co., Liverpool, U.K.

Sodha, M.S. et al., 1987, *Solar Crop Drying*, Vols. I and II, CRC Press, Boca Raton, FL.

Strumillo, C. and T. Kudra, 1987, *Drying Principles, Applications and Design*, Gordan & Breach Sci. Pub., New York, p. 466.

Subramanian, V. and R. Jambunathan, 1980, Traditional methods of processing of sorghum and pearl millet grains in India, *Rep. Int. Assoc. Cereal Chem.*, 10, 115–118.

Sulc, D.R. et al., 1982, Utilization of secondary raw materials in food production, *Hrana I Ishrana*, 257(10), 199.

Swaminathan, K.R., 1980, Project report for using agricultural waste for paper-board making, College of Agricultural Engineering, TNAU, Coimbatore, India.

Thompson, R.A. and G.H. Foster, 1963, Stress cracks and breakage in artificially dried corn, Marketing Research Report No. 631, USDA-AMS, Washington, DC.

Thompson, R.A. and G.W. Issas, 1967, Porosity determination of grains and seeds with air comparison pyenometer, *Trans. ASAE*, 10, 693–696.

Thompson, J.F., F.G. Mitchell, T.R. Rumsey, R.F. Kasmire, and C.H. Crisosto, 1998, *Commercial Cooling of Fruits, Vegetables and Flowers*, DANR Publication 21567, University of California, Davis, CA.

Thomson, H.J. and C.K. Shedd, 1954, Equilibrium moisture and heat of vaporization of shelled corn and wheat, *Agric. Eng.*, 35, 786–788.

Tillman, D.A., 1982, *Wood as an Energy Source*, McGraw-Hill, New York.

Togrul, I.T. and D. Pehlivan, 2003, Modeling of drying kinetics of single apricot, *J. Food Eng.*, 58, 23–32.

Toshizo, B., 1965, *Drying of Rice in Japan*, Institute of Agr. Mach., Japan.

Treybal, R.E., 1955, *Mass Transfer Operations*, McGraw-Hill, New York.

Turkan, A. et al., 2007, Changes in the drying characteristics water activity values of selected pistachio cultivars during hot air drying, *J. Food Proc. Eng.*, 30(5), 607–624.

USDA, 1952, Drying shelled corn and small grain with unheated air, *Leaflet*, 332, 1–10.

USDA, 1959, Research on conditioning and storage of rough and milled rice, Review through 1958, ARS 20–27, USDA, Washington, DC.

USDA, 1962, Drying shelled corn and small grain with heated air, *Leaflet*, 331, 1–12.

Valentas, K.J. et al., 1997, *HB of Food Engineering Practice*, CRC Press, New York.

Van Arsdel, W.B. et al. (Eds.), 1973, *Food Dehydration*, Vol. 1, 347pp, Vol. II, 529pp, AVI Publishing Co., Westport, CT.

Viraktamath, C.S. and H.S.R. Desikachar, 1971, Inactivation of M phase in rice bran in Indian rice-mills, *J. Food Sci. Technol.*, 8(2), 70–74.

Wang, C.Y. and R.P. Singh, 1978, A single layer drying for rough rice, ASAE Paper No. 78-300, ASAE, St. Joseph, MI.

Wasserman, T. et al., 1956a, Drying characteristics of western rice, Part I. Equal moisture removal and constant drying air temperature in all phases, *Rice J.*, 59(3), 12–16.

Wasserman, T. et al., 1956b, Drying characteristics of western rice, *Rice J.*, 59(4), 41–45.

Wasserman, T. et al., 1957, Commercial drying of western rice, *Cereal Sci. Today*, 2(9), 251 254.

Watson, S.A., 1976a, Manufacture of corn and milo starches, *Starch: Chemistry and Technology*, Vol. II, Whistler, R.L. and E.F. Paschall (Eds.), Academic Press, New York.

Webster, F.H., 1986, *Oats: Chemistry and Technology*, AACC, St. Paul, MN.

Weinecke, L.A., O.L. Brekke, and E.L. Griffin, Jr., 1963, Corn dry milling: Effect of Beall degerminator tail-gate configuration on product streams, *Cereal Chem.*, 40, 575–581.

Witcombe, J.R. (Ed.), 1986, *Proceedings of Pearl Millet Workshop*, ICRISAT, April 7–11, 1986, Hyderabad, India.

Wolf, W. et al., 1985, *Sorption Isotherms and Water Activity of Food Materials*, Elsevier, Amsterdam, the Netherlands, p. 239.

Yokochi, K., 1972, Rice bran processing for the production of rice bran oil and rice bran protein meal, UNIDO Publication, New York.

Zoebel, H.F. et al., 1986, Starch gelatinization: X-ray diffraction study, *Cereal Chem.*, 65(6), 443–446.

Index

Printed and bound by CPI Group (UK) Ltd, Croydon, CR0 4YY

21/10/2024

01777107-0018